Earth and Environmental Sciences Library

The *Earth and Environmental Sciences Library* (EESL) is a multidisciplinary book series that showcases innovative approaches and in-depth reviews, aimed at enhancing the role of the Earth and Environmental Sciences communities. The series also offers valuable guidance for stakeholders, decision-makers, policymakers, international organizations, and NGOs.

Topics of interest include oceanography, the marine environment, atmospheric sciences, hydrology, soil sciences, geophysics, geology, agriculture, environmental pollution, remote sensing, climate change, water resources, and natural resource management.

To further advance these critical areas, the *Earth and Environmental Sciences Library* invites contributions from the Earth and Environmental Sciences communities in the form of edited volumes, monographs, textbooks, conference proceedings, and more. The series fosters knowledge exchange and collaboration across disciplines, encouraging comprehensive discussions and solutions to pressing environmental challenges.

Pradeep Khyalia · Sangita Yadav
Editors

Microplastic Pollution in India and Its Environmental Impacts

Assessment of Ecosystems and Aquatic Environments, Impacts on Human Health, Management and Mitigation

Springer

Editors
Pradeep Khyalia
Department of Environmental Science
Maharshi Dayanand University
Rohtak, Haryana, India

Sangita Yadav
Department of Environmental Science
and Engineering
Guru Jambheshwar University of Science
and Technology
Hisar, Haryana, India

ISSN 2730-6674 ISSN 2730-6682 (electronic)
Earth and Environmental Sciences Library
ISBN 978-3-031-92532-0 ISBN 978-3-031-92530-6 (eBook)
https://doi.org/10.1007/978-3-031-92530-6

This Springer imprint is published by the registered company Springer Nature Switzerland AG
The registered company address is: Gewerbestrasse 11, 6330 Cham, Switzerland

Preface

Microplastic pollution has emerged as a pervasive and pressing environmental issue, posing significant global challenges to ecosystems, biodiversity, and human health. India, with its vast and diverse terrestrial and aquatic environments, is no exception to this crisis. Despite increasing awareness, the scale and impact of microplastic contamination across India remain underexplored and poorly understood. This book endeavors to bridge this knowledge gap by offering a comprehensive assessment of microplastic pollution in the Indian context. Microplastic Pollution in India and Its Environmental Impacts provides an in-depth analysis of microplastic prevalence across various ecosystems, including terrestrial, inland water, sedimentary, and marine environments. It also delves into the multifaceted health implications for ecosystems and humans, offering insights into potential monitoring indicators and ecological risk assessment methodologies tailored to India's unique landscape. The chapters in this book are structured to guide the reader through the intricate interplay between microplastic pollution and its environmental and health ramifications. Beginning with an introduction to microplastic pollution as an emerging concern in India, the text traverses diverse ecosystems, exploring the impacts and risks associated with this ubiquitous contaminant. The latter sections of the book focus on health-related aspects, highlighting the consequences for both ecological systems and human populations. The book also discusses indicators for monitoring pollution and provides a framework for ecological risk assessment to ensure practical applicability. It is a resource intended to inform and inspire initiatives for effective management and mitigation strategies in India. By shedding light on the complexity of microplastic pollution in India's ecosystems, this book seeks to contribute to a deeper understanding of its impacts and support the development of sustainable solutions.

Editors extend the deepest gratitude to everyone who has supported us throughout the journey of contributing to this book. This endeavor would not have been possible without the guidance, encouragement, and contributions of many.

We sincerely thank our mentors, colleagues, and peers for their valuable insights and unwavering support. Their constructive feedback and thought-provoking discussions have been instrumental in shaping our work. A special note of appreciation to

Gaurav Sharma for his support and encouragement, which have been invaluable in this endeavor. We also express our heartfelt gratitude to our families and friends for their patience, motivation, and belief in us, even during the most challenging times. Their encouragement has been our greatest strength. Lastly, we are grateful to the series editorial team and publishers for their guidance and for giving us the opportunity to participate in this significant work. To everyone who played a role in this journey—thank you! Your support has made this experience both rewarding and enriching.

Rohtak, India Pradeep Khyalia
Hisar, India Sangita Yadav

Contents

Microplastic Pollution in India: An Overview 1
Sangita Yadav, Gaurav Sharma, Pradeep Khyalia, and Shiksha Malik

Microplastic Pollution in India as an Emerging Concern 11
Sweta Kumari, Asha Gupta, Neha, Nisha, Ankur, and Sangita Yadav

Microplastic in the Terrestrial Ecosystem of India 43
Manisha Devi, Smita Chaudhry, Gajender Yadav, and R. K. Yadav

Microplastic in the Inland Water Ecosystem of India 63
Alka Rao, Shikha Kumari, Jitender Singh Laura, and Geeta Dhania

Microplastic in the Sediment Ecosystem of India 77
Divya Yadav, Sweta Kumari, Santosh Bhukal, Sumistha Das, and Sangita Yadav

Microplastics in the Marine Ecosystem of India 95
Soniya Kaushik, Manisha Saini, Jitender Singh Laura, Meenakshi Nandal, and Pradeep Khyalia

Health Impacts of Microplastics on the Indian Terrestrial Ecosystem 123
Pooja Yadav, Manju Rani, and Sudesh Chaudhary

Health Impact of Microplastic on Indian Aquatic System 143
Shikha Kumari, Alka Rao, Manjeet Kaur, and Geeta Dhania

Human Health Impact of Microplastics 165
Manjulata Yadav, Soumya Chatterjee, and Pankaj Kumar

Potential Bioindicators of Microplastics for Monitoring Microplastic Pollution 193
Vikram Mor, Wazir Singh, and Pradeep Khyalia

Assessment of Potential Ecological Risk of Microplastics in India 207
Manish Pant, Naveen Singhal, and Jabrinder Singh

Microplastic Pollution in India: An Overview

Sangita Yadav, Gaurav Sharma, Pradeep Khyalia, and Shiksha Malik

Abstract Microplastics (MPs) are a major environmental problem in the world's oceans, and their presence in the Indian continent as a well-known emerging contaminant has attracted a lot of attention in recent years. The chapter provides a detailed overview of India's geographical, demographic, and environmental attributes. India is the seventh largest country in the world, with diverse physical features, from the Indian Ocean to the Himalayas. It is the most populated country, recently surpassing China, and comprises multiple states and territories with varying population densities. The country is also characterized by numerous ecosystems, including terrestrial and marine biogeographic regions. The growing issue of plastic pollution globally and in India, emphasizing the shortcomings in existing policies to manage the increasing plastic waste. India is a major producer and consumer of plastics, contributing significantly to the global plastic waste problem. Overall, India's ecological and environmental challenges are intertwined with its demographic and industrial growth, requiring comprehensive policy interventions to address issues such as population pressure and plastic pollution.

1 Introduction

India is located north of the equator between latitudes $8°4'$ north and $37°6'$ north and longitudes $68°7'$ east to $97°25'$ east. With a total size of 3,287,263 square kilometres (1,269,219 square miles), it is the seventh largest country in the world. India's length

S. Yadav (✉)
Department of Environmental Science and Engineering, Guru Jambheshwar University of Science and Technology, Hisar, India
e-mail: yadavsangitaevs@gmail.com

G. Sharma
Department of Applied Psychology, Guru Jambheshwar University of Science and Technology, Hisar, India

P. Khyalia · S. Malik
Department of Environmental Science, Maharshi Dayanand University, Rohtak, India

P. Khyalia and S. Yadav (eds.), *Microplastic Pollution in India and Its Environmental Impacts*, Earth and Environmental Sciences Library,
https://doi.org/10.1007/978-3-031-92530-6_1

is 2,933 km (1,822 mi) from east to west and 3,214 km (1,997 mi) from north to south. Its shoreline is 7,516.6 km (4,671 mi), and its land border is 15,200 km (9,445 km) (Annual Report 2016–17, Ministry of Home Affairs; India Yearbook, 2007, Publications Division, Ministry of Information and Broadcasting, Govt. of India). In addition to the mainland, India is home to a few island groups. India's physical characteristics are incredibly diverse; the Indian Ocean in the south, the vast Indian Desert in the west, and the majestic Himalayas in the north. India has been divided into six different physiographic divisions as a result of these features: the Northern Plain, Peninsular Plateau, Indian Desert, Islands, Northern and North-eastern Mountain, and Coastal Plains. Each of these divisions has its own features.

India is the most populated nation with one-sixth of the world's population. With a population of 1,425,775,850 as of the end of April 2023, India has surpassed China as the nation with the highest population in the world, according to UN projections (The Economic Times, 2023; United Nations Report, 2023). India is home to a staggering 17.78% of the world's population despite making up only 2.4% of the planet's surface (Statistic Time, India population, 2024), most populated states of India are shown in Fig. 1. With a projected population of 23.89 crore in 2024, Uttar Pradesh is the most populated state in India. Nearly 17% of the nation's population resides in Uttar Pradesh. With about 129 million residents, Bihar has surpassed Maharashtra as the second-largest state in terms of population. With a population of 2.19 crore, Delhi is India's largest union territory. The five most populated states in the nation are Madhya Pradesh, West Bengal, Maharashtra, Uttar Pradesh, and Bihar, where about half of the population resides. Seventy-four percent of the nation's population lives in the ten most populous states. Just 1% of the nation's population lives in the 12 least populated states or UTs, and 10% of the nation's population lives in the 21 least populous states or UTs. Based on population, Lakshadweep (69,000) is the smallest union territory in India, and Sikkim (697,000) is the smallest state (Statistics Times, Population of Indian states, 2024).

The OECD's first Global Plastics Outlook, released ahead of UN negotiations on global action to reduce plastic waste, reveals that policies to stop plastic from leaking into the environment are failing as growing populations and incomes fuel an unrelenting increase in the amount of plastic used and discarded. According to a recent OECD analysis, the amount of plastic trash produced worldwide has doubled in the previous 20 years, with just 9% of it being properly recycled and the majority ending up in landfills, being burned, or leaking into the environment (OECD, 2022). Modern society is characterized by plastic pollution, which scientists from all continents have observed to be pervasive. The deep sea, rainwater, the highest point of Mount Everest, and even human placenta and blood have all been shown to be filled with plastic particles. It is present in the water we drink, is consumed by plankton, causes injury to innumerable marine creatures, and annually starves or chokes millions of birds. Although it is far more difficult to determine the whole extent of the harm caused by microplastics, a new assessment clearly shows that plastics are having an ecological impact. In light of this situation, it makes sense to conclude that plastic pollution directly impedes global objectives for sustainable development (Stoett et al., 2024).

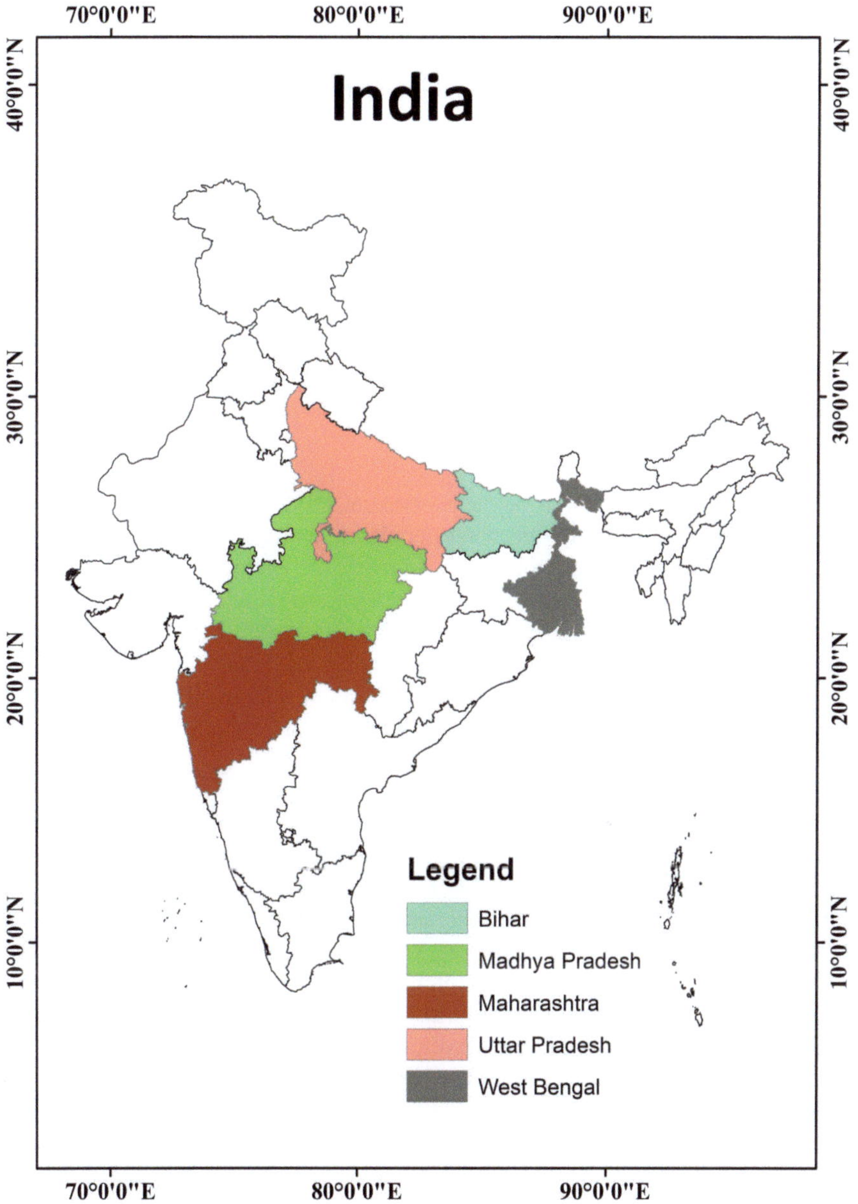

Fig. 1 Most populated states of India

E.A. Poe's "Maelström" and spiral symbolism both allude to the need for a profound ecological-psychological shift in our lopsidedly logos-dominated society. Plastic has come to represent our desire for immortality because it is a material that is nearly non-biodegradable and that humanity has created. Humanity must abandon its ecologically and psychologically harmful consumerist mindset and embrace a less resource-intensive way of living to prevent what Jung called a "catastrophic enantiodromia." Recognizing the depths of our ecological-psychological crisis and taking action based on knowledge rather than giving in to paralysing fear or denial (Bucher, 2022). This chapter discusses the location and population distribution over different regions of India with plastic pollution.

2 Major Ecosystem of India

Environmental balance depends on the ecosystem. The ecosystem and environment are impacted by the constantly growing population and the exploitation of nature. Abiotic influences include things like rain, temperature, light, wind, humidity, and so forth; edaphic factors include soil, pH, topography, minerals, and so forth; and organic compounds include producers, consumers, and decomposers. The food chain, which consists of producers, primary, secondary, and tertiary consumers, and decomposers, makes up an ecosystem. The food web is consists of food chains. Terrestrial and marine ecology are the two categories of ecosystems. Forests, grasslands, deserts, and tundra habitats are examples of terrestrial ecosystems. The aquatic environment also includes freshwater and marine ecology. Ten biogeographic regions in India are the coastal region, Andaman and Nicobar Islands, Deccan Peninsula, Gangetic Plains, Northeast India, Western Ghats, Himalayas, Indian Desert, and semi-arid regions of India (Induria & Sahu, 2023). One of the world's many biogeographical provinces is India. Subtropical, tropical, arctic, and temperate climates are among the nation's many climatic zones. Different ecosystem types have been formed by the different climatic zones, ranging from scorching deserts (Rajasthan) to humid tropical regions (Western Ghats), mild coastlines, and frigid deserts (Ladakh). National parks, wildlife sanctuaries, conservation reserves, and community reserves are all part of the nation's extensive network of protected areas, which span 1,71,921 km^2 (5.03%) of the total land area (Dar et al., 2022).

Of all the terrestrial ecoregions in the world, 45 are located wholly or partially inside India's borders. The Indo-Malayan and Palearctic biogeographic realms comprise these ecoregions. The ten biomes that they belong to are tropical and subtropical coniferous forests, tropical and subtropical dry broadleaf forests, tropical and subtropical grasslands, savannas and shrublands, mangroves, flood-infested grasslands and savannas, tropical and subtropical moist broadleaf forests, and tropical and subtropical mixed forests. No biome or biogeographic realm includes the ecoregion Rock and Ice (Dinerstein et al., 2017). The term "a large area encompassing one or more freshwater systems with a distinct assemblage of natural freshwater communities and species" refers to the world's freshwater ecoregions. A specific ecoregion's freshwater species, dynamics, and environmental conditions are more comparable to one another than to those of nearby ecoregions, and as a whole, they constitute a conservation unit. In India, there are fourteen distinct freshwater ecoregions: the Upper Indus, the Indus Himalayan Foothills, the Lower and Middle Indus, the Ganges Himalayan Foothills, the Middle Brahmaputra, the Chin Hills–Arakan Coast, the Ganges Delta and Plain, the Narmada–Tapti, the Northern and Southern Deccan Plateau, the Southeastern and Western Ghats, and the Andaman and Nicobar Islands (Abell et al., 2008). Throughout the oceans and seas, marine ecoregions have been presented. The Western Indo-Pacific marine domain includes the waters of India. This comprises six marine ecoregions and four provinces: the Andaman Province, which includes the Andaman and Nicobar Islands; the Bay of Bengal Province, which includes Eastern India and the Northern Bay of Bengal; the Central Indian Ocean Islands Province; and the West and South Indian Shelf Province, which includes Western India, South India, and Sri Lanka (Spalding et al., 2007).

3 Main Producers of MPs in India

India's plastics industry is expanding quickly, and the states of Gujarat, Maharashtra, Madhya Pradesh, Daman and Diu, Chhattisgarh, and Dadra and Nagar Haveli are the biggest consumers of plastics in Western India (47%). Overall, India consumes over 11 kg of plastic per person year (CSE, 2019), and as a big consumer, it produces about 26 million metric tons of plastic garbage annually (Law et al., 2020), therefore playing a significant role in the community that generates plastic garbage. Many scientists are investigating the occurrence of MPs in different parts of the Indian environment, which are eventually produced by the weathering and fragmentation of this plastic trash (Vaid et al., 2021). The trend of MP research in Indian surroundings has only recently attracted the scientific world's attention. Since freshwater systems, terrestrial systems, the atmosphere, and human consumables are all equally vital to us, there is a substantial paucity of research interventions in these areas, which calls for more attention.

4 MP Pollution in India

Plastics can linger for years in various environmental compartments after being discarded and exposed to the physical and chemical influences that dismantle their plastic structure to smaller pieces (Rose et al., 2023). In the maritime environment, microplastic pollution is regarded as a severe problem. The Punnakayal estuary on India's southeast coast was found to be contaminated with up to 19.9 MPs per L in research by Selvam et al. (2021), suggesting that MPs could be channelled from inland sources to the Gulf of Mannar. The movement of plastic waste from crowded regions to the ocean through South Juhu Creek was calculated in a different study by Manickavasagam et al. (2020). The study found that macroplastic and megaplastics, which fragment over time and eventually turn into MPs, made up a significant amount of the carried plastic debris. This is a significant problem that has to be addressed. These beaches have MPs from a variety of man-made and natural sources. While natural sources include surface and wind currents, aeolian processes, runoff, and riverine transport, anthropogenic sources include fishing, tourism, recreational, religious, port, and industrial activities, as well as poorly managed plastic garbage and untreated wastewater discharges (Maharana et al., 2020; Robin et al., 2020). Even though freshwater is a source of drinking water, very little research has been done on freshwater microplastics (MPs), even though MPs are a widespread and harmful environmental stressor in the world's oceans.

One of the most essential things for living things to exist and survive is air. Airborne MPs are one of the several atmospheric pollutants that have recently come to light as contaminants of concern. MPs pose a serious risk to human health since they can enter the body directly. Dris et al. (2016) conducted preliminary research on airborne MPs and discovered that MP fibers can become airborne following their disintegration from their source. Synthetic textiles continue to be the main source of MPs in the air (Dris et al., 2016), but it's also crucial to take into account slow releases from streets, landfills, and incomplete rubbish incineration (Liu et al., 2019). Air pollution is an issue that affects people on a local, national, and international level. In rapidly developing nations like China and India, industrialization and urbanization are essential components of a nation's infrastructure and economic development. However, they also bring with them a number of problems, including an increase in the number of people living in urban areas, a rise in infrastructure and industrialization, and an increase in vehicle traffic, all of which contribute to an increase in atmospheric near-surface pollution (Martin et al., 2019). Unfortunately, India's poor air quality is no longer limited to metropolitan areas because high pollution levels are also seen in semi-rural and/or rural areas (Dey et al., 2020). The reason for this is because local emissions and advected pollution from far-off sources combine to form air pollution across a region (Tan et al., 2022). Mishra et al. (2023) showed how emissions from nearby and/or distant primary biomass burning throughout the night cause haze formation in Delhi by accelerating the growth of nanoparticles.

5 Attitudes and Behaviour Pertaining to Plastic Uses

The extent to which a population's awareness (or lack thereof) and the attitudes that come from that awareness contribute to plastic pollution is neither straightforward nor simple to measure. However, it is logical to believe that a community's awareness level, and hence its attitude and conduct, play a significant influence in assessing the severity of the plastic pollution problem. Even while there has been a noticeable increase in the problems associated with plastic pollution in recent years, and consequently in public knowledge of these issues, there appears to be a disparity in people's capacity to connect the issues at hand to their own unique behaviors.

People's connection with plastics also appears to be complicated because, despite the apparent adverse environmental effects, they cannot stop using the material (Menzel et al., 2021). Environmental awareness is often influenced by sociodemographic characteristics like age, gender, and level of education (Soares et al., 2021). Compared to men, women are more likely to undertake environmentally friendly behaviors (Charitou et al., 2021). Increased awareness also sparked a greater desire to act environmentally responsibly. People believed that stopping the sale of reusable bags, enforcing additional tariffs, and providing free reusable bags may be crucial steps in preventing pollution. Lack of awareness combined with social media access may potentially be an issue because, even though many participants had internet access, they rarely visited websites that addressed environmental issues (Hammami et al., 2017). Inadequate infrastructure for garbage management, disposal, and recycling combined with a lack of public awareness (Dumbili & Henderson, 2020).

Although finding a solution to the serious problem of plastic pollution is not easy, it can be agreed that spreading awareness of the issue may have some beneficial effects, and encouraging students to think critically about the issue may be crucial to the development of workable solutions. It is concerning to see that, despite India's 7,500 km of coastline, there are either few laws governing marine plastic pollution or few studies assessing public awareness of the problem. Women were shown to be more conscious than men and more inclined to embrace environmentally friendly behaviors. The degree of awareness was also significantly and directly correlated with educational background and field of study. Although there was sufficient awareness of plastic contamination, there was little understanding of microplastics specifically. The current problem of plastic/microplastic pollution may be mitigated by educating people about it through mass media and structured education, strictly enforcing laws, and substituting eco-friendly plastics for conventional ones (Dowarah et al., 2022).

Human behavior can be changed, particularly by minimising plastic waste, to ensure environmental sustainability. This study examined whether millennials' Willingness to Reduce Plastic Waste (WRPW), which measures their propensity to avoid or minimise the purchase of single-use plastic tableware, food packaged in plastic, and plastic water bottles, is influenced by their natural environmental orientation, environmental knowledge, environmental concern, and environmental attitude (Solekah et al., 2022).

6 Organization of This Book

This book's objective is to outline the most current occurrences in the field of research on microplastics as pollutants in various environments. This book has eleven chapters, majorly divided into five parts, which discuss microplastics and their presence in the Indian continent as a well-known emerging contaminant. Part I: Microplastic Pollution and Monitoring consists of 3 chapters, which talk about the location and geography of India, microplastic and its emergence and noticeable effect on the Indian continent. After this, potential indicators of microplastics for monitoring microplastic pollution are discussed. Part II: Impacts on Ecosystems, consist of 4 chapters. The following four chapters discuss the impacts of microplastic on terrestrial, inland waters, sediment, and marine ecosystems, as well as their presence and the effects caused by their presence in the respective environments. Part III: Impacts on Soil and Aquatic Ecosystem consists of 2 chapters. These talk about the health impacts on terrestrial and aquatic ecosystems, respectively. Part IV: Impacts on Food Security and Human Health, consists of two chapters. Microplastic enters the food chain, and the food web of humans is discussed, further discussing the health effects caused to humans because of their presence. Part V: Ecological Risk Assessment, Management, and Regulations consists of the last two chapters of the book and discusses the assessment of potential ecological risk assessment of microplastics in India and the methodologies used for the assessment.

7 Conclusion

The chapter discusses the significant environmental challenge posed by microplastics (MPs) in India, focusing on the country's geographical, demographic, and ecological aspects. India, a major producer and consumer of plastics, faces intertwined challenges of plastic pollution and rapid industrial growth. The chapter reviews the status of MP pollution, identifying shortcomings in current policies and emphasising the need for comprehensive interventions to address these issues. It highlights research on MPs across various ecosystems in India, noting insufficient attention to freshwater and terrestrial environments. The chapter underscores the relationship between population density, industrial activities, and environmental impact, drawing attention to the health risks associated with MP exposure. The chapter calls for increased awareness, better waste management strategies, and educational initiatives to mitigate plastic pollution and its ecological effects.

References

Abell, R., Thieme, M. L., Revenga, C., Bryer, M., Kottelat, M., Bogutskaya, N., Coad, B., Mandrak, N., Balderas, S. C., Bussing, W., Stiassny, M. L., & Petry, P. (2008). Freshwater ecoregions of the world: A new map of biogeographic units for freshwater biodiversity conservation. *BioScience, 58*(5), 403–414. https://doi.org/10.1641/B580507

Annual Report 2016–17. Ministry of Home Affairs. Wayback Machine. Accessed 14 January 2025

Bucher, S. (2022). The spiral of plastic pollution: A compensatory urge from the collective unconscious for an ecological-psychological transformation of civilization. *Journal of Analytical Psychology, 67*(5), 1386–1409. https://doi.org/10.1111/1468-5922.12861

Charitou, A., Aga-Spyridopoulou, R. N., Mylona, Z., Beck, R., McLellan, F., & Addamo, A. M. (2021). Investigating the knowledge and attitude of the Greek public towards marine plastic pollution and the EU single-use plastics directive. *Marine Pollution Bulletin, 166*, 112182. https://doi.org/10.1016/j.marpolbul.2021.112182

CSE. (2019). The problem of plastics. Retrieved January 15, 2025, from https://cdn.cseindia.org/attachments/0.57139300_1570431848_Factsheet1.pdf

Dar, S. A., Dar, S. A., & Nabi, M. (2022). Conservation of Biodiversity in India: Current status and future strategies. In *Towards sustainable natural resources: Monitoring and managing ecosystem biodiversity* (pp. 195–214). Springer International Publishing. https://doi.org/10.1007/978-3-031-06443-2_11

Dey, S., Purohit, B., Balyan, P., Dixit, K., Bali, K., Kumar, A., Imam, F., Chowdhury, S., Ganguly, D., Gargava, P., Shukla, V. K. (2020). A satellite-based high-resolution (1-km) ambient PM2.5 database for India over two decades (2000–2019): Applications for air quality management. *Remote Sensing, 12*(23), 3872. https://doi.org/10.3390/rs12233872

Dinerstein, E., Olson, D., Joshi, A., Vynne, C., Burgess, N. D., Wikramanayake, E., Hahn, N., Palminteri, S., Hedao, P., Noss, R., Hansen, M., & Saleem, M. (2017). An ecoregion-based approach to protecting half the terrestrial realm. *BioScience, 67*(6), 534–545. https://doi.org/10.1093/biosci/bix014

Dowarah, K., Duarah, H., & Devipriya, S. P. (2022). A preliminary survey to assess the awareness, attitudes/behaviours, and opinions pertaining to plastic and microplastic pollution among students in India. *Marine Policy, 144*, 105220. https://doi.org/10.1016/j.marpol.2022.105220

Dris, R., Gasperi, J., Saad, M., Mirande, C., & Tassin, B. (2016). Synthetic fibers in atmospheric fallout: A source of microplastics in the environment? *Marine Pollution Bulletin, 104*(1–2), 290–293. https://doi.org/10.1016/j.marpolbul.2016.01.006

Dumbili, E., & Henderson, L. (2020). The challenge of plastic pollution in Nigeria. In *Plastic waste and recycling* (pp. 569–583). Academic Press. https://doi.org/10.1016/B978-0-12-817880-5.00022-0

Hammami, M. B. A., Mohammed, E. Q., Hashem, A. M., Al-Khafaji, M. A., Alqahtani, F., Alzaabi, S., & Dash, N. (2017). Survey on awareness and attitudes of secondary school students regarding plastic pollution: Implications for environmental education and public health in Sharjah city, UAE. *Environmental Science and Pollution Research, 24*, 20626–20633. https://doi.org/10.1007/s11356-017-9625-x

India Yearbook 2007. (2007). Publications Division, Ministry of Information & Broadcasting, Govt. of India. ISBN: 978-81-230-1423-4. Accessed 14 January 2025.

Induria, S. K., & Sahu, K. K. (2023) Ecosystem of India, functions, biogeographic regions and importance. In *Role of applied science in social implication.* Probecell Press.

Law, K. L., Starr, N., Siegler, T. R., Jambeck, J. R., Mallos, N. J., & Leonard, G. H. (2020). The United States' contribution of plastic waste to land and ocean. *Science Advances, 6*(44), eabd0288. https://doi.org/10.1126/sciadv.abd0288

Liu, K., Wang, X., Fang, T., Xu, P., Zhu, L., & Li, D. (2019). Source and potential risk assessment of suspended atmospheric microplastics in Shanghai. *Science of the Total Environment, 675*, 462–471. https://doi.org/10.1016/j.scitotenv.2019.04.110

Maharana, D., Saha, M., Dar, J. Y., Rathore, C., Sreepada, R. A., Xu, X. R., Koongolla, J. B., & Li, H. X. (2020). Assessment of micro and macroplastics along the west coast of India: Abundance, distribution, polymer type and toxicity. *Chemosphere, 246*, 125708. https://doi.org/10.1016/j.chemosphere.2019.125708

Manickavasagam, S., Kumar, S., Kumar, K., Bhuvaneswari, G. R., Paul, T., & Shukla, S. P. (2020). Quantitative assessment of influx and efflux of marine debris in a water channel of South Juhu Creek, Mumbai, India. *Regional Studies in Marine Science, 34*, 101095. https://doi.org/10.1016/j.rsma.2020.101095

Martin, R. V., Brauer, M., van Donkelaar, A., Shaddick, G., Narain, U., & Dey, S. (2019). No one knows which city has the highest concentration of fine particulate matter. *Atmospheric Environment: X, 3*, 100040. https://doi.org/10.1016/j.aeaoa.2019.100040

Menzel, C., Brom, J., & Heidbreder, L. M. (2021). Explicitly and implicitly measured valence and risk attitudes towards plastic packaging, plastic waste, and microplastic in a German sample. *Sustainable Production and Consumption, 28*, 1422–1432. https://doi.org/10.1016/j.spc.2021.08.016

Mishra, S., Tripathi, S. N., Kanawade, V. P., Haslett, S. L., Dada, L., Ciarelli, G., Kumar, V., Singh, A., Bhattu, D., Rastogi, N., Daellenbach, K. R., Prevot, A. S. (2023). Rapid night-time nanoparticle growth in Delhi driven by biomass-burning emissions. *Nature Geoscience, 16*(3), 224–230. https://doi.org/10.1038/s41561-023-01138-x

OECD. (2022). Plastic pollution is growing relentlessly as waste management and recycling fall short, says OECD. In *Plastic pollution is growing relentlessly as waste management and recycling fall short, says OECD*. OECD. Accessed 14 January 2025.

Robin, R. S., Karthik, R., Purvaja, R., Ganguly, D., Anandavelu, I., Mugilarasan, M., & Ramesh, R. (2020). Holistic assessment of microplastics in various coastal environmental matrices, southwest coast of India. *Science of the Total Environment, 703*, 134947. https://doi.org/10.1016/j.scitotenv.2019.134947

Rose, P. K., Yadav, S., Kataria, N., & Khoo, K. S. (2023). Microplastics and nanoplastics in the terrestrial food chain: Uptake, translocation, trophic transfer, ecotoxicology, and human health risk. *TrAC Trends in Analytical Chemistry, 167*, 117249. https://doi.org/10.1016/j.trac.2023.117249

Selvam, S., Jesuraja, K., Venkatramanan, S., Roy, P. D., & Kumari, V. J. (2021). Hazardous microplastic characteristics and its role as a vector of heavy metal in groundwater and surface water of coastal south India. *Journal of Hazardous Materials, 402*, 123786. https://doi.org/10.1016/j.jhazmat.2020.123786

Soares, J., Miguel, I., Venâncio, C., Lopes, I., & Oliveira, M. (2021). Public views on plastic pollution: Knowledge, perceived impacts, and pro-environmental behaviours. *Journal of Hazardous Materials, 412*, 125227. https://doi.org/10.1016/j.jhazmat.2021.125227

Solekah, N. A., Handriana, T., & Usman, I. (2022). Millennials' deals with plastic: The effect of natural environmental orientation, environmental knowledge, and environmental concern on willingness to reduce plastic waste. *Journal of Consumer Science*, *7*(2), 115–133. https://doi.org/10.29244/jcs.7.2.116-134

Spalding, M. D., Fox, H. E., Allen, G. R., Davidson, N., Ferdaña, Z. A., Finlayson, M. A. X., Halpern, B.S., Jorge, M. A., Lombana, A. L., Lourie, S. A., Martin, K. D., & Robertson, J. (2007). Marine ecoregions of the world: A bioregionalization of coastal and shelf areas. *BioScience, 57*(7), 573–583. https://doi.org/10.1641/B570707

Statistic Times. (2024). *India population 2024*. . Retrieved January 14, 2025 from StatisticsTimes.com

Statistics Times. (2024). *Population of Indian states 2024*. Retrieved January 14, 2025 from StatisticsTimes.com

Stoett, P., Scrich, V. M., Elliff, C. I., Andrade, M. M., Grilli, N. D. M., & Turra, A. (2024). Global plastic pollution, sustainable development, and plastic justice. *World Development, 184*, 106756. https://doi.org/10.1016/j.worlddev.2024.106756

Tan, Q., Ge, B., Xu, X., Gan, L., Yang, W., Chen, X., Pan, X., Wang, W., Li, J., & Wang, Z. (2022). Increasing impacts of the relative contributions of regional transport on air pollution in Beijing: Observational evidence. *Environmental Pollution, 292*, 118407. https://doi.org/10.1016/j.envpol.2021.118407

The Economic Times. (2023). India population: India overtakes China to become world's most populous nation with with 1.428 billion people: UN - The Economic Times. ISSN 0013-0389. (Accessed on 14-01-2025).

United Nations. (2023). *India to overtake China as world's most populous country in April 2023, United Nations projects*. United Nations. Accessed 14 January 2025

Vaid, M., Mehra, K., & Gupta, A. (2021). Microplastics as contaminants in Indian environment: A review. *Environmental Science and Pollution Research*, 1–28. https://doi.org/10.1007/s11356-021-16827-6

Microplastic Pollution in India as an Emerging Concern

Sweta Kumari, Asha Gupta, Neha, Nisha, Ankur, and Sangita Yadav

Abstract A novel class of environmental contaminants, known as microplastics (MPs), have emerged as a result of increased plastics manufacture and usage. MPs are less than 5 mm-sized plastic fragments that come from both primary and secondary sources. Despite the fact that secondary MPs are shaped when bigger particles are fragmented and MP size range is where initial MPs are formed. Though MP pollution is a worldwide issue, India has little research on it. This chapter's objective is to provide the latest information about MP contamination in Indian soil, rivers, and the atmosphere. In order to support the objectives of emerging pollutant management, it evaluates the body of scientific knowledge that is currently accessible, finds gaps, and indicates future research paths. In order to further enhance the management of MP in Indian contexts, the assessment also suggests areas for improvement and offers future research topics.

Keywords Environmnetal contamination · Environment · India · Microplastics · MP pollution · Sources

1 Introduction

The numerous new threats, such as plastic waste, that pose a lethal threat to both human health and the ecosystem are worrying. Interest is majorly in the enormous production and its disposal (Vaid et al., 2021). Exponential growth in production was initially seen in the 1970s, after which it has not seen any downfall, making 'The Earth' a plastic planet. Plastic has become so indispensable that it is choking the globe to incompetence and death. Anything that we touch is plastic, from household to commercial level. The estimated production of plastic initially in the 1950s was 1.5 million tons, which increased to 359 million tons in 2018 and 380 million tons in 2022. According to the World Population Review, India is ranked 1st among the

S. Kumari · A. Gupta · Neha · Nisha · Ankur · S. Yadav (✉)
Guru Jambheshwar University of Science and Technology, Hisar, India
e-mail: yadavsangitaevs@gmail.com

P. Khyalia and S. Yadav (eds.), *Microplastic Pollution in India and Its Environmental Impacts*, Earth and Environmental Sciences Library,
https://doi.org/10.1007/978-3-031-92530-6_2

top 10 countries with the most plastic pollution, producing around 12,994,100 tons of plastic waste. According to CPCB, out of the total collected plastic waste, only 60% is recycled, and rest is dumped in open, which paves the way in the aquatic and terrestrial environment in the form of small plastic fragments due to natural degradation. The breakage of plastic into small fragments is due to its exposure to UV rays playing the role of a catalyst in photooxidation. When plastic gets exposed to UV rays, it becomes brittle and more prone to breakage, resulting in smaller fragments known as microplastics (MPs) (Akdogan & Gven, 2019). Carpenter et al. (1972) published the first study on microplastics. The tiniest plastic particles served as the basis for the investigation (Carpenter et al., 1972). Plastics are more precisely defined on the basis of their size. They basically fall into four categories: (i) Microplastics (<0.5 cm) (ii) Mesoplastics (0.5–5 cm) (iii) Macroplastics (5–50 cm) and (iv) Megaplastics (>50 cm) (Veerasingam et al., 2020a, 2020b). Microplastics are mainly classified as primary and secondary MPs. Primary MPs are defined as fragments that are intentionally blown up in the environment by industries that manufacture plastic (Yadav et al., 2023). MPs can be found in the extrusion and grinding of the raw materials used to make personal care products like toothpaste, face cleanser, handwash, and other cosmetic items that are recently being used on a larger scale by humans (Prakash et al., 2020; Veerasingam et al., 2020a, 2020b). Secondary MPs are mainly imported into the environment by the process of the fragmentation and anthropogenic activities of large plastics that are dumped in the open. For example, food packaging, fishing gear, synthetic textiles, cosmetics, car tyres, etc., make for secondary MPs (Prakash et al., 2020; Vaid et al., 2021). Microplastics and their other category, nanoplastics, are attracting the attention of researchers these days due to the damage they are causing to the harmony of the planet "The Earth". MPs are found in different environmental compartments (see Fig. 1). In the lithosphere, animals, plants and microbes are the transportation medium for MPs. Microplastics are disseminated and transported by animals (MPs) in soils by ingesting and eating them. Precipitation events may drive MP deposition in the atmosphere, with wind activity affecting the migration of synthetic polymer particles and MP concentration in wind-eroded sediments. MPs' size and form may also impact their transportation through the atmosphere (Zhang et al., 2020).

The possible toxicity and durability of MPs serve as a growing danger to ecosystems and the surroundings. Despite their small size, MPs are absorbed by fish, barnacles, mussels, lugworms, and amphipods. MPs retain the properties of plastics and contain chemical additives like pigments, colorants, processing chemicals, antioxidants, and phthalate plasticizers. They can harm species like salt-marsh grasses, mussels, corals, zooplankton, primary tropic levels, and keystone organisms. MPs can be consumed by low trophic suspension, filter feeders, detritivores, and planktivores due to their size resembling sediments and planktonic species.

The United Nations (UN) has highlighted the negative effects of plastic pollution on marine life and the livelihoods of 3 billion people relying on the oceans. The UN's 2021 SDGs report reveals that plastic pollution, along with factors like ocean warming, eutrophication, acidification, and fishery collapse, threaten ocean sustainability. Dead zones, water regions with insufficient oxygen to sustain marine

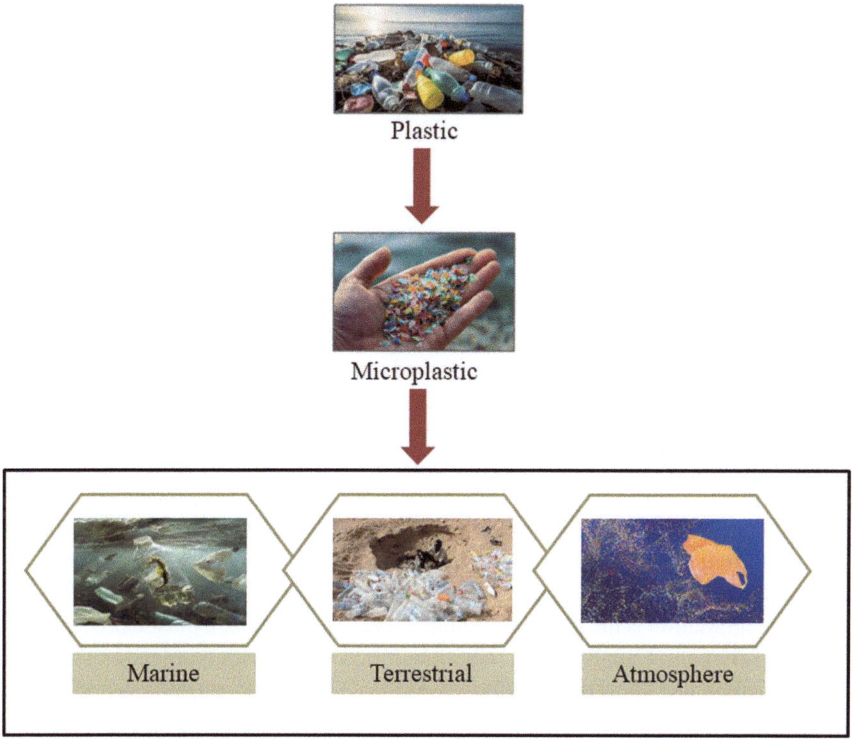

Fig. 1 Microplastic in different environmental compartments including. **a** Marine, **b** terrestrial and **c** atmosphere

life, have grown from 400 in 2008 to 700 in 2019, making over half of maritime critical biodiversity regions vulnerable. The UN Environmental Assembly approved resolutions in 2019 to address marine plastic litter and microplastics, aiming to limit their release, offer substitutes, and reverse their impacts. This is part of the 2030 goal for sustainable development.

This chapter will address microplastics as a growing environmental problem in India, including their impact and potential hazards. This section provides a general overview of microplastics by discussing their prevalence in various ecosystems, distribution across India, and laws and regulations.

2 Microplastic Pollution in the Environment

Standard plastics are often fairly resistant to deterioration. Depending on the polymers' characteristics, the projected lifespan of plastics is hundreds or even thousands of years, as the surrounding natural environment. Even though it happens extremely slowly, environmental weathering nonetheless leads to the degradation of plastics, which leads to the occurrence of MPs in the environment (Zhang et al., 2021). The shattering of macroplastic debris is a key source of these synthetic polymers, in keeping with the study on the frequency of MPs in lakes of India (Vaid et al., 2021). Surface water and sediment components of MP experiments were carried out in Indian lakes using nylon plankton nets for surface waters and plankton nets for sediment samples. The breakup of macroplastic debris was shown to be a key source of MPs during a recent investigation into the occurrence of these polymers in Indian lakes (Ganesan et al., 2019; Gopinath et al., 2020; Manikanda Bharath et al., 2020; Sruthy & Ramasamy, 2017). Additional variables that may be at blame include the principal MPs that leak from personal care items or companies, the MPs that are transported by rivers, and dry deposition and rainfall through runoff. The Meghna, Ganges and Brahmaputra rivers are believed to release one to three billion MPs each day in Bay of Bengal. While some MPs do not completely enter the oceans, some do collect in the sediments, which functions as a significant MP sink. Indian rivers are contaminated with MPs due to the disintegration of macroplastic waste, such as ropes, materials for packaging of plastic, fisher nets, synthetic textiles, wrappers and pipes. Different hydrodynamic and meteorological parameters direct these artificial polymers to distinct regions of the water section.

Photodegradation is the most important mechanism via which plastic breaks down in the environment. Usually, solar radiation initiates processes that lead to free radicals mediating the photodegradation of plastics. The main culprits are medium-energy UV-A (315–400 nm) and high-energy UV-B (290–315 nm) irradiation (Liu et al., 2019). Esters, aldehydes, and propyl end groups are produced as a result of the light-induced polymer breakdown (Manzoor et al., 2022). "Thermal degradation" refers to the disintegration of polymers brought on by input of energy at elevated temperatures. Plastics may undergo thermo-oxidative reactions when temperatures are high. Polymer chains can break and produce radicals when exposed to heat (Zhang et al., 2021), combining with oxygen to create hydroperoxide and splitting into alkoxy radicals and free hydroxyl. Heat degradation can result in molecule enlargement and reduction (Crawford & Quinn, 2016). The thermal characteristics of plastics and the presence of oxygen affect the temperature needed for thermal deterioration. Mechanical degradation occurs when plastics break down due to external forces, such as collisions with rocks and abrasion from waves and wind. In aquatic settings, thawing and freezing can cause polymer breakdown (Pal et al., 2018). The mechanical characteristics of the plastic determine how external aggressors behave. Fracture strain, also known as elongation at break, measures how well a plastic product can withstand shape changes without developing cracks. In the presence of additional tensile stresses, polymers are more likely to fracture when their elongation is lower at break values. Polymer chain scission inevitably occurs in plastics that are continuously stressed.

Tire wear, brake wear, and road wear particles are mostly produced by mechanical forces (Wagner et al., 2018). Tire interactions with road surfaces, brake pads, and disks cause frictional stresses, causing tire fragments to tear off the rubber surface. This continuous stress causes material fatigue and causes tire treads to bulge, causing prolonged stretching. The material's limiting strength may be reached by abrasion with the road surface, leading to elongated rubber particles, scratching, and micro-cutting (Frank Sommer et al., 2018).

3 Sources of Microplastic

Several investigations identified the following as examples of anthropogenic activities: run-off, riverine transport, aeolian processes, tourism, fishing, recreational, industrial activities, port and religious; on the other hand, natural sources include aeolian, wind currents, and surface processes (Maharana et al., 2020; Robin et al., 2020; Tiwari et al., 2019). The two basic sources of plastic particles that contribute to the presence of microplastics (MPs) in the environment are referred to as primary and secondary sources, respectively. The sources of microplastic pollutants include plastic pellets, paint, washing wastewater, sewage sludge, artificial grass, rubberised roads in urban areas, and tyre wear from vehicles. Plastic pellets, granular polymers, are used to create various plastic items with a consistent form and 2–4 mm diameter. They are processed, transported, and stored in semifinished goods. The primary ingredients of plastic include petroleum and coal. When MPs are reduced to minute particles, they become microbeads. Many personal care products use synthetic pigments for cosmetic purposes, such as cleaning, whitening, exfoliating, and the removal of dead skin. These pigments may be easily substituted with microbeads. Meanwhile, secondary sources include large-scale plastic wastes from farming, fishing, and other sources, as well as city trash like plastic bottles and bags. Because of the quick rise in the number of cars on the planet, tyre wear on automobiles is thought to be among the significant sources of environmental microplastic among them. However, surprisingly little research is now accessible about the prevalence of rubber in the environment.

4 Analysis of MPs

Microplastics' prevalence in the seas, bioavailability, and capacity to transport hazardous substances have made them a significant worldwide environmental concern in recent years. It is becoming more and more common practice to analyse microplastics in environmental and experimental materials. Microplastics are present everywhere in the ocean, from the top to the bottom, from the coast to the vast ocean, and from subtropical to polar waters (Kang et al., 2015). As they get smaller, they become more bioavailable to zooplankton and other tiny aquatic animals (Shim et al., 2017). Microplastics that are ingested have been documented across a wide range of taxa and trophic levels (Song et al., 2014). Fundamental scientific information on the pollution state, hotspots of level, historical trends, and the exposition and species fate may be obtained by observing microplastics in various biotic and abiotic environmental matrices (De Witte et al., 2014). Reliable and consistent sampling and analytical techniques are necessary for microplastics monitoring investigations. The two main processes in microplastic analysis are physical (such as microscopy) and chemical (such as spectroscopy) characterizations, which are used to confirm the presence of plastics. There are benefits and drawbacks to every technique and combination (Shim et al., 2017). Techniques like scanning electron microscopy (SEM), Fourier transform infrared spectroscopy (including micro-FTIR and ATR-FTIR) and Raman spectroscopy help grade microplastics from environmental samples by identifying shape, color, morphology, structure, and chemical composition which are crucial in determining environmental plastic pollution.

4.1 Visual Identification

Large microplastics (1–5 mm) were usually discovered on beaches and in surface waters prior to the term "microplastics" becoming widely used (McDermid & McMullen, 2004). Traditionally, forceps and the human eye were used for sorting and identification. While colourful plastic pieces and pre-production resin pellets are easily recognised with the unaided eye, many tiny plastics can only be detected visually (Heo et al., 2013). Sorting, however, may miss minute particles in high strandline samples including large concentrations of confusing polymers or interfering inorganic and organic components (Shim et al., 2017). For both specialists and untrained volunteers, visual sorting and identification of big microplastics is a quick, straightforward, and simple process (Hidalgo-Ruz & Thiel, 2013).

4.2 Microscopy

A popular technique for spotting microplastics in the hundreds of micron range, as seen in neuston net samples is stereo-microscopy. In order to distinguish unclear, plastic-like particles, it offers comprehensive surface texture and structural information (Nel & Froneman, 2015). It is challenging to classify as plastics sub-hundred micron particles (less than 100 μm) that lack colour and a conventional form (Song et al., 2015). Biogenic components from sediment and neuston net samples can impede microscopic viewing, and sediment samples with poorly separated light particles can also cause problems with microscopic identification (Shim et al., 2017).

Images of plastic-like particles at high magnification may be obtained using scanning electron microscopy (SEM), which boasts exceptional clarity. Organic and microplastic particles can be distinguished from one another by using high-resolution photos of the particles' surfaces (Cooper & Corcoran, 2010). The elemental makeup of the same item is revealed by further investigation using energy-dispersive X-ray spectroscopy (EDS) (Vianello et al., 2013). When distinguishing carbon-dominant polymers from inorganic particles, the elemental makeup of the particles might be helpful (Shim et al., 2017). There is a limit to how many samples can be processed by a SEM-EDS because of its high cost and significant labour requirements for sample preparation and inspection. SEM cannot be used to identify polymers by their colour. More surface characterisation and elemental analysis of particular plastic particles are advised using this approach.

Using an optical microscope, also known as a light microscope, to enlarge pictures of tiny objects through the use of lenses and visible light is known as optical microscopy. A micrograph can be produced from the picture of an optical microscope using a standard, photosensitive camera. This approach relies on visual identification made possible by a microscope (Wirnkor et al., 2019). Large microplastics and possible microplastics from other organic or inorganic material in the sample residues may be effectively separated using visual sorting (Hidalgo-Ruz & Thiel, 2013), while tiny microplastic particles can be effectively separated using a dissection microscope (Morét-Ferguson et al., 2010). Wang et al. (2017) claim that optical microscopy makes it possible to quantify the size ranges of particles and fibre lengths contained in a sample, as well as to classify particles or fibre kinds based on their morphology.

4.3 Polymer Identification

Spectroscopy is crucial for identifying plastics and the polymers used in their manufacture for particles smaller than 1 mm (Hanvey et al., 2017). FTIR serves as a commonly employed technique to detect microplastic samples with a high spectral resolution data collection rate. It is the most common method for identifying polymers from sediment samples (Lusher et al., 2013). A combination of optical microscopy (micro-FTIR) and FTIR has been used for polymer identification. FTIR spectroscopy has numerous advantages in plastic identification, including its minimally disruptive character, which permits specimens to be examined without affecting them (Song et al., 2015). Micro-FTIR (μ-FTIR) is a technique that must be used to identify tiny microplastics. It works by switching between the IR probe and object lens to image micro-sized plastic-like particles in order to validate their identity spectroscopically on a single platform (Shim et al., 2017). Transmission (Ugolini et al., 2013), reflectance (Ng & Obbard, 2006) and attenuated total reflectance (ATR) modes (Song et al., 2014, 2015) exist in the FTIR microplastics analysis. Furthermore, from uneven microplastic surfaces, the ATR mode yields steady spectra, but in the reflectance mode, the spectra are unstable. Theoretically, microplastics as tiny as the ATR probe's IR beam aperture diameter (10 μm) can be detected. μ-ATR-FTIR, which combines microscopy and spectroscopy for chemical validation, is a helpful technique for detecting microplastics from environmental materials (Shim et al., 2017). Raman spectroscopy is a type of spectroscopy that relies on the sample interaction and the resulting change in photons in light (monochromatic) (Hanvey et al., 2017). It may be used to determine the polymer type of plastic by analysing its structural properties. Raman spectroscopy was only applied in two investigations, where Raman micro-spectroscopy was utilised in conjunction with it to detect and quantify plastics in sediment. This enables the quantification of particles in the mm range using a non-destructive identification instrument (Allen et al., 1999). Tiny microplastics, measuring only a few μm can be identified using Raman spectroscopy because of the lower laser beam diameter compared to FTIR (Cole et al., 2011). The advantage of using non-contact Raman spectroscopy analysis is that the microplastic samples are kept whole for potential future examination. Microplastics in samples of zooplankton may be identified using confocal microscopy in conjunction with Raman spectroscopy (Tagg et al., 2015).

4.4 Thermal Analysis

Differential scanning calorimetry (DSC), a thermo-analytical technique, has been used for microplastic identification. DSC quantifies alterations in the chemical and physical characteristics, based on polymer's heat stability (Tsukame et al., 1997). Though it can help detect main microplastics like polyethylene microbeads, it necessitates materials to serve as references to determine polymer kinds (Majewsky et al., 2016). DSC combined with Thermogravimetry analysis (TGA) identified polypropylene and polyethylene, but could not identify polyamide (PA), polyvinylchloride (PVC), polyester (PES), polyurethane (PU) and polyethylene terephthalate (PET) because phase shift signals are overlapping. In conjunction with solid-phase (SPE) extraction and thermal desorption gas chromatography-mass spectrometry (TDS-GC-MS), TGA offers a larger sample size and a higher resolution compared to DSC. PE was quantitatively detected using this approach after being introduced into mussel and soil samples (Shim et al., 2017).

Another technique for analysing thermally broken-down gas from polymers is pyro-GC-MS. Reference pyrograms of known polymer samples were contrasted with the pyrograms produced from the sample. Small sample particles of plastic were isolated and subjected to GC-MS analysis after being pyrolised at a temperature greater than that of TGA (e.g., 700 °C) (Fabbri et al., 2000). Bulk examination of silt and solid suspended particles revealed the existence of PS, PVC, polyvinyl acetate (PVA), styrene-butadiene rubber (SBR) and poly(acrylonitrile butadiene styrene) (ABS). The segregated possible plastic particles from sediment samples were identified by Pyro-GC-MS analysis as PP, PE, PS, PVC, PA, chlorinated or chlorosulfonated PE and PET (Dekiff et al., 2014).

5 Microplastic in India

India, which is mapped in southern Asia, has a coastline that stretches 7,500 km and is home to union territories and 13 coastline-rich states. The Arabian Sea and the Bay of Bengal encircle it on both the east and west. Coastline of India is embraced by an immense variety of environments, including mangrove woodlands, coral reefs, seagrass beds, sand ridges, mudflats, wetlands, and sandy and rough shores (Daniel et al., 2020a, 2020b). The distribution of MPs in different environmental domains is compiled in Table 1. Open disposal contaminates surface water due to the unrestrained flow of leachate and material. A noticeable effect explicitly influencing the oceans and seas internationally is mostly plastic trash to blame for marine debris. India's plastics export and import trade has expanded radically throughout recent times to satisfy the needs of the dramatically increasing population. This trend of increased accumulation of microplastic in the marine environment has attracted the research community to study the present scenario created by MPs. The rest of the chapter goes on to detail the sorts of MPs identified, their sources and the proportion of MPs gathered from research undertaken in coastal and other areas of India.

5.1 Northern and Central Parts of India

Jammu and Kashmir, Delhi, Uttarakhand, Uttar Pradesh, Punjab, Himachal Pradesh, and Haryana, are just a few of the populous states that make up northern and central India. The Indus, Ganges, and Brahmaputra rivers, which are the three largest freshwater rivers, traverse northern and central India.

A considerable quantity of human pollution, including plastic garbage, damages the holy Ganges River and its several tributaries (Chauhan et al., 2021; Xu et al., 2020). Six different varieties of particulates of plastic, including HDPE, polyethylene terephthalate (PET), low-density polyethylene, polystyrene, polyvinyl chloride, and polypropylene were found to be present among the Ganges River's easternmost deposits (Napper et al., 2021; Singh et al., 2021). Sediment and water sample analysis was done according to the National Oceanic Atmospheric Administration's (NOAA) standard methodology (Lusher et al., 2017). Particles were sorted into five size groups: 4–5 mm, 3–5 mm, 2–3 mm, 1–2 mm, and <1 mm using methods such as SEM and EDX (Napper et al., 2021). Water and sediment samples from five major cities along the Ganga River were examined for microplastics (MPs). The ranges for the sediment and water samples were 17–36 items/kg dw and 10–45 mg/kg dw, respectively, and 380–684 items/1000 m^3 and 143–340 mg/1000 m^3, respectively. Film-shaped MPs and white color were predominant, with 2.5–5 mm MPs contributing more mass and quantity. A commonly used plastic type was discovered to be polyethylene (Singh et al., 2021). The five types of MP particles—fragments, foam, pellets, films, and threads/fibers—that were found in samples totaled over 955. Common MPs were identified as PET, PVC, and nylon polymers. It is possible that

Table 1 Occurrence of MPs in different Indian environments

Location	Ecosystem	Shape	Type of polymer	Concentration	Refs.
Kerala, Thiruvananthapuram District	Soil	Flakes, fragments, filaments, sheets, foams, and fibers	LDPE, HDPE, PET, PP, PS	799 ± 0.09 pieces/kg	Krishna et al. (2024)
Bhopal	Soil	Fragment, filament, foam	PE, PP, PVC, PET, PVAc	752.5–6.36 particles/kg	Singh et al. (2023)
Nagpur	Soil	Fragments	Propylene polyethylene polyamide	8.54×10^4 to 8.51×10^5 particles/kg	Prajapati et al. (2023)
Kerala Mangalavanam	Soil	Fragments	PP, LDPE, HDPE, PVC, PE	933 ± 564 particles/kg	Kannankai et al. (2022)
Kerala (Padinjarekkara, Mahe, Munakkal, Koyilandy, Varkala, Azheekkal, Poovar, Veli)	Sediments	Fragment, fibre, foam	ABS, PA PP,, PS, RY, PET, PU, PVF alkyd, PVC, CE	40.7 items/m^2 PE,	Robin et al. (2020)
Tamil Nadu (Silver Beach)	Sediments	Pellets and fiber	NY, PE, PVC	204 items/kg	Vidyasakar et al. (2020)
Tamil Nadu (Rameswaram Island)	Sediments	Foam, fiber, film, fragment	PE, PP, PET, PA, CP, PU, PEST, PS, PVA, PVC	55–259 items/kg	Jeyasanta et al. (2020a)
Tamil Nadu (Tuticorin)	Sediments	Fibre, film, foam, fragment	PS, PE, NY, PVC, PET, PP	25–83 items/m^2	Jeyasanta et al. (2020b)
Tamil Nadu (Tuticorin)	Water	Fragment, fibre, film	PE, PP	12.14–31.05 items/L	Rajaram et al. (2021)
Tamil Nadu (Kanyakumari)	Sediments	Fibre and fragments	–	6860 items/kg	Sundar et al. (2020)

(continued)

Table 1 (continued)

Location	Ecosystem	Shape	Type of polymer	Concentration	Refs.
Andaman and Nicobar Islands	Water	Pellet, fragment, fibres	Ethylene vinyl alcohol, surlyn ionomer, PPS, PEI, PIP, acrylic, acrylonitrile, NY, PVC, EVA, PU	0.93×10^{-3} items/L	Goswami et al. (2020)
Andaman and Nicobar Islands	Sediments	Fragment, fibre, pellets	Acrylonitrile, surlyn ionomer, acrylic, PEI, PPS, EVA ethylene vinyl alcohol, NY, PU, PIP	45.17 items/kg	Goswami et al. (2020)
Tamil Nadu (Chennai)	Sediment	Pellet	PE, PP	304 items (March) 896 items (November)	Suman et al. (2020)
Andaman and Nicobar Archipelago	Sediments	Film, irregular, pellets, filament	NY, PE, PP, PVC, PS, others	–	Krishnakumar et al. (2020)
South Andaman Islands	Sediment	Fragments and fibre	Spherule poly(dimer acid-coalkyl polyamine), melamine, poly(perfuoroethylene oxide), polysulfde, polybutadiene, PVB, poly(butadiene-acrylonitrile acrylic acid), PVC, epoxy epichlorohydrin, ABS, nylon 6	414.35 items/kg	Patchaiyappan et al. (2020)
Maharashtra (Juhu, Aksa, Girgaon, Dadar)	Sediments	Fragment, fbre, flm, pellet	PE, PP, others	43.6–346.0 items/m^2	Maharana et al. (2020)
Goa (Colva, Calangute, Vagator)	Sediments	Fragment, fbre, flm, pellet	PE, PP, others	17.0–95.6 items/m^2	Maharana et al. (2020)

(continued)

Table 1 (continued)

Location	Ecosystem	Shape	Type of polymer	Concentration	Refs.
Karnataka (Devbagh, Karwar, Kasarkod)	Sediment	Fragment, fiber, film, pellet	PE, PP, others	21.0–155.3 items/m^2	Maharana et al. (2020)
Tamil Nadu (Chennai)	Water	Fiber and fragments	PA	11 items/L	Ganesan et al. (2019)
Pondicherry (Puducherry)	Sediment	Fiber, fragment, pellet, foam flm/sheet	EAA copolymer, PP, PS, HDPE, PVC, LDPE, PVK, CA	720.30 items/kg	Dowarah and Devipriya (2019)
Lakshadweep (Tinnakara)	Sediment	Pellet	–	603 items	Mugilarasan et al. (2017)
Goa (Colva, Calangute, Galgibaga Keri, Mobor and Vagator beaches)	Sediments	Pellet	PE, PP	1345 items (January) 1655 items (June)	Veerasingam et al. (2016)
Maharashtra (Mumbai)	Sediment	Pellet	PE, PP, others	–	Ogata et al. (2009)
Gujarat (Alang-Sosiya ship-yard)	Sediments	Fragment	PU, NY, PEST, PS	81.43 mg/kg	Reddy et al. (2004, 2006)

the fibres came from religious rites in the Ganges because they were less than 5 mm and showed biofilm development. MPs were discovered in the sediments of Anchar Lake in the Kashmir Valley, Northwest Himalaya, Portion of (1.4%) PET, (1.4%) PS, (0.9%) PVC, and (0.7%) PP were the next most common polymers after polyamide (Neelavannan et al., 2022).

5.2 Southern Part of India

The five largest states Kerala, Tamil Nadu, Karnataka, Telangana, and Andhra Pradesh, make up the southern area of India, which is home to roughly 20% of the country's population. There are two states where a substantial number of research studies on MPs are conducted, Tamil Nadu and Kerala. Rivers, lakes, and mangroves were utilised in populated states for a variety of reasons, including irrigation, agriculture, power generation, and recreational activities. Plastic waste dumped carelessly in rivers and lakes are seriously disrupting the local ecology and food chain is contaminated (Kaviarasan et al., 2022).

Research on microplastics (MPs) pollution on sandy beaches in Tamil Nadu, India, has shown that beaches near rivers have a higher MPs abundance than those affected by fishing and tourism. Plastic particles, line/fibres, and foam are the largest types of debris found. Estuaries are the main hub for MPs occurrence, with mangroves also present. Tourism activities are responsible for MPs residues throughout the shoreline (Karthik et al., 2018). Sruthy and Ramasamy's research on Vembanad Lake sediments revealed that the materials collected included films, fragments, fiber/line, and foam, pellets made of PS, HDPE, and LDPE. LDPE was the most common polymer, used in various products like milk cartons, household goods, plastic bags, and films. In addition, it was reported that samples that were taken from eight sandy beaches in the Tuticorin District of Tamil Nadu, including Vellapati, Tharuvaikulam, Inigonaga, Thirespuram, Muthu Nagar, Harbour Beach, Manappad beaches and Thiruchendur. Additionally, the properties and average concentrations of the plastic particles classified as macro (>2.5 cm), meso (5 mm–2.5 cm), and micro (5 mm) were documented (Jeyasanta et al., 2020a, 2020b). Fishermen's nets, synthetic apparel, and single-use plastics including bags, PET bottles and straws were among the recovered MPs from surface sediments and water. The MPs varied in size, colour, and composition (Fibre > fragment > foam > pellet), and they were mostly composed of nylon (39%), PE (23%), PS (19%), PP (15%), and PVC (4%). Using density separation and flotation, MPs were separated according to the National Oceanic and Atmospheric Administration (NOAA) methodology. The dimensions of MPs' were found to be in the region of 0.3–2 mm. Sediment and water samples included microparticles of the most prevalent polymers, including PS, HDPE, PP, LDPE and with the morphology of (37.9%) fibers, (27%) fragments, (24%) films and (11.1%) pellets (Fries et al., 2013). The quantity, traits, origins, and biological hazards of MPs in surface sediments and water samples collected from the Gulf of Mannar's coral reef ecosystem, India, were also assessed (Patterson et al., 2022). The investigation came to the conclusion that polymer fibers 2 mm predominated the MP contamination whereas fragments more than 3 mm predominated in sediments. The most frequent polymers found throughout the investigation were polyethylene (PE) and polypropylene (PP).

MPs were recently found in salt samples taken from several parts of India. In salts of food grade made from Tuticorin (Thoothukudi) borewell water and seawater samples, the presence of MPs was examined for their frequency, physical characteristics, elemental composition, surface morphology and polymer composition (Selvam et al., 2020). There was a broad spectrum of salt concentrations (3515–7240 MPs/kg), with fiber (87%) being the predominant form. Few studies have looked at the quantity and mobility of MPs in groundwater, which has a direct impact on human health because rural communities depend on it for drinking reasons. Along with cloth fibers, nets for fishing, natural or semi-synthetic chemical colours and ropes, common polymers including PVC, PP and polyamide were frequently discovered in groundwater. Untreated direct well water is consumed by a sizable portion of the rural Indian population, subjecting to heavy metal ion and MPs poisoning.

5.3 Western Part of India

Gujarat, Maharashtra and Goa state's maritime habitats, as well as Rajasthan's desert ecology, make up the Western Indian area. A little body of research has been done on the issue of microplastic contamination in the saltpans, rivers, seas, and shellfish in the states of west India (see Fig. 4). Food wrappers were the most abundant plastic item found between January and March 2020, ropes, plastic straws, fishnets, cups, cigarettes and tiny plastic fragments were next in the row. The colours of the objects were arranged in the following order: white > transparent > brown > blue > Grey > red > green > orange > yellow > black (Chinglenthoiba et al., 2023). Gurjar (2022) looked at the MPs' ability to be transported as harmful carriers detecting dissolved impurities in the water next to the coastline of Mumbai. In sediment samples, the average MP abundance was 9630 items/kg dw, compared to 372 ± 143 items/L for water samples. The investigation also highlighted the MPs' ability to bind to tissues in fish taken off the Mumbai coast. Less than 250 mm in size, the MPs found in the samples were Malabar sole fish and Bombay duck. They came in seven different colour varieties: brown, red, blue, black, yellow, green and translucent. Sand samples taken from six Goa beaches at the high tide of the South-West (SW) and North-East (NE) monsoons were used to study the MPs presence along the coastline of Goa. The study analyzed surface modification, morphology, and degradation of microplastics (MPs) in samples of sand taken during the NE and SW monsoons. Plastic deterioration was accelerated by physical friction and UV light, and the ecosystem's nature was revealed via weathering and biofouling (Veerasingam et al., 2016).

5.4 East and Northeastern Part of India

India's east and north-east feature distinctive habitats and vegetation. The major states with big populations include Jharkhand, Odisha, West Bengal, and Bihar. The large Brahmaputra River flows through the eight N–E states, which contain a crucial ecology of the Himalayan mountains. The Hooghly River in West Bengal (Mukhopadhyay & Chakraborty, 2021) and the Brahmaputra River in Arunachal Pradesh were the subjects of the majority of the studies on MPs from the area. A few observations from Odisha and Chennai beaches indicated MPs were present and were linked to human activities (Patchaiyappan et al., 2021a, 2021b). An estimated 227.94 ± 91.37 microplastics were found in every 100 grammes of street dust sample of Chennai coastline. The mean microplastic density in the nine stations sampled alongside the coastal stretch of Odisha is 258.7 ± 90.0 particles/kg of sediment at beach. Out of all the sites, Abhayachandpur had the lowest microplastic density (153.3 ± 27.3 particles/kg), whereas Swargadwara had the highest (378.3 ± 39.7 particles/kg of beach sediment). The Ganges eventually receives the plastic trash after being cut into tiny bits by wind erosion, UV light and tidal currents. Tsering et al. (2021) investigated a variety of sediment samples from the Brahmaputra River in India's Arunachal Pradesh and Assam. Microplastics with a smaller size range (20–150 µm) were found to be more abundant in the Indus and Brahmaputra River, with a total of 525–1752 MP/kg in the Indus River and 531–3485 MP/kg in the Brahmaputra River. The characteristics and geographic MP's prevalence in the Ganges river basin water and sediment on the riverbank were examined, and they were linked to other factors, such as proximity to nature reserves, industrial zones and population density (Sarkar et al., 2020a, 2020b). FT-IR spectroscopy and an optical microscope were used to analyse MP samples from beach sediments and water from five Ganges River locations for polymer amount, shape, size, type and colour. MPs were classified into three sizes: 10–5 mm, 5 mm–850 µm, and 850–60 µm. Most macroparticles were PE and PET, coming from bottles and food packaging made of plastic. Nylon fibers PP, PS and PE from industrial-generated waste were among the microparticles (Saha et al., 2021).

6 Impact of Microplastic on Different Ecosystems

Despite a large number of ecotoxicological studies already completed, the true environmental and ecological risks associated with microplastics remain controversial because the level and attributes of microplastics implied for experimentation in the lab do not match those observed in the field (Zhang et al., 2021). The available scientific data indicates the contact with microplastics can cause an extensive array of toxins, ranging from physical ingestion, energy metabolism disruptions, altered liver physiology, and/or synergistic and antagonistic effects with additional hydrophobic organic pollutants, among other things, going up in trophic levels from lower ones.

Therefore, it is imperative to pursue risk assessment and legislative action in response to microplastic buildup and the detrimental impacts it causes. Research has shown that microplastics obstruct algae's ability to eat (Cole et al., 2013), reduced fertility and higher copepod morbidity (Lee et al., 2013), cause weight loss and poor feed intake in plankton, and boost the bioaccumulation of polychlorinated biphenyls (PCBs) linked with plastic.

6.1 Impact on Aquatic Ecosystem

Unknowingly, various natural and anthropogenic activities are harming the aquatic environment, and MPs are among them. It is also the most prevalent environmental issue that will have an impact on biodiversity preservation (Du et al., 2021). Particle size reduction increases the risk of microplastic injury, but its exact impacts are still unknown (Law & Thompson, 2014). Nevertheless, because of their frequent biofouling and ingestion, microplastics are likely to change their density and buoyancy, making them accessible to different species at different depths in the water column. When aquatic species eat MP, it bioaccumulates and exhibits particle or chemical toxicity as well as combination effects that alter the metabolism, feeding habits, development, and reproduction of the species, with varying levels of toxicological concerns (Sundar et al., 2020). For instance, polystyrene concentration decreased the algal chlorophyll (Hazeem et al., 2020), whereas *Daphnia magna* showed reproductive cycle disruption (Issac & Kandasubramanian, 2021). Organisms residing in their respective biological niche are exposed to the toxic, synthetic polymers which are ingested by them and making them highly vulnerable to health complications. Marine biota can experience oxidative stress, reduction infiltration competence and reduced immunity inflammation in tissues, pseudo-satiation and impaired digestive tract (Dowarah et al., 2020). Plastic waste production and distribution in marine environments are increasing over time, which accelerates the accumulation of plastic debris on the ocean's surface and seabed (Kavitha & Kandasubramanian, 2020; Rastogi & Kandasubramanian, 2020). Due to their buoyant and persistent nature, microplastics are universal contaminant in marine environments, providing a pathway through which contaminants are transferred to aquatic life (Rodrigues et al., 2019). Unambiguous proof of biomagnification is not obtained because many marine animals rapidly absorb them and primarily excrete those (Cózar et al., 2014). Nevertheless, the consequences of MP uptakes include altered behavior, developmental problems, and decreased food intake. According to Martin et al. (2020), around the globe, about 700 creatures of water have suffered as a result of the influx of microplastics, including penguins, sea turtles, and a range of crustaceans. However, because most victims remain undiscovered across the vast oceans, the situation caused by microplastics becomes less serious (Pabortsava & Lampitt, 2020). According to Thushari and Senevirathna (2020), the ongoing, huge creation and distribution of plastics into marine environment exacerbate the already-polluted medium. Impact

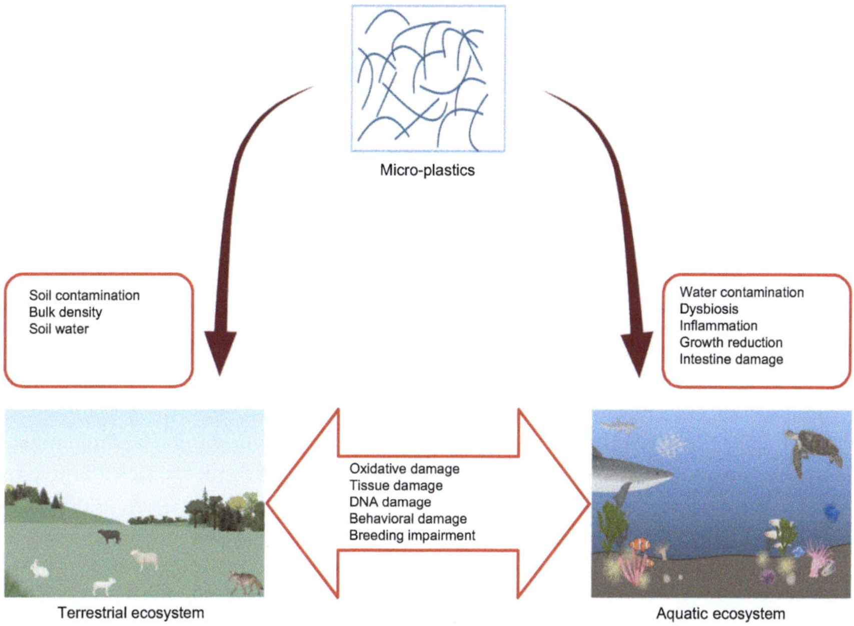

Fig. 2 Impact of MPs on terrestrial and aquatic environment

of MPs on aquatic and terrestrial ecosystem has been shown in Fig. 2. Microplastics' varied impacts and small size make them an ideal home for microorganisms in growth (Yang et al., 2020). The concentration of dangerous organic pollutants, such as DDT, polybrominated diphenyl ethers, and other additives incorporated into water during manufacture, can be increased by microplastics due to their ease of accumulation and release (Campanale et al., 2020; Rajhans et al., 2019; Thakur & Kandasubramanian, 2019).

6.2 Impact on Terrestrial Ecosystem

An important part of terrestrial ecosystems under heavy pollution strain are the soils. The amount of evidence indicating soil contamination by microplastics is growing, and this may have an impact on the function and biodiversity of the soil (de Souza Machado, 2018a). Carbon sequestration, biogeochemical cycling, and biodiversity enhancement are just a few of the many functions that soils facilitate. Comparable researches in aquatic ecosystems have provided a better mechanistic knowledge of the impacts of new stresses from the ecosystem, including microplastics, on modifications in soil function. To understand these consequences, soil biophysical parameters may be crucial (Rillig, 2012). Nonetheless, it is fair to anticipate a nearly everlasting increase in microplastic on land contamination at the existence of human-relevant time frames and the management of pollution (de Souza Machado, 2018b). In general, MPs are less dense than comparable-sized dirt particles. This has the potential to improve soil aeration, which might facilitate root insertion significantly. Conversely, microfibers have the potential to grasp young roots and impede seedling development. Microfibers (MPs) can alter soil properties by reducing or increasing cohesiveness between elements of soil that form conglomerates. 72% of MPs in soil participate in aggregate formation, according to Zhang and Liu (2018).

The impact of MPs on terrestrial creatures is only comprehensible in light of the information that is now available regarding plastic manufacture, use, and disposal. MPs come in a variety of sizes, but those with a diameter of smaller than one millimetre have a higher chance of being consumed by soil-dwelling organisms (Rillig, 2012). The terrestrial ecology is at risk from MP contamination due to their risky characteristics that could result in cancer, endocrine disturbance, and mutation (Prakash et al., 2020). It has been suggested that due to their similar feeding strategies, the exposure route and potential negative effects of MPs on species that live in soil and sediment are likely to be similar (Prakash et al., 2020). Two species of earthworms—*Eisenia andrei* and *Lumbricus terrestris*—have been used in the experiment to expose MPs in a lab setting, and MP ingestion has been documented (Rodriguez-Seijo et al., 2017). One of the key elements influencing the uptake of any particle is the mouth-to-particle size ratio, which should be consumed because smaller particles have a higher likelihood of being ingested by soil organisms. Rodriguez-Seijo et al. (2017) introduced PE particles to *E. Andrei* at various concentrations in order to test potential negative impacts on fatalities, development rate, and reproduction, etc. Polyethylene particles (MPs) have been found to act as a vector for pesticide transfer in agricultural fields and bind with various hydrophobic organic chemicals (HOCs). Different chemicals may accumulate in plants and animals as an aftermath of MPs drainage, such as *E. fetida*, via bioaccumulation of PBDE exposed to *E. fetida*. Studies have shown that MP fragments of human-generated trash result high levels of biogenic absorption rates from the ground surface to soil walls dug underneath, causing MPs to leak as a result of circulation of groundwater, potentially exposing soil organisms (Prakash et al., 2020).

Lozano and Rillig (2020) found that MPs can impact the predominance of plant genera, with rampant species of Europe like *Calamagrostis* and allelopathic species taking over in a soil brimming with microfibers. Nevertheless, despite reduced plant matter, plant species like Holcus may have robust community architecture. Polymers (PFs) also induce aggregation. Polyester fibers (PFs) induce soil particle aggregation through heaping, increasing water-holding capacity and aeration (Zhang et al., 2019). In sandy terrain, nevertheless, they are not involved in the creation of clumps. Water-holding capacity and soil consolidation are unaffected via additional microplastic particles such as polyacrylic fibres, fragments or beads (Lehmann et al., 2019). The adhesion mechanism of soil granules is immediately impacted by the integration of PLA and HDPE, resulting in sturdy soil gravel. This structure is ideal for plant root growth and development as it holds moisture efficiently. However, after being exposed to LDPE, Lolium perenne's root development was shown to decrease, albeit this is yet unknown. More investigation is required to comprehend the impact of microplastic on soil–plant systems (Boots et al., 2019). Soundarya and Sujatha (2023) conducted a pilot-scale study to explore the impacts of MPs (glitters) in the seed germination of *Vigna radiata* (Mung bean). Plant components were observed to be discoloured by the glitter's dye, and the cross-section of the root verified the penetration of glitters. Moreover, in a different study by Maity et al. (2020), the effects of MP exposure were assessed in *Allium cepa* (onion). This study found cytogenotoxic effects in *A. cepa* caused by polystyrene MPs, showing the presence of MPs in our terrestrial ecosystems. Germination is a crucial stage in plant life, starting with water absorption and metabolic responses. Microplastics have been found to decrease seed germination rates in plants like *Lepidium sativum* and polystyrene MPs exposed to herbaceous plants at different concentrations and particle sizes. However, at increased ratios of polystyrene MPs were germination indicators like germination percentage, vigor, and germination index reduced in rice seedlings under stress (Kumari et al., 2022). MPs are causing significant environmental damage, but research on terrestrial environments is limited in India. Scaling up research is crucial to better understand the situation and adopt necessary management actions.

7 Microplastic Waste Management and Regulation in India

India consumes and disposes of plastics massively due to its enormous population. By restricting its usage, presents a chance to significantly minimises the production of plastic garbage. The amount of secondary MPs released into the environment will immediately decrease with a decrease in plastic trash. In India, there are several laws that control the usage of plastics and trash management. The government's 2016 plastic waste management rules include guidelines for plastic recycling, the imminent demise of multilayer non-reusable bags, and instructions for municipal and rural organisations (Patil et al., 2021). The Environment (protection) Act of 1986 (EPA, 1986), The Plastics (manufacture, usage, and waste management) Rules of 2009 (The Plastics Rules 2009), and The Plastic Waste (management and handling) Rules of 2011 (The Plastic Waste Rules 2011) are a few of the laws in India that regulate the management of plastic waste. To stop the growing plastic pollution load, the Hazardous and Other Wastes (management and transboundary movement) Rules, 2016, have barred the importation of polymers trash into India. Tragically, the actual implementation of these regulations is not perfect. In India, waste management and its disposal are largely disregarded; approximately 90% of trash is disposed of outdoors (Banerjee et al., 2014). Effective waste-gathering, sorting, and management facilities are still lacking in India (Patil et al., 2021). Nonetheless, there is still a significant window of opportunity for reducing the plastic problem, which may be achieved by properly enforcing laws and regulations, encouraging garbage separation, raising public awareness, organising a trained workforce, etc. A fantastic alternative is to limit the usage of plastic packaging in typical consumer goods. Innovation is crucial for disposing of plastic waste in India, as large capital investments are difficult to make. Incineration is an easy method but it releases toxic gases. Economizing waste disposal through recycling and reuse is essential. Technological implementations, such as road construction, cement kiln co-processing, RDF oil conversion, plasma pyrolysis, and green recycling, are needed for safer disposal (Patil et al., 2021). India's corporates have implemented initiatives to reduce environmental impact. Reliance Industries converts over 2 billion PET plastic bottles annually to polyester yarn, reducing CO_2 release and landfill space. PET bags and T-shirts reduce water usage, pesticide usage, and carbon footprint by 1400 L, 50%, and 32%, respectively. In 2015, the government mandated the inclusion of plastic waste in road construction, as it offers better water resistance, no potholes, and environmental conservation benefits. Diesel must be recycled in order to get the necessary qualities, but fuel made with the combined pyrolysis of plastic waste technique is equivalent to fossil fuels in characteristics and may be utilised as motor fuel or for transit right away (Thahir et al., 2019). The creation and usage of bioplastics can further lessen the environmental effects, consumption of carbon footprint, pollution load, and petrochemical resources.

By raising public awareness via educational initiatives, plastic dumping, which is otherwise challenging to control, has decreased to some extent. The litter weight/m^2/year decreased by 76% and 8%, respectively, as a result of beach tidy up efforts at Panambur and Thanneerbhavi beaches (Patil et al., 2021). Secondary MP incidence may be decreased by source reduction, cleaning, and remediation techniques (Wu et al., 2017). Therefore, the application of the "design for environment concept," creation of products with extra significance, improved use towards the end of life, adoption of environmentally friendly remedies, and creative options as a result of technical developments can enhance how the ecology is doing and offset MP pollution (Bhattacharya et al., 2018). Limited public awareness and road waste littering remain key challenges in India. Strict regulations and enforcement are needed for proper waste management. Straightforward elimination and handling of microplastics (MPs) from India's environment is scarce, on the other hand, practical approaches to dealing with polluted regions may be developed.

8 Potential Ecological Risk Assessment of Microplastics in India

Many marine and land species, including molluscs (Naidu, 2019), prawns (Daniel et al., 2020a, 2020b), fishes (Kumar et al., 2018), and crabes (Piarulli et al., 2019), ingest MPs, which people could ingest through food chains. As a result, it would provide serious threats to both human health and the biota (Pan et al., 2021). MPs are made up from a range of polymeric materials that are created from a monomeric chain during a polymerization process, which may also involve dangerous additives and unreacted monomers. The sun's heat and sunshine lead to the weathering of MPs as well as the discharge of harmful chemicals. More than half of plastics' chemical components are detrimental, based on a model of hazard rating constructed from the

Globally Harmonised System of classification and labelling of chemicals (Ranjani et al., 2021).

Potential ecological risk index (PERI), Pollution load index (PLI) and Polymer hazard index (PHI) metrics are employed to gauge the environmental danger of microplastics (MPs) in sediments from both land and sea. The total danger of MP pollution in India was classified as Hazard category IV–V based on PHI values. PHI values were used to classify the total danger of pollution by MP in India as Hazard categories IV–V. A noteworthy pattern of MP contamination is shown in MPs PHI values of marine and territorial sediments. Since MPs are present and have high hazard score, such as PA and PS, the Keralan, Tamil Nadu, and Maharashtra coasts, for instance, possess elevated PHI levels (>1000). Although PE, PET, and PP's hazard ratings are somewhat fewer than those of PS and PA, they should nevertheless be taken into account when calculating risk. The chemical toxicity of the MP should also not be disregarded, even if the concentration is modest (Ranjani et al., 2021). According to Pan et al. (2021), PLI is used to assess the level of MP contamination. PLI findings indicate that the sediments along the west coast of india (WCI) are both terrestrial and marine and are substantially polluted with MPs, whereas less pollution is seen in the sediments along India's east coast. Since PLI was determined by dividing the quantity of quantified MPs by the preceding value, it seems that PLI is unaffected by the chemical composition of MPs. Localized human endeavors, such as industrialization rate, population density, financial progress activities, and marine-related pursuits (fishing and shipping), influence the quantity of MPs present in the marine ecosystem (Jang et al., 2020; Pan et al., 2021). The hazard score and the prevalence of MP polymers are used to calculate PERI. For instance, Tuticorin has the greatest PERI since it has a high PA and PS abundance. On the other hand, the lack of PS and PA in other regions only along the Indian coast led to a small (PERI 150) to medium (150–300) ecological risk. Although no discernible relationship was found between MP (and PLI) and PHI abundance, a higher MP abundance might pose an ecological concern. The cumulative utilization of PERI, PLI and PHI in this work offered a preliminary evaluation of the ecological risk brought on the Indian coast's marine and terrestrial sediments contaminated with MP (Ranjani et al., 2021).

9 A Case Study on Patna City, Bihar, India

The Indian subcontinent's atmospheric MPs deposition has never before been examined, as in this study. This analysis tries to deepen our comprehension of the accumulation of MPs in peri-urban and urban settings. Additionally, the connection between climatic variables and the allocation of MPs has been researched (Napper et al., 2023). The possible origins and regional scope of MPs transport to the investigated place have been identified using the backward air trajectory information. Two locations were chosen for the study one in the urban setting a residential area near the Patna airport and the other in the peri-urban setting near IIT Patna (Kumar, 2017) which was monitored during the monsoon period for five months. Since precipitation is a more significant factor in MPs' transit and deposition than dry deposition, wet atmospheric repercussions were the focus of our investigation into MPs. Parashar and Hait (2022) collected triple copies of samples on the roofing of both locations at an altitude of around 18 m above ground. Samples were collected for a total of 63 days, resulting in a collection of 378 samples from two sites, including replicates. The study found that the average count of microplastics (MPs) in peri-urban and urban areas was 1320.4 ± 126 MPs/m^2/day and 1959.6 ± 205 MPs, respectively. The most common forms were spheres and microbeads, followed by fibers and pieces. The most common polymers were PET and PP, making up around 26% of urban and 27% of peri-urban regions. The distribution of MPs was correlated with wind velocity and rainfall intensity. MPs could originate from nearby and far-off sources, such as building sites, automobile emissions, and improper waste disposal. Smaller MPs from land-based sources could move across vast distances. The distribution of MPs was also correlated with their source and climatic factors.

10 Conclusion

India, one of the world's top manufacturers of plastic trash, is steadily stepping up its microplastics study. The extent of India's contribution to MP pollution worldwide is yet insufficient. This study aims to comprehend the information gaps about MP contamination in several environmental compartments of India. The south-east coast, where the state of Tamil Nadu has conducted the most research, has been the primary focus of a significant majority of MP studies in this dataset. Additionally, there have been more studies done on sediment contamination of marine systems with MP. In comparison to the lakes and groundwater counterparts, the river segment of freshwater systems has the most studies. However, the scientific community has to focus on gaining a thorough knowledge of the origins, paths, and destiny of MPs inside these environments. MPs have expanded to include human consumable items, such as tap water, drinking water, seafood and salt. In India, salt, drinking water, and seafood have been found to be affected. Worldwide MPs are detected in various things of food and drink. It's uncertain how common MPs are in Indian drink and food items, and their toxicity has not been well investigated. It is crucial to investigate MPs' ability to interact with other pollutants and their health impacts on biota interacting upon inhalation or consumption. The rising manufacturing and consumer appetite for plastics will cause an upsurge in MP contamination, potentially causing environmental damage. To address this, research should be scaled up at both field and laboratory scales, focusing on various environmental matrices, including toxicities, marine, terrestrial, freshwater, atmosphere, and human consumables. Creating appropriate methods for MP extraction from these matrices should be the focus of innovation.

11 Recommendations

To effectively mitigate the escalating issue of microplastic (MP) pollution, a multidimensional strategy involving researchers, policymakers, and stakeholders is imperative. Long-term, systematic monitoring programs should be implemented across major aquatic systems, including rivers, lakes, estuaries, coral reefs, and groundwater reserves in under-explored regions, employing standardized methodologies such as NOAA protocols to ensure data uniformity and cross-regional comparability. Comprehensive source attribution studies are critical, focusing on identifying and quantifying contributions from industrial effluents, urban runoff, discarded fishing gear, tourism-driven waste, and culturally linked plastic usage. Public outreach initiatives must prioritize raising awareness within vulnerable populations, particularly in rural and riverine areas, emphasizing the ecological and health ramifications of MP contamination while advocating for sustainable alternatives to single-use plastics and responsible waste disposal practices. Policymakers are urged to enforce stringent regulatory frameworks targeting the reduction of single-use plastics, incentivize the adoption of biodegradable and sustainable materials, and mandate the integration of advanced microplastic filtration technologies in wastewater treatment infrastructure. Strategic investments are required to establish robust waste collection and recycling systems in proximity to aquatic ecosystems, alongside providing incentives for the adoption of environmentally conscious practices in coastal and fishing communities. Concurrently, research should intensify its focus on elucidating the long-term ecological impacts of MPs on sensitive biomes, including coral reefs, mangrove ecosystems, and Himalayan reserves, as well as assessing potential health risks posed by MP contamination in groundwater and salt. Cross-sectoral collaboration among local governments, non-governmental organizations (NGOs), industries, and international stakeholders is indispensable, particularly for addressing transboundary MP pollution in shared hydrological systems like the Ganges and Brahmaputra basins. Designating ecologically vulnerable areas as "MP-Free Zones" and incorporating microplastic mitigation strategies into conservation frameworks will further facilitate ecosystem resilience and sustainable development.

References

Akdogan, Z., & Guven, B. (2019). Microplastics in the environment: A critical review of currentunderstanding and identification of future research needs. *Environmental Pollution, 5*, 113011.

Allen, V., Kalivas, J. H., & Rodriguez, R. G. (1999). Post-consumer plastic identification using Raman spectroscopy. *Applied Spectroscopy, 53*(6), 672–681.

Banerjee, T., Srivastava, R. K., & Hung, Y. T. (2014). Chapter 17: Plastics waste management in India: An integrated solid waste management approach. In *Handbook of environment and waste management: Land and groundwater pollution control* (pp. 1029–1060).

Bhattacharya, R. R. N., Chandrasekhar, K., Roy, P., & Khan, A. (2018). Challenges and opportunities: Plastic waste management in India.

Boots, B., Russell, C. W., & Green, D. S. (2019). Effects of microplastics in soil ecosystems: Above and below ground. *Environmental Science & Technology, 53*(19), 11496–11506.

Campanale, C., Massarelli, C., Savino, I., Locaputo, V., & Uricchio, V. F. (2020). A detailed review study on potential effects of microplastics and additives of concern on human health. *International Journal of Environmental Research and Public Health, 17*(4), 1212.

Carpenter, E. J., Anderson, S. J., Harvey, G. R., Miklas, H. P., & Peck, B. B. (1972). Polystyrene spherules in coastal waters. *Science, 175*, 749–750.

Chauhan, J. S., Semwal, D., Nainwal, M., Badola, N., & Thapliyal, P. (2021). Investigation of microplastic pollution in river Alaknanda stretch of Uttarakhand. *Environment, Development and Sustainability, 23*, 16819.

Chinglenthoiba, C., Pukhrambam, B., Chanu, K. T., Devi, K. S., Meitei, N. J., Devika, Y., & Valiyaveettil, S. (2023). A review on microplastic pollution research in India. *Regional Studies in Marine Science, 58*, 102777.

Cole, M., Lindeque, P., Fileman, E., Halsband, C., Goodhead, R., Moger, J., & Galloway, T. S. (2013). Microplastic ingestion by zooplankton. *Environmental Science & Technology, 47*(12), 6646–6655.

Cole, M., Lindeque, P., Halsband, C., & Galloway, T. S. (2011). Microplastics as contaminants in the marine environment: A review. *Marine Pollution Bulletin, 62*(12), 2588–2597.

Cooper, D. A., & Corcoran, P. L. (2010). Effects of mechanical and chemical processes on the degradation of plastic beach debris on the island of Kauai, Hawaii. *Marine Pollution Bulletin, 60*(5), 650–654.

Cózar, A., Echevarría, F., González-Gordillo, J. I., Irigoien, X., Úbeda, B., Hernández-León, S., Palma, Á. T., Navarro, S., García-de-Lomas, J., Ruiz, A., Fernández-de-Puelles, M. L., & Duarte, C. M. (2014). Plastic debris in the open ocean. *Proceedings of the National Academy of Sciences, 111*(28), 10239–10244.

Crawford, C. B., & Quinn, B. (2016). *Microplastic pollutants*. Elsevier.

Daniel, D. B., Ashraf, P. M., & Thomas, S. N. (2020a). Abundance, characteristics and seasonal variation of microplastics in Indian white shrimps (Fenneropenaeus indicus) from coastal waters off Cochin, Kerala, India. *Science of the Total Environment, 737*, 139839.

Daniel, D. B., Thomas, S. N., & Thomson, K. T. (2020b). Assessment of fishing-related plastic debris along the beaches in Kerala Coast, India. *Marine Pollution Bulletin, 150*, 110696.

de Souza Machado, A. A., Lau, C. W., Till, J., Kloas, W., Lehmann, A., Becker, R., & Rillig, M. C. (2018a). Impacts of microplastics on the soil biophysical environment. *Environmental Science & Technology, 52*(17), 9656–9665.

de Souza Machado, A. A., Kloas, W., Zarfl, C., Hempel, S., & Rillig, M. C. (2018b). Microplastics as an emerging threat to terrestrial ecosystems. *Global Change Biology, 24*(4), 1405−1416.

De Witte, B., Devriese, L., Bekaert, K., Hoffman, S., Vandermeersch, G., Cooreman, K., & Robbens, J. (2014). Quality assessment of the blue mussel (Mytilus edulis): Comparison between commercial and wild types. *Marine Pollution Bulletin, 85*(1), 146–155.

Dekiff, J. H., Remy, D., Klasmeier, J., & Fries, E. (2014). Occurrence and spatial distribution of microplastics in sediments from Norderney. *Environmental Pollution, 186*, 248–256.

Dowarah, K., & Devipriya, S. P. (2019). Microplastic prevalence in the beaches of Puducherry, India and its correlation with fishing and tourism/recreational activities. *Marine Pollution Bulletin, 148*, 123–133.

Dowarah, K., Patchaiyappan, A., Thirunavukkarasu, C., Jayakumar, S., & Devipriya, S. P. (2020). Quantifcation of microplastics using Nile red in two bivalve species Perna viridis and Meretrix meretrix from three estuaries in Pondicherry, India and microplastic uptake by local communities through bivalve diet. *Marine Pollution Bulletin, 153*, 110982.

Du, S., Zhu, R., Cai, Y., Xu, N., Yap, P. S., Zhang, Y., He, Y., & Zhang, Y. (2021). Environmental fate and impacts of microplastics in aquatic ecosystems: A review. *RSC Advances, 11*(26), 15762–15784.

Fabbri, D., Tartari, D., & Trombini, C. (2000). Analysis of poly(vinyl chloride) and other polymers in sediments and suspended matter of a coastal lagoon by pyrolysis-gas chromatography-mass spectrometry. *Analytica Chimica Acta, 413*(1–2), 3–11.

Frank Sommer, V. D., Baum, A., Sauer, J., Gilge, J., Maschowski, C., & Giere, R. (2018). Tire abrasion as a major source of microplastics in the environment. *Aerosol and Air Quality Research, 18*, 2014–2028.

Fries, E., Dekiff, J. H., Willmeyer, J., Nuelle, M. T., Ebert, M., & Remy, D. (2013). Identification of polymer types and additives in marine microplastic particles using pyrolysis-GC/MS and scanning electron microscopy. *Environmental Science: Processes & Impacts, 15*(10), 1949–1956.

Ganesan, M., Nallathambi, G., & Srinivasalu, S. (2019). Fate and transportof microplastics from water sources. *Current Science, 117*, 1874–1879.

Gopinath, K., Seshachalam, S., Neelavannan, K., Anburaj, V., Rachel, M., Ravi, S., Bharath, M., & Achyuthan, H. (2020). Quantification of microplasticin Red Hills lake of Chennai city, Tamil Nadu, India. *Environmental Science and Pollution Research, 27*, 33297–33306.

Goswami, P., Vinithkumar, N. V., & Dharani, G. (2020). First evidence of microplastics bioaccumulation by marine organisms in the Port Blair Bay Andaman Islands. *Marine Pollution Bulletin, 155*, 111163.

Gurjar, U. R., Xavier, K. M., Shukla, S. P., Jaiswar, A. K., Deshmukhe, G., & Nayak, B. B. (2022). Microplastic pollution in coastal ecosystem off Mumbai coast, India. *Chemosphere, 288*, 132484.

Hanvey, J. S., Lewis, P. J., Lavers, J. L., Crosbie, N. D., Pozo, K., & Clarke, B. O. (2017). A review of analytical techniques for quantifying microplastics in sediments. *Analytical Methods, 9*(9), 1369–1383.

Hazeem, L. J., Yesilay, G., Bououdina, M., Perna, S., Cetin, D., Suludere, Z., Barras, A., & Boukherroub, R. (2020). Investigation of the toxic effects of different polystyrene micro-and nanoplastics on microalgae Chlorella vulgaris by analysis of cell viability, pigment content, oxidative stress and ultrastructural changes. *Marine Pollution Bulletin, 156*, 111278.

Heo, N. W., Hong, S. H., Han, G. M., Hong, S., Lee, J., Song, Y. K., Jang, M., & Shim, W. J. (2013). Distribution of small plastic debris in cross-section and high strandline on Heungnam beach, South Korea. *Ocean Science Journal, 48*, 225–233.

Hidalgo-Ruz, V., & Thiel, M. (2013). Distribution and abundance of small plastic debris on beaches in the SE Pacific (Chile): A study supported by a citizen science project. *Marine Environmental Research, 87*, 12–18.

Issac, M. N., & Kandasubramanian, B. (2021). Effect of microplastics in water and aquatic systems. *Environmental Science and Pollution Research, 28*, 19544–19562.

Jang, M., Shim, W. J., Cho, Y., Han, G. M., Song, Y. K., & Hong, S. H. (2020). A close relationship between microplastic contamination and coastal area use pattern. *Water Research, 171*, 115400.

Jeyasanta, K. I., Patterson, J., Grimsditch, G., & Edward, J. P. (2020a). Occurrence and characteristics of microplastics in the coral reef, sea grass and near shore habitats of Rameswaram Island, India. *Marine Pollution Bulletin, 160*, 111674.

Jeyasanta, K. I., Sathish, N., Patterson, J., & Edward, J. P. (2020b). Macro-, meso- and microplastic debris in the beaches of Tuticorin district, Southeast coast of India. *Marine Pollution Bulletin, 154*, 111055.

Kang, J. H., Kwon, O. Y., Lee, K. W., Song, Y. K., & Shim, W. J. (2015). Marine neustonic microplastics around the southeastern coast of Korea. *Marine Pollution Bulletin, 96*(1–2), 304–312.

Kannankai, M. P., Alex, R. K., Muralidharan, V. V., Nazeerkhan, N. P., Radhakrishnan, A., & Devipriya, S. P. (2022). Urban mangrove ecosystems are under severe threat from microplastic pollution: A case study from Mangalavanam, Kerala, India. *Environmental Science and Pollution Research, 29*(53), 80568.

Karthik, R., Robin, R. S., Purvaja, R., Ganguly, D., Anandavelu, I., Raghuraman, R., Hariharan, G., Ramakrishna, A., & Ramesh, R. (2018). Microplastics along the beaches of southeast coast of India. *Science of the Total Environment, 645*, 1388–1399.

Kaviarasan, T., Dhineka, K., Sambandam, M., Sivadas, S. K., Sivyer, D., Hoehn, D., Pradhan, U., Mishra, P., & Murthy, M. R. (2022). Impact of multiple beach activities on litter and microplastic composition, distribution, and characterization along the southeast coast of India. *Ocean & Coastal Management, 223*, 106177.

Kavitha, V. U., & Kandasubramanian, B. (2020). Tannins for wastewater treatment.

Krishna, A. B., Madhu, M., & Jayadev, A. (2024). Investigation of microplastics and microplastic communities in selected river and lake basin soils of Thiruvananthapuram District, Kerala, India. *Environmental Monitoring and Assessment, 196*(1), 66.

Krishnakumar, S., Anbalagan, S., Kasilingam, K., Smrithi, P., Anbazhagi, S., & Srinivasalu, S. (2020). Assessment of plastic debris in remote islands of the Andaman and Nicobar Archipelago, India. *Marine Pollution Bulletin, 151*, 110841.

Kumar, R., Singh, R., & Sharma, P. (2021, December). Evidence of urbanization on microplastic pollution in the sub-surface ecosystem of River Ganga near Patna, Bihar (India). In *AGU fall meeting abstracts* (Vol. 2021, p. H11G-03).

Kumar, D. (2017). River Ganges—Historical, cultural and socioeconomic attributes. *Aquatic Ecosystem Health & Management, 20*(1–2), 8–20.

Kumar, V. E., Ravikumar, G., & Jeyasanta, K. I. (2018). Occurrence of microplastics in fishes from two landing sites in Tuticorin, South east coast of India. *Marine Pollution Bulletin, 135*, 889–894.

Kumari, A., Rajput, V. D., Mandzhieva, S. S., Rajput, S., Minkina, T., Kaur, R., Sushkova, S., Kumari, P., Ranjan, A., Kalinitchenko, V. P., & Glinushkin, A. P. (2022). Microplastic pollution: An emerging threat to terrestrial plants and insights into its remediation strategies. *Plants, 11*(3), 340.

Law, K. L., Thompson, R. C. (2014). Microplastics in the seas. *Science (80-), 345*, 144–145.

Lee, K. W., Shim, W. J., Kwon, O. Y., & Kang, J. H. (2013). Size-dependent effects of micro polystyrene particles in the marine copepod Tigriopus japonicus. *Environmental Science & Technology, 47*(19), 11278–11283.

Lehmann, A., Fitschen, K., & Rillig, M. C. (2019). Abiotic and biotic factors influencing the effect of microplastic on soil aggregation. *Soil Systems, 3*(1), 21.

Liu, K., Wang, X. H., Wei, N., Song, Z. Y., & Li, D. J. (2019). Accurate quantification and transport estimation of suspended atmospheric microplastics in megacities: Implications for human health. *Environment International, 132*, 10.

Lozano, Y. M., & Rillig, M. C. (2020). Effects of microplastic fibers and drought on plant communities. *Environmental Science & Technology, 54*(10), 6166–6173.

Lusher, A. L., Mchugh, M., & Thompson, R. C. (2013). Occurrence of microplastics in the gastrointestinal tract of pelagic and demersal fish from the English channel. *Marine Pollution Bulletin, 67*(1–2), 94–99.

Lusher, A. L., Welden, N. A., Sobral, P., & Cole, M. (2017). Sampling, isolating and identifying microplastics ingested by fish and invertebrates. *Analytical Methods, 9*, 1346.

Maharana, D., Saha, M., Dar, J. Y., Rathore, C., Sreepada, R. A., Xu, X.-R., Koongolla, J. B., & Li, H.-X. (2020). Assessment of plastics along the west coast of India: Abundance, distribution, polymer type and toxicity. *Chemosphere, 246*, 125708.

Maity, S., Chatterjee, A., Guchhait, R., De, S., & Pramanick, K. (2020). Cytogenotoxicpotential of a hazardous material, polystyrene microparticleson *Allium cepa* L. *Journal of Hazardous Materials, 385*, 121560.

Majewsky, M., Bitter, H., Eiche, E., & Horn, H. (2016). Determination of microplastic polyethylene (PE) and polypropylene (PP) in environmental samples using thermal analysis (TGA-DSC). *Science of the Total Environment, 568*, 507–511.

Manikanda Bharath, K., Srinivasalu, S., Natesan, U., Ayyamperumal, R., Kalam, N., Anbalagan, S., Sujatha, K., & Alagarasan, C. (2020). Microplasticsas an emerging threat to the freshwater ecosystems of Veeranam lake in South India: A multidimensional approach. *Chemosphere, 264*, 128502.

Manzoor, S., Naqash, N., Rashid, G., & Singh, R. (2022). Plastic material degradation and formation of microplastic in the environment: A review. *Materials Today: Proceedings, 56*, 3254–3260.

Martin, C., Baalkhuyur, F., Valluzzi, L., Saderne, V., Cusack, M., Almahasheer, H., Krishnakumar, P. K., Rabaoui, L., Qurban, M. A., Arias-Ortiz, A., Masqué, P., & Duarte, C. M. (2020). Exponential increase of plastic burial in mangrove sediments as a major plastic sink. *Science Advances, 6*(44), eaaz5593.

McDermid, K. J., & McMullen, T. L. (2004). Quantitative analysis of small-plastic debris on beaches in the Hawaiian archipelago. *Marine Pollution Bulletin, 48*(7–8), 790–794.

Morét-Ferguson, S., Law, K. L., Proskurowski, G., Murphy, E. K., Peacock, E. E., & Reddy, C. M. (2010). The size, mass, and composition of plastic debris in the western North Atlantic Ocean. *Marine Pollution Bulletin, 60*(10), 1873–1878.

Mugilarasan, M., Venkatachalapathy, R., Sharmila, N., & Gurumoorthi, K. (2017). Occurrence of microplastic resin pellets from Chennai and Tinnakkara Island: Towards the establishment of background level for plastic pollution.

Mukhopadhyay, M., & Chakraborty, P. (2021). Plasticizers and bisphenol A: Emerging organic pollutants along the lower stretch of River Ganga, north-east coast of the Bay of Bengal. *Environmental Pollution, 276*, 116697.

Naidu, S. A. (2019). Preliminary study and first evidence of presence of microplastics and colorants in green mussel, Perna viridis (Linnaeus, 1758), from southeast coast of India. *Marine Pollution Bulletin, 140*, 416–422.

Napper, I. E., Baroth, A., Barrett, A. C., Bhola, S., Chowdhury, G. W., DaviesBFR, D. E. M., Kumar, S., Nelms, S. E., Niloy, M. N. H., Nishat, B., Maddalene, T., Thompson, R. C., & Koldewey, H. (2021). The abundanceand characteristics of microplastics in surface water in the transboundary Ganges River. *Environmental Pollution, 274*, 116348.

Napper, I. E., Parker-Jurd, F. N. F., Wright, S. L., & Thompson, R. C. (2023). Examining the release of synthetic microfibres to the environment via two major pathways: Atmospheric deposition and treated wastewater effluent. *Science of the Total Environment, 857*, 159317.

Neelavannan, K., Sen, I. S., Lone, A. M., & Gopinath, K. (2022). Microplastics in the high-altitude himalayas: Assessment of microplastic contamination in freshwater lake sediments, Northwest Himalaya (India). *Chemosphere, 290*, 133354.

Nel, H. A., & Froneman, P. W. (2015). A quantitative analysis of microplastic pollution along the south-eastern coastline of South Africa. *Marine Pollution Bulletin, 101*(1), 274–279.

Ng, K. L., & Obbard, J. P. (2006). Prevalence of microplastics in Singapore's coastal marine environment. *Marine Pollution Bulletin, 52*(7), 761–767.

Ogata, Y., Takada, H., Mizukawa, K., Hirai, H., Iwasa, S., Endo, S., Mato, Y., Saha, M., Okuda, K., Nakashima, A., Murakami, M., & Thompson, R. C. (2009). International Pellet Watch: Global monitoring of persistent organic pollutants (POPs) in coastal waters. 1. Initial phase data on PCBs, DDTs, and HCHs. *Marine Pollution Bulletin, 58*(10), 1437–1446.

Pabortsava, K., & Lampitt, R. S. (2020). High concentrations of plastic hidden beneath the surface of the Atlantic Ocean. *Nature Communications, 11*(1), 4073.

Pal, P., Pandey, J. P., & Sen, G. (2018). In: V. K. Thakur (Ed.), *Biopolymer grafting* (pp. 153–203). Elsevier.

Pan, Z., Liu, Q., Jiang, R., Li, W., Sun, X., Lin, H., Jiang, S., & Huang, H. (2021). Microplastic pollution and ecological risk assessment in an estuarine environment: The Dongshan Bay of China. *Chemosphere, 262*, 127876.

Parashar, N., & Hait, S. (2022). Occurrence and removal of microplastics in a hybrid growth sewage treatment plant from Bihar, India: A preliminary study. *Journal of Cleaner Production, 376*, 134295.

Patchaiyappan, A., Ahmed, S. Z., Dowarah, K., Jayakumar, S., & Devipriya, S. P. (2020). Occurrence, distribution and composition of microplastics in the sediments of South Andaman beaches. *Marine Pollution Bulletin, 156*, 111227.

Patchaiyappan, A., Dowarah, K., Ahmed, S. Z., Prabakaran, M., Jayakumar, S., Thirunavukkarasu, C., & Devipriya, S. P. (2021a). Prevalence and characteristics of microplastics present in the street dust collected from Chennai metropolitan city, India. *Chemosphere, 269*, 128757.

Patchaiyappan, A., ZakiAhmed, S., Dowarah, K., Khadanga, S. S., Singh, T., Jayakumar, S., Thirunavukkarasu, C., & Devipriya, S. P. (2021b). Prevalence of microplastics in the sediments of Odisha beaches, southeastern coast of India. *Marine Pollution Bulletin, 167*, 112265.

Patil, S., Bafana, A., Naoghare, P. K., Krishnamurthi, K., & Sivanesan, S. (2021). Environmental prevalence, fate, impacts, and mitigation of microplastics—A critical review on present understanding and future research scope. *Environmental Science and Pollution Research, 28*(5), 4951–4974.

Patterson, J., Jeyasanta, K. I., Laju, R. L., Booth, A. M., Sathish, N., & Edward, J. P. (2022). Microplastic in the coral reef environments of the Gulf of Mannar, India—Characteristics, distributions, sources and ecological risks. *Environmental Pollution, 298*, 118848.

Piarulli, S., Scapinello, S., Comandini, P., Magnusson, K., Granberg, M., Wong, J. X., Sciutto, G., Prati, S., Mazzeo, R., Booth, A. M., & Airoldi, L. (2019). Microplastic in wild populations of the omnivorous crab Carcinus aestuarii: A review and a regional-scale test of extraction methods, including microfibres. *Environmental Pollution, 251*, 117–127.

Prajapati, A., Jadhao, P. S., Khare, A., Vaidya, A. N., Kumar, A. R., & Das, S. (2023). Characterization of microplastics and associated heavy metals in urban soils affected by anthropogenic littering: Distribution, spatial variation, and influence of soil properties. *Soil and Sediment Contamination: An International Journal*, 1–29.

Prakash, V., Dwivedi, S., Gautam, K., Seth, M., & Anbumani, S. (2020). Occurrence and ecotoxicological effects of microplastics on aquatic and terrestrial ecosystems. In *Microplastics in terrestrial environments: Emerging contaminants and major challenges* (pp. 223–243).

Rajaram, R., Ganeshkumar, A., Vinothkumar, S., & Arun, G. (2021). Ecological risk assessment of toxic metals contamination in Tuticorin coast of Gulf of Mannar, Southern India. *Chemistry and Ecology, 37*(2), 132–148.

Rajhans, A., Gore, P. M., Siddique, S. K., & Kandasubramanian, B. (2019). Ionimprinted nanofibers of PVDF/1-butyl-3-methylimidazolium tetrafluoroborate for dynamic recovery of europium (III) ions from mimicked effluent. *Journal of Environmental Chemical Engineering, 7*, 103068.

Ranjani, M., Veerasingam, S., Venkatachalapathy, R., Mugilarasan, M., Bagaev, A., Mukhanov, V., & Vethamony, P. J. M. P. B. (2021). Assessment of potential ecological risk of microplastics in the coastal sediments of India: A meta-analysis. *Marine Pollution Bulletin, 163*, 111969.

Rastogi, S., & Kandasubramanian, B. (2020). Application of electrospun materials in water treatment. In: Inamuddin, R. Boddula, M. I. Ahamed, & A. M. Asiri (Eds.), *Electrospun materials and their allied applications* (1st ed., pp. 151–183). Wiley.

Reddy, M. S., Basha, S., Adimurthy, S., & Ramachandraiah, G. (2006). Descriptionof the small plastics fragments in marine sediments along the Alang-Sosiya ship-breaking yard, India. *Estuarine, Coastal and Shelf Science, 68*, 656–660.

Reddy, M. S., Basha, S., Kumar, V. S., Joshi, H. V., & Ramachandraiah, G. (2004). Distribution, enrichment and accumulation of heavy metals in coastal sediments of Alang-Sosiya ship scrapping yard, India. *Marine Pollution Bulletin, 48*(11–12), 1055–1059.

Rillig, M. C. (2012). Microplastic in terrestrial ecosystems and the soil? *Environmental Science and Technology, 46*, 6453–6454.

Rillig, M. C., Ziersch, L., & Hempel, S. (2017). Microplastic transport in soil by earthworms. *Science and Reports, 7*, 1362.

Robin, R. S., Karthik, R., Purvaja, R., Ganguly, D., Anandavelu, I., Mugilarasan, M., & Ramesh, R. (2020). Holistic assessment of microplastics in various coastal environmental matrices, southwest coast of India. *Science of the Total Environment, 703*, 134947.

Rodrigues, J. P., Duarte, A. C., Santos-Echeandía, J., & Rocha-Santos, T. (2019). Significance of interactions between microplastics and POPs in the marine environment: A critical overview. *TrAC, Trends in Analytical Chemistry, 111*, 252–260.

Rodriguez-Seijo, A., Lourenço, J., Rocha-Santos, T. A., Da Costa, J., Duarte, A. C., Vala, H., Pereira, R. (2017). Histopathological and molecular effects of microplastics in Eisenia andrei Bouché. *Environmental Pollution, 220*, 495–503.
Rodríguez-Seijo, A., da Costa, J. P., Rocha-Santos, T., Duarte, A. C., & Pereira, R. (2018). Oxidative stress, energy metabolism and molecular responses of earthworms (Eisenia fetida) exposed to low-density polyethylene microplastics. *Environmental Science and Pollution Research, 25*, 33599–33610.
Saha, M., Naik, A., Desai, A., Nanajkar, M., Rathore, C., Kumar, M., & Gupta, P. (2021). Microplastics in seafood as an emerging threat to marine environment: A case study in Goa, west coast of India. *Chemosphere, 270*, 129359.
Sarkar, D. J., Sarkar, S. D., Manna, R. K., Samanta, S., & Das, B. K. (2020a). Microplastics pollution: An emerging threat to freshwater aquatic ecosystem of India. *Journal of Inland Fisheries Society of India, 52*(1), 05–15.
Sarkar, D., Das Sarkar, S., Manna, R., & Das, B. (2020b). Microplastic pollution: Threat to Indian river system. *Everyman's Science, 54*(6), 378–384.
Selvam, S., Manisha, A., Venkatramanan, S., Chung, S. Y., & Paramasivam, C. R. (2020). Microplastic presence in commercial marine sea salts: A baseline study along Tuticorin Coastal salt pan stations, Gulf of Mannar, South India. *Marine Pollution Bulletin, 150*, 110675.
Shim, W. J., Hong, S. H., & Eo, S. E. (2017). Identification methods in microplastic analysis: A review. *Analytical Methods, 9*, 1384–1391.
Singh, N., Mondal, A., Bagri, A., Tiwari, E., Khandelwal, N., Monikh, F. A., & Darbha, G. K. (2021). Characteristics and spatial distribution of microplastics in the lower Ganga River water and sediment. *Marine Pollution Bulletin, 163*, 111960.
Singh, S., Chakma, S., Alawa, B., Kalyanasundaram, M., & Diwan, V. (2023). Identification, characterization, and implications of microplastics in soil—A case study of Bhopal, central India. *Journal of Hazardous Materials Advances, 9*, 100225.
Song, Y. K., Hong, S. H., Jang, M., Han, G. M., Rani, M., Lee, J., & Shim, W. J. (2015). A comparison of microscopic and spectroscopic identification methods for analysis of microplastics in environmental samples. *Marine Pollution Bulletin, 93*, 202–209.
Song, Y. K., Hong, S. H., Jang, M., Kang, J. H., Kwon, O. Y., Han, G. M., & Shim, W. J. (2014). Large accumulation of micro-sized synthetic polymer particles in the sea surface microlayer. *Environmental Science & Technology, 48*(16), 9014–9021.
Soundarya, M., & Sujatha, K. (2023). A pilot-scale study on microplastics on the growth of Vigna radiata (mung bean). *Biochemical & Cellular Archives, 23*(1).
Sruthy, S., & Ramasamy, E. V. (2017). Microplastic pollution in Vembanad Lake, Kerala, India: The frst report of microplastics in lake and estuarine sediments in India. *Environmental Pollution, 222*, 315–322.
Suman, T. Y., Li, W. G., Alif, S., Faris, V. R. P., Amarnath, D. J., Ma, J. G., & Pei, D. S. (2020). Characterization of petroleum-based plastics and their absorbed trace metals from the sediments of the Marina Beach in Chennai, India. *Environmental Sciences Europe, 32*, 1–10.
Sundar, S., Chokkalingam, L., Roy, P. D., & Usha, T. (2020). Estimation of microplastics in sediments at the southernmost coast of India (Kanyakumari). *Environmental Science and Pollution Research, 28*, 18495–18500.
Tagg, A. S., Sapp, M., Harrison, J. P., & Ojeda, J. J. (2015). Identification and quantification of microplastics in wastewater using focal plane array-based reflectance micro-FT-IR Imaging. *Analytical Chemistry, 87*, 6032–6040.
Thahir, R., Altway, A., & Juliastuti, S. R. (2019). Production of liquid fuel from plastic waste using integrated pyrolysis method with refinery distillation bubble cap plate column. *Energy Reports, 5*, 70–77.
Thakur, K., & Kandasubramanian, B. (2019). Graphene and graphene oxidebased composites for removal of organic pollutants: A review. *Journal of Chemical and Engineering Data, 64*, 833–867.

The Environment (Protection) Act. (1986). Govt. of India. https://www.indiacode.nic.in/bitstream/123456789/6196/1/the_environment_protection_act%2C1986.pdf. Assessed on 24 September 2024.

Thushari, G. G. N., & Senevirathna, J. D. M. (2020). Plastic pollution in the marine environment. *Heliyon, 6*(8).

Tiwari, M., Rathod, T. D., Ajmal, P. Y., Bhangare, R. C., & Sahu, S. K. (2019). Distribution and characterization of microplastics in beach sand from three different Indian coastal environments. *Marine Pollution Bulletin, 140*, 262–273.

Tsering, T., Sillanpää, M., Sillanpää, M., Viitala, M., & Reinikainen, S. P. (2021). Microplastics pollution in the Brahmaputra River and the Indus River of the Indian Himalaya. *Science of the Total Environment, 789*, 147968.

Tsukame, T., Ehara, Y., Shimizu, Y., Kutsuzawa, M., Saitoh, H., & Shibasaki, Y. (1997). Characterization of microstructure of polyethylenes by differential scanning calorimetry. *Thermochimica Acta, 299*(1–2), 27–32.

Ugolini, A., Ungherese, G., Ciofini, M., Lapucci, A., & Camaiti, M. (2013). Microplastic debris in sandhoppers. *Estuarine, Coastal and Shelf Science, 129*, 19–22.

Vaid, M., Mehra, K., & Gupta, A. (2021). Microplastics as contaminants in Indian environment: A review. *Environmental Science and Pollution Research*, 1–28.

Veerasingam, S., Ranjani, M., Venkatachalapathy, R., Bagaev, A., Mukhanov, V., Litvinyuk, D., Verzhevskaia, L., Guganathan, L., Vethamony, P. (2020a). Microplastics in different environmental compartments in India: Analytical methods, distribution, associated contaminants and research needs, *Trends in Analytical Chemistry.*

Veerasingam, S., Ranjani, M., Venkatachalapathy, R., Bagaev, A., Mukhanov, V., Litvinyuk, D., Verzhevskaia, L., Guganathan, L., & Vethamony, P. (2020b). Microplastics in different environmental compartments in India: Analytical methods, distribution, associated contaminants and research needs. *TrAC Trends in Analytical Chemistry, 133*, 116071.

Veerasingam, S., Saha, M., Suneel, V., Vethamony, P., Rodrigues, A. C., Bhattacharyya, S., & Naik, B. G. (2016). Characteristics, seasonal distribution and surface degradation features of microplastic pellets along the Goa coast, India. *Chemosphere, 159*, 496–505.

Vianello, A., Boldrin, A., Guerriero, P., Moschino, V., Rella, R., Sturaro, A., & Da Ros, L. (2013). Microplastic particles in sediments of Lagoon of Venice, Italy: First observations on occurrence, spatial patterns and identification. *Estuarine, Coastal and Shelf Science, 130*, 54–61.

Vidyasakar, A., Krishnakumar, S., Kasilingam, K., Neelavannan, K., Bharathi, V. A., Godson, P. S., Prabha, K., & Magesh, N. S. (2020). Characterization and distribution of microplastics and plastic debris along Silver Beach, Southern India. *Marine Pollution Bulletin, 158*, 111421.

Vidyasakar, A., Neelavannan, K., Krishnakumar, S., Prabaharan, G., Priyanka, T. S. A., Magesh, N. S., Godson, P. S., & Srinivasalu, S. (2018). Macrodebris and microplastic distribution in the beaches of Rameswaram Coral Island, Gulf of Mannar, southeast coast of India: A first report. *Marine Pollution Bulletin, 137*, 610–616.

Wagner, S., Hüffer, T., Klockner, P., Wehrhahn, M., Hofmann, T., & Reemtsma, T. (2018). Tire wear particles in the aquatic environment—A review on generation, analysis, occurrence, fate and effects. *Water Research, 139*, 83–100.

Wang, W., Ge, J., & Yu, X. (2020). Bioavailability and toxicity of microplastics to fish species: A review. *Ecotoxicology and Environmental Safety, 189*, 109913.

Wang, Z. M., Wagner, J., Ghosal, S., Bedi, G., & Wall, S. (2017). SEM/EDS and optical microscopy analyses of microplastics in ocean trawl and fish guts. *Science of the Total Environment, 603*, 616–626.

Wirnkor, V. A., Ebere, E. C., & Ngozi, V. E. (2019). Microplastics, an emerging concern: A review of analytical techniques for detecting and quantifying microplatics. *Analytical Methods in Environmental Chemistry Journal, 2*, 13–30.

Wu, W. M., Yang, J., & Criddle, C. S. (2017). Microplastics pollution and reduction strategies. *Frontiers of Environmental Science & Engineering, 11*(1), 6.

Xu, Y., Chan, F. K. S., He, J., Johnson, M., Gibbins, C., Kay, P., Stanton, T., Xu, Y., Li, G., Feng, M., Paramor, O., Yu, X., & Zhu, Y. G. (2020). A critical review of microplastic pollution in urban freshwater environments and legislative progress in China: Recommendations and insights. *Critical Reviews in Environment Science and Technology, 3*, 1.

Yadav, S., Kataria, N., Khyalia, P., Rose, P. K., Mukherjee, S., Sabherwal, H., Chai, W. S., Rajendran, S., Jiang, J. J., & Khoo, K. S. (2023). Recent analytical techniques, and potential eco-toxicological impacts of textile fibrous microplastics (FMPs) and associated contaminates: A review. *Chemosphere, 326*, 138495.

Yang, Y., Liu, W., Zhang, Z., et al. (2020). Microplastics provide new microbial niches in aquatic environments. *Applied Microbiology and Biotechnology, 104*, 6501–6511.

Zhang, G. S., & Liu, Y. F. (2018). The distribution of microplastics in soil aggregate fractions in southwestern China. *Science of the Total Environment, 642*, 12–20.

Zhang, K., Hamidian, A. H., Tubić, A., Zhang, Y., Fang, J. K., Wu, C., & Lam, P. K. (2021). Understanding plastic degradation and microplastic formation in the environment: A review. *Environmental Pollution, 274*, 116554.

Zhang, M., Zhao, Y., Qin, X., Jia, W., Chai, L., Huang, M., & Huang, Y. (2019). Microplastics from mulching film is a distinct habitat for bacteria in farmland soil. *Science of the Total Environment, 688*, 470–478.

Zhang, Z., Kang, S., Allen, S., Allen, D., Gao, T., & Sillanpaa, M. (2020). Atmospheric microplastics: A review on current status and perspectives. *Earth-Science Reviews, 203*, 103118.

Microplastic in the Terrestrial Ecosystem of India

Manisha Devi, Smita Chaudhry, Gajender Yadav, and R. K. Yadav

Abstract Microplastics, identified as plastic fragments that are less than five millimeters in size, have emerged as a major environmental pollutant across various ecosystems globally, including the terrestrial landscapes of India. These particles come from a variety of origins, such as agricultural operations (via plastic mulch and irrigation systems), urban runoff, incorrect trash disposal, industrial discharges, and decomposition of bigger plastic detritus. A total of seventy-five papers (from 2010 to 2024) pertaining to MPs were found, and their production, distribution, and sources in the Indian terrestrial environment were thoroughly examined. Furthermore, the production of plastics and microplastics worldwide as well as the state of information regarding MPs pollution in the terrestrial ecosystem of India are examined. The major cause of MPs contamination in agricultural soil and sediments is the widespread use of controlled release fertilizers, compost, plastic mulching and plastic greenhouses. Our research on the terrestrial ecosystem of India has revealed a broad prevalence of microplastics in soils, especially in urban and agricultural settings. Fragments were the most frequently discovered MPs type, followed by fibers and films. The physical characteristics of soil are changed by microplastics, which may have an impact on microbial populations and plant growth. We also emphasized the harmful effects of MPs on plants, including decreased weight and vegetative development. The problem of microplastic contamination in India is made worse by the country's fast urbanization, poor waste management systems, heavy reliance on plastic in agriculture, and paucity of scientific research. Addressing this issue requires a comprehensive understanding of the sources, distribution, and ecological consequences of microplastics in the terrestrial environment. This chapter underscores the urgent need for an integrated approach combining scientific research, public awareness, and regulatory measures to mitigate the growing threat of microplastics to India's terrestrial ecosystem.

M. Devi (✉) · S. Chaudhry
Institute of Environmental Studies, Kurukshetra University, Kurukshetra, Haryana, India
e-mail: manisharavish001@gmail.com

M. Devi · G. Yadav · R. K. Yadav
Central Soil Salinity Research Institute (ICAR), Karnal, Haryana, India
e-mail: rk.yadav@icar.gov.in

P. Khyalia and S. Yadav (eds.), *Microplastic Pollution in India and Its Environmental Impacts*, Earth and Environmental Sciences Library,
https://doi.org/10.1007/978-3-031-92530-6_3

Keywords Plastic pollution · Microplastic · Soil environment · Terrestrial ecosystem

1 Introduction

Currently, we are living in the "Plastic Age." These days, it is not easy to find an ordinary item that is not made of plastic. The modern lifestyle makes it impossible to go a day without using some form of plastic. It is widely used in different industries because of its unique qualities, which include low cost, mobility, durability, and light weight (Surendran et al., 2023). The widespread manufacture of plastics began in the 1950s. Since then, plastics output has expanded dramatically, from 1.5 Mt (million tonnes) annually to 374.8 Mt by 2019. The COVID-19 epidemic caused a brief halt in output in 2020 at 375.5 Mt, but it resumed to 390.7 Mt in 2022 (Bouaicha et al., 2022; Plastics Europe, 2022). With worldwide production predicted to reach 940 Mt by 2040, this trend is anticipated to continue (Surendran et al., 2023). China holds a leading position in the manufacture and distribution of plastic products, accounting for 32% of the global production volume (Fig. 1). It is followed by the North American Free Trade Agreement (NAFTA) at 19%. Europe produces 15%, while the rest of Asia accounts for 17% (Theroundup.org, 2024).

Every day, India produces 26,000 tonnes of plastic garbage, which amounts to 3.46 million tonnes annually, second only to the USA and the European Union. The three most widely used polymers—polypropylene, polyethylene, and polyvinyl chloride—account for 75% of the waste produced. Packaging accounts for 59% of consumption, with building and construction (13%) and agricultural following closely behind (9%). If current trends continue, India's consumption of plastic is anticipated to rise to roughly 39 Mt by 2025 and 70.5 Mt by 2035 (OECD, 2019). According to data from the NITI Aayog (SDG Index, 2021), Goa generates the most plastic garbage in India i.e., 21 tonnes per year per 1,000 people. Among Union Territories, Delhi generates the highest amount, at 11.49 tonnes per year per 1,000 people. Household waste contains a significant amount of plastic, of which only 50–80% is collected, and just 40% is separated for resource recovery. India's reported plastic recycling rate is 60–70%, but this only applies to valuable plastics (CPCB, 2020; Kapur-Bakshi et al., 2021). In 2019, only 8% of the 24.1 million tonnes of plastic consumed in India were recycled (Dhodapkar et al., 2023). The residual plastic is generally burnt or dumped, allowing it to enter various environmental compartments and finally end up in the food chain. (Dhodapkar et al., 2023; Issac & Kandasubramanian, 2021).

Plastic waste undergoes weathering and fragmentation, which ultimately induces the formation of microplastics (MPs). According to their origins, plastic fragments lesser than 5 mm in diameter are referred to as microplastics (Xu et al., 2020). They fall into two different categories: principal and secondary. Chemical agencies and manufacturers create primary MPs for various consumer goods, including skincare products, cosmetics, and dermal exfoliators (Crawford & Quinn, 2017). Conversely, secondary MPs come from the disintegration of bigger plastic products such as

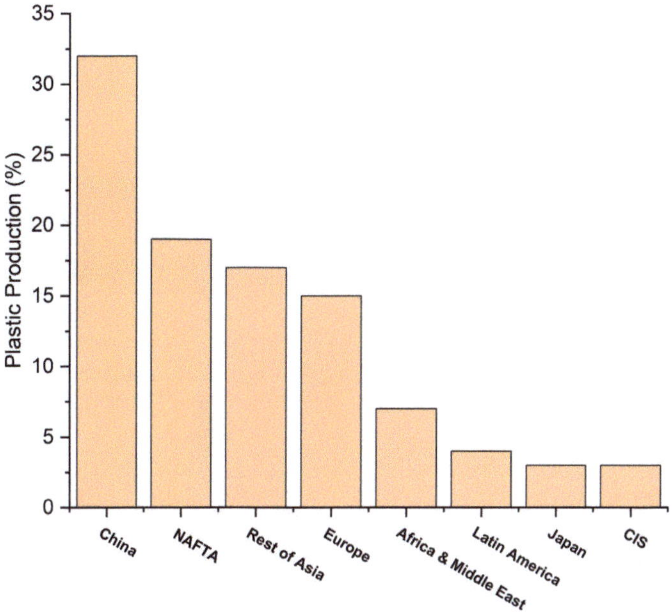

Fig. 1 Distribution of global plastic production for geographic regions [NAFTA-North American Free Trade Agreement, CIS—Commonwealth of Independent States). *Source* https://theroundup.org/plastic-waste-statistics/

paints, cosmetics, automobile tires, fishing gear, food packaging, synthetic textiles, and plastic bottles (Barboza & Gimenez, 2015). MPs can be classified as fibres, foam, pellets, films, and fragments based on their structural properties (Yadav et al., 2023).

Many experts are researching the presence of MPs in numerous components of the Indian environment. MPs research in India has progressed since 2010 but requires more comprehensive studies due to the various injuriousness associated with this emerging pollutant (Ajith et al., 2020). When microplastics enter aquatic and terrestrial settings, they have the potential to cause harm to humans and ecosystems via trophic exchange (Kataria et al., 2024). These MPs break down quickly in water, soil, and air currents because of their minimal density, tiny size, and resilience to natural biological degradation. Additionally, metals such as lead and mercury, agricultural pesticides, organic pollutants and atmospheric pollution or deposition are just a few of the contaminants that can be carried by microplastics and end up in the environment (Fig. 2). When mixed with other contaminants, the MPs are ingested by the organisms and may be hazardous to the health of both peoples and animals (Austen et al., 2022). MPs have the potential to interfere with the activities of species found in terrestrial systems, including fungi, plant pollinators, and soil-dwelling invertebrates (de Souza Machado et al., 2018; Madhav et al., 2020). The physical, chemical, and biological aspects of soil may alter as a result of their increased bioavailability, potentially having an impact on terrestrial vegetation (Bi et al., 2020). Moreover,

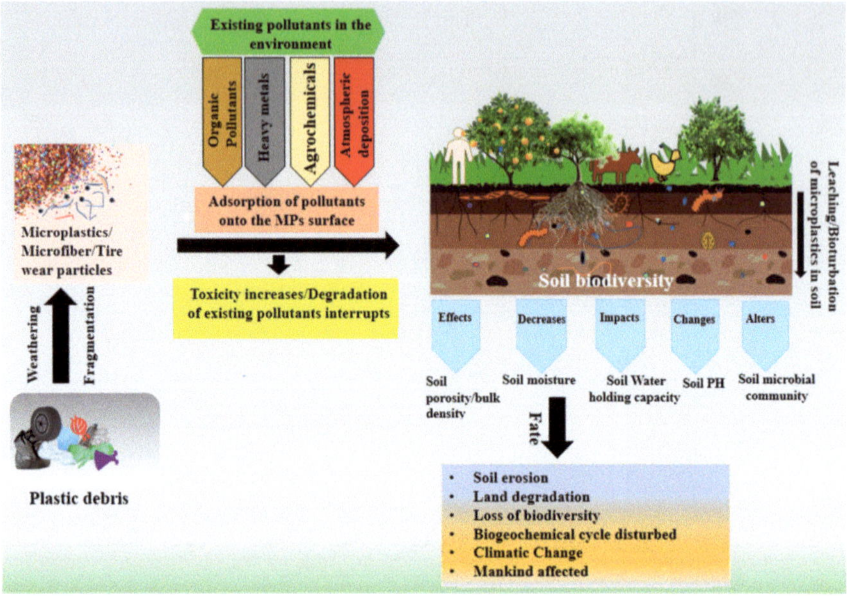

Fig. 2 Microplastic interactions with existing soil pollutants, as well as their effects on other ecological functions (Rai et al., 2023)

MPs may potentially be transferred trophically or accumulate through bioaccumulation. Ragusa et al. (2021) found that MPs can pass through the human placenta and may use respiratory or gastrointestinal systems as potential routes.

Plastic production and consumption have increased exponentially, making it a significant concern for waste management and needs effective monitoring. There are limited reports on microplastics and their impact in various environments. Research on microplastics has primarily been undertaken in Australia, Europe and the United States of America. India possesses a restricted dataset regarding this matter (Bhattacharya & Khare, 2019). The scientific community has recently shown interest in researching microplastics in Indian environments. Most studies in India were undertaken in Southern India (including Indian Ocean, Andaman and Nicobar Islands and some southern states), Western India, which includes specifically Maharashtra and Gujarat and Eastern India This study aims to gather pertinent data regarding MP pollution in Indian terrestrial ecosystems. Although a considerable amount of research has been carried out on plastic and microplastic pollution in marine ecosystems, there is a lack of information on this topic in terrestrial ecosystems.

2 Origins of Microplastics in Terrestrial Environment

The present text pertains to the potential sources of microplastic contamination in terrestrial ecosystem. The issue of microplastic contamination in the terrestrial setting remains a matter of significant concern for several environmental and health reasons. Microplastics can enter the soil ecosystem through countless routes, including agricultural practices, waste discarding, atmospheric deposition, and sewage sludge application (Corradini et al., 2019). Furthermore, the transportation industry significantly contributes to microplastic emissions in the soil ecosystem (Cao et al., 2021), mostly through tire tread and brake friction, as well as road markings. Kole et al. (2017) estimated that 0.81 kg of microplastics per capita per year are emitted from the abrasion of car tires on the road. Soil has accumulated micro-pollutants over decades as shown in Fig. 3, posing environmental and soil quality risks.

2.1 *Sludge from Sewage System*

Sludge from the sewage system is the primary cause of MPs found in the environment, particularly fiber microplastics. These microplastics come from various places, including home garbage and industrial effluents. Li et al. (2018) and Mason et al. (2016) have reported that approximately 80–90% of plastic fragments infiltrating the soil and the environment come from sewage sludge. The primary sources of these particles include domestic, industrial and agriculture. De Falco et al. (2019) identified microfibers in wastewater produced by the laundering of synthetic textiles. The

Fig. 3 Microplastic's major souces in the terrestrial environment (Campanale et al., 2022)

number of microfibers ranged from 124 to 308 mg/kg of fabric washed, implying 640,000–1,500,000 microfibers per wash cycle, depending on garment type. This study emphasizes the need for more investigation and proactive measures to lower microfiber pollution.

It is noteworthy that even household washing machines can generate secondary microplastic filaments, subsequently infiltrating agricultural land via water treatment facilities. On average, top-loading washing machines release seven times more microfibers per garment than front-loaders. The output ranges from 1471 to 2121 microfibers for each garment (Hartline et al., 2016).

2.2 *Plastic Mulching in Agriculture Sector*

He et al. (2018) report plastic mulching as a widespread agricultural practice that aims to increase crop yields and quality. Polyethylene (PE), the most prevalent material used in plastic mulching, does not completely break down (Tian et al., 2022). It has come to attention that plastic mulches, which are frequently composed of oxo-plastics and marketed as single-use plastics, present a challenge in terms of their biodegradability. These oxo-plastics, much like their traditional plastic counterparts, persist in the soil for long period due to their limited capacity for biodegradation. As a result, the practice of plastic mulching can contribute to the increase of MPs pollution in soil (European Commission 2018). Almost 80% of global plastic mulch film consumption comes from three Asian countries (Chinese, Japanese, and Korean). Plastic film mulching is extensively used in China's agriculture, consuming roughly 1.5 million tons annually. China does not recycle more than 40% of this plastic film (Yan et al., 2017). Regretfully, there aren't many differences in the circumstances in other nations. Plastic mulching is responsible for 33–56% of microplastics observed in soil at depths ranging from 0 to 100 cm. The topsoil (0–10 cm) includes an average of 8885 particles/kg, and the deep subsurface (80–100 cm) carries an estimate of 2899 particles/kg (Li et al., 2022). Ding et al. (2020) found fragmented plastics and MPs in soil from farming with plastic mulch. Furthermore, another research informed that the concentration of MPs was 571 pieces per kg in mulched soil, compared to 263 pieces per kg in non-mulched soil (Zhou et al., 2020). The quantity of MPs in soil was discovered to vary in accordance with the intensity of mulching. Huang et al. (2020) found that the number of MPs in soil from plastic mulch rose with the length of mulching; after 5, 15, and 24 years, the average amounts were 80.3 ± 49.3, 308 ± 138.1, and 1075.6 ± 346.8 pieces per kg, respectively. The most prevalent polymers in the ecosystem are polyethylene (PE), polyvinyl chloride (PVC), and polyethylene terephthalate (PET). When plastic mulch was utilized, the concentration of phthalic acid esters (PAEs) in the farming soil ranged from 1.8 to 3.5 mg kg^{-1} (Shi et al., 2019).

2.3 *Plastic Greenhouses and Controlled Released Fertilizers from Agro-Sector*

Poly houses and greenhouses are used in agriculture for protected cultivation. Plastics are extensively used in these structures because of their low price and durability. However, this leads to plastic waste accumulation in the soil, causing MPs pollution, which is an important environmental issue. During mulching, plastic film is buried in the soil, where it degrades into MPs due to factors such as physical damage during intercultural operations, biochemical/biodegradation, and allied mechanisms (Huang et al., 2020; Steinmetz et al., 2016). The study by Wang et al. (2022a, 2022b) found that the distribution of MPs in discarded greenhouses, typical greenhouses and normal greenhouses was 2215.56 ± 1549.86 items kg^{-1}, 891.11 ± 316.71 items kg^{-1}, and 632.50 ± 566.93 items kg^{-1}, respectively. Controlled-release fertilisers (CRF) are a type of fertiliser that releases nutrients to the soil in a controlled manner. This leads to an increase in nutrient utilisation efficiency and reduces losses due to runoff and leaching. As a result, agro technocrats recommend using fewer fertilisers per unit area. Many CRFs are coated in polymers, slowly releasing nitrogen, phosphorus and potassium. However, this coating also breaks down, causing microplastics contaminating the terrestrial environment (Surendran et al., 2023). This is a result of the encapsulating polymer used in CRF, which takes a long time to completely decompose and stays in the soil.

2.4 *Microbeads from Pharmaceutical and Cosmetic Sector*

Micro-beads are little particles of solid plastic, measuring between 10 μm to 1 mm. Polyethylene is the most common substance utilized to make these microbeads, however occasionally polyethylene terephthalate, nylon, and polypropylene (PP) are also utilized (Shahul Hamid et al., 2018; Yuan et al., 2022). Numerous goods, including makeup, personal hygiene, and household cleaning supplies, contain microbeads. These goods have microbeads added to them to give them a rough texture since they act as an exfoliant, or abrasive. These microbeads are typically found in cosmetics and personal care items, including face washes, hand soaps, face creams, and other items. After usage, they are promptly flushed in the household sewage system. Primary treatment screening eventually penetrates the water-soil environment and is typically unsuccessful due to its relatively modest size. One of the main components of microbeads, polyacrylamide, a polymer of acrylamide, has been discovered to be a major source of toxicity in experimental subjects. It is apparently utilized in 110 distinct cosmetics in different amount ranging from 0.05 to 2.8% (Miraj et al., 2019). A recent study predicted that MPs productions from cosmetics and personal care products were between 4.25 and 12 tons annually in

Estonia itself, which was greater than the fibres released from cloth washers (3.52–11.24 tons per year), emphasizing the significance of cosmetics as a major contributor of microplastic in the terrestrial ecosystem. (Kukkola et al., 2024). Because of the harmful consequences of microbeads, a number of countries, including the United Kingdom, Canada, Ireland, and the Netherlands, have recently established laws prohibiting their usage in cosmetics and personal hygiene products. (Surendran et al., 2023).

3 Transportation of Microplastic in Terrestrial Ecosystem

Two significant causes of land pollution are the inappropriate dumping of plastic trash and ineffective waste management (Prokić et al., 2019). The majority of plastic garbage is generated and disposed of on the ground and current literature suggests that soil is a more significant repository of microplastics than marine basins (He et al., 2018). According to Machado et al. (2018), there may be 4–23 times as much MP contamination on land as in the ocean. Subsoil's low oxygen, reduced temperature and limited UV contact slow the deterioration of polymers. These factors cause prolonged photo-oxidative fragmentation, which in turn acts as a sink for MPs in soil (Duis & Coors, 2016). MPs entered into the soil ecosystem can go under the photo- and thermal-initiated oxidative breakdown and bioturbation. Surface coarseness and oxygen-containing groups in microplastics can improve their sorption and mobility in soil, leading to their pathway in aquifers and groundwater (Ren et al., 2021).

Several factors affect the vertical and horizontal distribution of MPs in soil, including soil aggregates, soil cracking, soil microorganisms and biota and other land preparation activities (tillage, ploughing, etc.) used by farmers. Figure 4 gives a complete overview of the origins of microplastics, their transit pathways, and their interactions within ecosystems. Abiotic (such as water or wind) and biotic (such as human activity, floras, or soil creatures) variables can also cause MPs to travel vertically or horizontally (You et al., 2022). The plastics' properties (such as form, size, type, density, etc.) also significantly impact the transport processes. Soil particles are better able to capture plastic fibres, whereas spherical plastic particles have a better chance of passing through the pores in the soil. Higher-density polymers often travel to deeper layers of soil, whereas lower-density polymers are carried horizontally by wind or water close to the soil surface (Zhang et al., 2023). Microplastics can travel by aeolian means, especially in dry and semi-arid areas. Smaller particles, less than 100 μm in size, can be deposited on soils by dry or wet accumulation methods (P'erez-Reveron et al., 2022; Rezaei et al., 2022). Denser and lighter plastic particles may seep deeper into the soil layer as a result of infiltrating water (O'Connor et al., 2019; Zhang et al., 2023).

Plastics may become stuck in plant leaves, stalks, and shoots during surface runoff, limiting their movement (Han et al., 2022). Furthermore, vegetation greatly influences the way plastics are dispersed downward in the soil at the root level. Primary and secondary roots increase the soil's porosity when pore spaces fill with water,

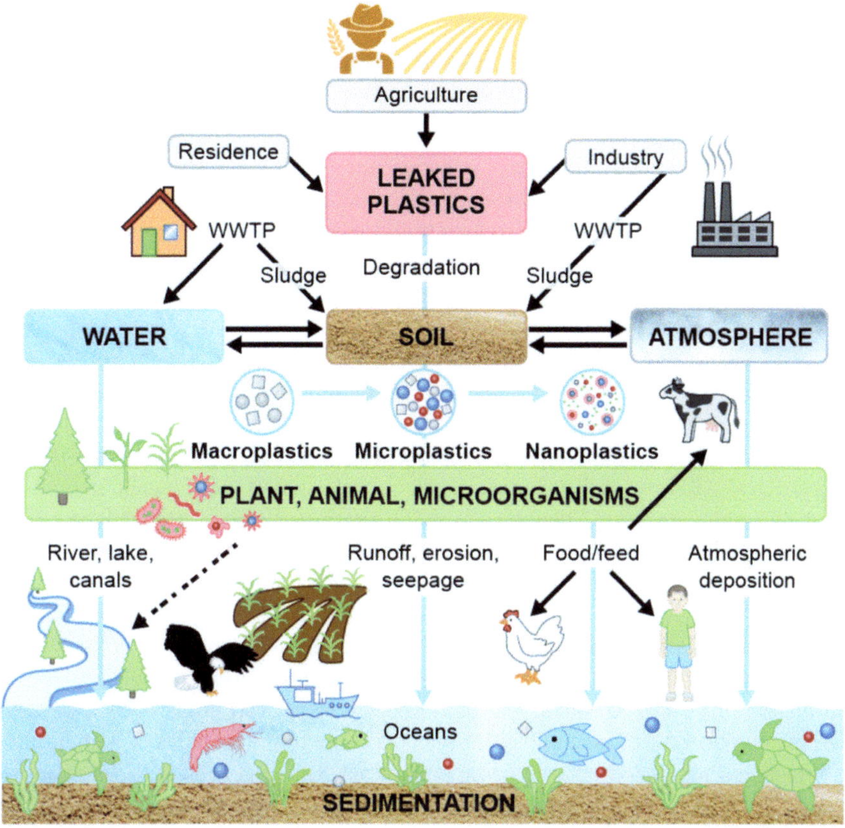

Fig. 4 Overview of the origins of MPs, their transit pathways and their interactions within envrionment (Roy et al., 2022)

which causes lighter plastic particles to rise to the surface. Plastics in the soil layer are typically retained by plants with tertiary roots (Li et al., 2021). The movement of (micro) plastics within soils is also facilitated by soil faunas, like earthworms, burrowing mammals, and springtails (Bouaicha et al., 2022; Dissanayake et al., 2022; Lwanga et al., 2017). Conversely, conventional farming practices (like ploughing, raking, tilling and watering) and bioturbation by creatures that lives in the soil are the main causes of plastic' short-distance mobility (Perez-Reveron et al., 2022; Zhu et al., 2019).

4 The Present State of Research on Microplastics in the Terrestrial Ecosystem of India

With India ranking among the world's top manufacturers of plastic waste, the country is progressively accelerating its microplastics study. This study aims to pinpoint the knowledge and information gaps around MP contamination in India's terrestrial environment. As we use more plastic products daily, MPs seep into our terrestrial systems due to the objects' aging and eventual fragmentation. These MPs may come from various sources, such as wrongly handled plastic garbage, tire wear and tear, broken building and construction supplies, and aeolian dispersion of MPs from large-scale producers (Patchaiyappan et al., 2020). Table 1 provides a summary of the MPs found in Indian terrestrial ecosystem. MPs in outdoor house dust and indoor street dust samples were investigated in Chennai (Patchaiyappan et al., 2020) and Patna (Zhang et al., 2020). A notable large quantity of MPs was discovered, demonstrating the particles' widespread distribution. The concentration of microplastic in outdoor street dust samples from Chennai was 2279.4 items/kg, while indoor house dust samples from Patna had 12 μg/g (free terephthalic acid-based MPs), 2000 μg/g (PET-based MPs), 0.71 μg/g (free bisphenol A-based MPs), and 20 μg/g (polycarbonate-based MPs). In diverse age groups (adults, adolescents, kids and infants), the assessed intake of microplastics generating from the dust samples is 770–10,000 ng/kg bw/day (polyethylene terephthalate (PET)-based MPs) and 0.88–11.00 ng/kg bw/day (polycarbonate (PC)-based MPs) (Zhang et al., 2020). Moreover, it has been suggested that, in comparison to consumables, human exposure to the particles of MP through this channel is significantly more common. For instance, ingesting sea salt can result in 0–36,135 MP particles per person year, whereas consuming mussels can result in an intake of 123 MP particles per person. However, recent investigations (Catarino et al., 2018; Peixoto et al., 2019) have suggested that dust exposure in homes can cause an enormous exposure of 13,731–68,415 particles per person.

4.1 Microplastic in Soil

The majority of microplastics studies in India focus on coastal regions (Amrutha & Warrier, 2020; Karthik et al., 2018). Yet, there is limited information on microplastics in the soil, despite their potentially serious implications for soil health and crop production. It's crucial to remember that because microplastics are so tiny, they frequently gather in the roots region, where they alter the chemical and microbiological characteristics. Given the significance and necessity of research in the soil matrix, here are some efforts made to gather some information on microplastic studies in the Indian soil environment.

A study was performed by Sarkar et al. (2022) on the soil of Eastern India. Soil sampling was carried out in all forms of land cover throughout the city of Cooch Behar in West Bengal, India. This included housing areas, transport areas, market places,

Table 1 Presence of microplastic in the Indian terrestrial ecosystem

States	Site specification	Sampling technique	Microplastic concentration	Microplastics's types	Reference
Tamil Nadu (Chennai)	Dust from streets (outdoor)	Samples were collected using a paint brush and sterile paper in the quadrat method	2279.4 items/kg	Fragments, fibers	Patchaiyappan et al. (2020)
Bihar (Patna)	Dust from houses (Indoor)	Collection methods include vacuum cleaner or nylon brush sweeping	12 μg/g (free terephthalic acid–based MPs) 2000 μg/g (PET-based MPs) 0.71 μg/g (free bisphenol A-based MPs) 20 μg/g (polycarbonate-based MPs)	Fibers	Zhang et al. (2020)
West Bengal (Cooch Behar)	Soil of urban and peri-urban zones	Collection of surface soil samples using a stainless steel shovel	0.02 to 0.62 mg/g	Fibers	Sarkar et al. (2022)
Tamil Nadu (Chennai)	Dumping site soil	Samples were collected using metal trowel	1411 particles/kg	Fragments and films	Tun et al. (2022)
Madhya Pradesh (Bhopal)	Soil from tourism sites	The soil samples from the top 5 cm were dug up and collected using a stainless steel scoop and a metallic spade	180 ± 13.44 particles/kg	Fragments and filaments	Singh et al. (2023)

(continued)

Table 1 (continued)

States	Site specification	Sampling technique	Microplastic concentration	Microplastics's types	Reference
Himachal Pradesh (Una)	Urban, rural, and agricultural soil	Collection of samples targeting top 30 cm soil using a stainless steel shovel	0.014 particles in urban, 0.009 particles in rural and 0.024 particles in agricultural land per 150 g	Fragments, fibers and films	Dhiman et al. (2024)
Gujarat (Surat)	City's soil	Stainless steel shovels were used to gather the soil samples	103.2 ± 57.28 particles/100 g	Fibers	Bhavsar et al. (2024)

as well as suburban regions such as settlements, shops, markets, small-scale manufacturing and suburban croplands. The greatest amount of microplastics (>0.5 mg/g) was found in a tiny section of the municipal zone adjacent to the Cooch Behar train station. The nearby municipal and suburban regions also have a high concentration of MPs in the soil (0.2–0.5 mg/g). The remaining area of the city and its surroundings had a lower concentration of MPs in the soil (Sarkar et al., 2022). The results of another investigation on microplastic in Asian dumping site soils showed that the Indian soils had a average of 1411 particles/kg (Tun et al., 2022). Given that dumping site soils are heavily contaminated with a variety of waste products, it is evident that these soils have a significant concentration of microplastics. The investigation also revealed that the predominant polymers in the Indian soil samples from the dumpsite were polyethylene terephthalate, polypropylene, and polyethylene. This suggests that daily-used plastic goods are the most prevalent producer of MPs in Asian dumping site soils. MPs buildup in the soil has an immense effect on its physical features, particularly bulk density, permeability, and hydrological mechanism such as evaporation, retention, and infiltration. These changes, however, are not consistent and heavily reliant on variables such soil type, climate, MP quantity, parental material, and shape (Baho et al., 2021; Iqbal et al., 2021). Singh et al. (2023) documents the existence of MPs in the soil of Bhopal, Madhya Pradesh. A total of 752.5 ± 6.36 MP particles were detected in all 10 soil samples that were collected, according to the study. The maximum quantity recorded at any single location is 180 ± 13.44 particles/kg. Chemical composition research revealed that the particles were classified into five unique polymer classes: polyethylene, polypropylene, polyvinyl chloride, polyethylene terephthalate, and polyvinyl acetate. Dhiman et al. (2024) examined the MPs in soil samples taken from various locations in district Una, Himachal Pradesh, India (including urban, rural, and agricultural soil). The most common MPs type found was fragments followed by fibers and films. In a variety of sampling locations, the average concentration of microplastics was 0.014 particles/150 g in urban areas, 0.009 particles/150 g in rural areas, and 0.024 particles/150 g in agricultural land. Soil samples taken from farming region was found to be heavily polluted with microplastics (MPs). This contamination may have been caused by regular mulching that included plastic, pesticide packaging that contains microplastics, or environmental deposition Yoon et al. (2024). MPs are found in abundance in the soil of Surat, which is located in the state of Gujarat in India. The mean concentration of MPs in the soil was 103.2 ± 57.28 particles/100 g (Bhavsar et al., 2024). Polyethylene (70.19%), Polypropylene (17.53%), and Poly (ethylene terephthalate) (6.92%) polymers accounted for more than 90% of all MPs. The majority of MPs in the soil samples are fibers and pieces.

4.2 Microplastic in Plants

Microplastics represent major threats to soil ecosystems due to their abundance and persistence. The effect of microplastics on the soil-plant system varies according to their form, size, type, and concentration (Lozano et al., 2021), along with the soil type (Wang et al., 2022). The pollution caused by microplastics and its effects on plants are now a worldwide concern. However, few research have examined the consequences of microplastics on plant systems. This section examines the impact of microplastics on plant growth and performance. It also summarizes the tested microplastics and plant species used in previous investigations conducted in India.

Changmai et al. (2024) discovered that higher polyvinylchloride microplastic (PVC-MP) concentrations (2.5, 5,7.5, and 10% w/w) in the soil failed to result in phytotoxic (chlorosis or necrosis) signs, but did cause a dose-dependent decline in plant development indicators such as height, leaf surface area, diameter of the stem and the dry and fresh weights of the tomato. Furthermore, while the primary roots became longer, the quantity of secondary roots decreased. Moreover, PVC-MP increased the amount of reactive oxygen species (ROS) and lipid peroxidation in plants while significantly reducing the amounts of light-harvesting pigments such as carotenoids and chlorophylls. Nonetheless, concentration-dependent variations in photosynthetic markers and chlorophyll concenteration might occasionally occur (Azeem et al., 2021). The behavior of various PE-MP concentrations (0.25, 0.50, 0.75, and 1.00%) in two distinct soil types on blackgram seedling growth under controlled settings was examined in a different study conducted by Sahasa et al., (2022). The germination rate decreased in clay loam soil from 83.3 ± 3.08 (control) to 79.8 ± 0.03 (1.00% PE-MPs) and in loamy soil type from 87.1 ± 0.57 (control) to 73.1 ± 0.33 (1.00% PE-MPs). Similarly, root and shoot characteristics exhibited a marked decline along with rising microplastic concentrations. It's crucial to remember that a lot of these studies used artificially high plastic concentrations, which makes it difficult to reach firm findings. Studies using environmentally relevant plastic concentrations are therefore necessary. Chaudhary and Suthar (2023) examined the presence and attributes of MP in sewage sludge from three different cities (Haridwar, Dehradun and Rishikesh) along the Ganga River in India, as well as their harmful effects when the sludge is used for crop production. In laboratory experiments, sludge containing microplastics was found to have a significant toxic effect on *Vigna radiata*, leading to reduced vegetative growth and plant weight in setups where the sludge was used. Furthermore, fluorescence microscopy revealed the accumulation of microplastics in the tested seedlings.

Currently, MPs are encroaching into almost every aspect of our surroundings, but very little research has been done in India to assess the problem in terrestrial habitats. In order to better understand the current situation and take the required steps to address this issue as soon as feasible, it is critical to expand research investigations in this field.

In India, the problem of plastic pollution is becoming increasingly significant, and several initiatives are in place to lessen plastic footprints. For instance, in 2019, India became the first country to outlaw single-use plastics on board ships, demonstrating its earliest recognition of the problem. Sikkim became the first Indian state that prohibit single-use plastics in 1998, and by 2022, the entire country had also prohibited their usage (Pathak & Nichter, 2021; Wong et al., 2020). India banned the import of plastic trash in 2019 as part of the Swachh Bharat Mission. However, there is still a problem as these shipments are entering the country disguised as paper waste due to lack of proper screening at the ports. Despite its development potential, India's plastics industry continues to confront numerous hurdles, including the requirement for adequate disposal of waste and worries about severe adverse environmental effects.

5 Conclusion

Microplastic contamination is a global environmental issue, and it is particularly acute in India. The investigation on MPs in the terrestrial realms is still in its early stages but has developed steam in recent years. In this chapter, we have given a thorough overview of MP pollution's effects on terrestrial ecosystems. This chapter explores (a) the key sources of MP in terrestrial environment, (b) the routes by which microplastics enter in soil systems, and (c) the status of microplastic research in the Indian terrestrial environment at the moment. The information gathered from numerous studies suggests that although MPs can have certain positive impacts, such encouraging the growth of particular microbial species and enhancing soil aeration, their overall effects on the soil ecosystem are harmful. They have the ability to impact the soil environment directly or indirectly (for example, by adsorbing harmful compounds on their surface). There are a number of ways that MPs can get into the soil, but the most prevalent ones are related to agricultural activities like using compost or sewage sludge, as well as plastic mulching. India has fewer studies on MP contamination, with marine systems receiving more attention than freshwater and human consumption, and terrestrial environments receiving the least.

Therefore, it is crucial to expand laboratory and field research in India today, paying equal attention to all the elements of the various environmental matrices that include freshwater and marine systems, terrestrial systems, the atmosphere, human consumables, and related toxicities. Apart from that, additional research- ideally involving a variety of soil types, polymer kinds and test organisms is required to close these knowledge gaps. It is important to monitor and understand the interactions between MPs and soil biota through eco-toxicological studies. Enzyme activity assessments and other biomarker assays can reveal important information about how various organisms react to MPs and how efficiently they manage this novel stressor in terrestrial habitats. Evaluating studies in each of these ways will enable us to suggest an MP threshold level and design control measures to limit plastic usage and its eventual harm to the terrestrial habitat. Furthermore, stricter legislation, stronger legal actions, well-planned comprehensive waste management plans, and spontaneous public engagement are critical for raising awareness of terrestrial plastic pollution and mitigating the negative consequences of land-based plastics. Experts should evaluate the data in these manners to help minimize plastic usage and the damage it causes to the terrestrial ecology.

References

Ajith, N., Arumugam, S., Parthasarathy, S., Manupoori, S., & Janakiraman, S. (2020). Global distribution of microplastics and its impact on marine environment—a review. *Environmental Science and Pollution Research, 27*, 25970–25986. https://doi.org/10.1007/s11356-020-09015-5

Amrutha, K., & Warrier, A. K. (2020). The first report on the source-to-sink characterization of microplastic pollution from a riverine environment in tropical India. *Science of the Total Environment, 739*, Article 140377. https://doi.org/10.1016/j.scitotenv.2020.140377

Austen, K., MacLean, J., Balanzategui, D., & Hölker, F. (2022). Microplastic inclusion in birch tree roots. *Science of the Total Environment, 808*, Article 152085. https://doi.org/10.1016/j.scitotenv.2021.152085

Azeem, I., Adeel, M., Ahmad, M. A., Shakoor, N., Jiangcuo, G. D., Azeem, K., Ishfaq, M., Shakoor, A., Ayaz, M., Xu, M., & Rui, Y. (2021). Uptake and accumulation of nano/microplastics in plants: a critical review. *Nanomaterials, 11*(11), 2935. https://doi.org/10.3390/nano11112935

Baho, D. L., Bundschuh, M., & Futter, M. N. (2021). Microplastics in terrestrial ecosystems: Moving beyond the state of the art to minimize the risk of ecological surprise. *Global Change Biology, 27*(17), 3969–3986. https://doi.org/10.1111/gcb.15724

Barboza, L. G. A., & Gimenez, B. C. G. (2015). Microplastics in the marine environment: Current trends and future perspectives. *Marine Pollution Bulletin, 97*(1–2), 5–12. https://doi.org/10.1016/j.marpolbul.2015.06.008

Bhattacharya, A., & Khare, S. K. (2019). Microplastic pollution: An overview of current scenario, challenges, and research gaps. *Adv Biotech & Micro, 12*(3), 555836. https://doi.org/10.19080/AIBM.2019.12.555836

Bhavsar, P. S., Chovatiya, B. V., Kamble, S. B., & Gore, A. H. (2024). Extraction and analysis of microplastics in the soil of diamond city, Surat (Gujarat, India): Ecological risk, pollution indices, and greenness evaluation. *ACS Agricultural Science & Technology*. https://doi.org/10.1021/acsagscitech.4c00140

Bi, M., He, Q., & Chen, Y. (2020). What roles are terrestrial plants playing in global microplastic cycling? *Environmental Science and Technology, 54*, 5325–5327. https://doi.org/10.1021/acs.est.0c01009

Bouaicha, O., Mimmo, T., Tiziani, R., Praeg, N., Polidori, C., Lucini, L., Vigani, G., Terzano, R., Sanchez-Hernandez, J. C., Illmer, P., Cesco, S., & Borruso, L. (2022). Microplastics make their way into the soil and rhizosphere: A review of the ecological consequences. *Rhizosphere, 22*, 100542. https://doi.org/10.1016/j.rhisph.2022.100542

Campanale, C., Galafassi, S., Savino, I., Massarelli, C., Ancona, V., Volta, P., & Uricchio, V. F. (2022). Microplastics pollution in the terrestrial environments: Poorly known diffuse sources and implications for plants. Science of the Total Environment, *805*, 150431

Cao, L., Wu, D., Liu, P., Hu, W., Xu, L., Sun, Y., Wu, Q., Tian, K., Huang, B., Yoon, S. J., & Khim, J. S. (2021). Occurrence, distribution and affecting factors of microplastics in agricultural soils along the lower reaches of Yangtze River, China. *Science of the Total Environment*, *794*, 148694. https://doi.org/10.1016/j.scitotenv.2021.148694.

Catarino, A. I., Macchia, V., Sanderson, W. G., Thompson, R. C., & Henry, T. B. (2018). Low levels of microplastics (MP) in wild mussels indicate that MP ingestion by humans is minimal compared to exposure via household fibres fallout during a meal. *Environmental Pollution, 237*, 675–684. https://doi.org/10.1016/j.envpol.2018.02.069

Changmai, U., Sahana, S. K., Kumar, N., Borah, B., Chikkaputtaiah, C., Saikia, R., & Phukan, T. (2024). Impact of polyvinyl chloride (PVC) microplastic on growth, photosynthesis and nutrient uptake of Solanum lycopersicum L.(Tomato). *Environmental Pollution*, 123994. https://doi.org/10.1016/j.envpol.2024.123994

Chaudhary, M., & Suthar, S. (2023). Microplastic abundance and characterization in municipal sewage sludge from cities across upper Ganga River, India: Apprising microplastic uptakes and their toxicity in the plant during sludge application in agriculture. *Physics and Chemistry of the Earth, Parts a/b/c, 132*, Article 103468. https://doi.org/10.1016/j.pce.2023.103468

Corradini, F., Meza, P., Eguiluz, R., Casado, F., Huerta-Lwanga, E., & Geissen, V. (2019). Evidence of microplastic accumulation in agricultural soils from sewage sludge disposal. *Science of the Total Environment, 671*, 411–420. https://doi.org/10.1016/j.scitotenv.2019.03.368

CPCB. (2020). Annual report 2019–20 on implementation of plastic waste management rules, 2016, in annual report, central pollution control board, online, https://cpcb.nic.in/uploads/plasticwaste/Annual_Report_2019-20_PWM.pdf. Accessed 29 Sept 2022.

Crawford, C. B., & Quinn, B. (2016). Microplastic pollutants. *Elsevier Limited*. https://doi.org/10.1016/c2015-0-04315-5

Crawford, C.B., & Quinn, B. (2017). Plastic production, waste and legislation. In: Crawford CB, Quinn B (Eds.) *Microplastic pollutants* (pp. 39–56). Elsevier Science, Amsterdam. https://doi.org/10.1016/B978-0-12-809406-8.00003-7

CSE. (2019). The problem of plastics, https://cdn.cseindia.org/attachments/0.57139300_1570431848_Factsheet1.pdf. Accessed 30 Jan 2021.

De Falco, F., Di Pace, E., Cocca, M., & Avella, M. (2019). The contribution of washing processes of synthetic clothes to microplastic pollution. *Scientific Reports, 9*(1), 6633. https://doi.org/10.1038/s41598-019-43023-x

De Souza Machado, A. A., Kloas, W., Zarfl, C., Hempel, S., & Rillig, M. C. (2018). Microplastics as an emerging threat to terrestrial ecosystems. *Global Change Biology, 24*(4), 1405–1416. https://doi.org/10.1111/gcb.14020

Dhiman, S., Sharma, D., Kotia, N., & Sinha, R. (2024). Estimation of microplastics distribution in soil sample from District Una, Himachal Pradesh, India. *Journal of Toxicological Studies, 2*(1), 527–527. https://doi.org/10.59400/jts.v2i1.527

Dhodapkar, R., Bhattacharjya, S., Niazi, Z., Porter, N. B., Retamal, M., Sahajwalla, V., & Schandl, H. (2023). National Circular Economy Roadmap for Reducing Plastic Waste in India.

Ding, L., Zhang, S., Wang, X., Yang, X., Zhang, C., Qi, Y., & Guo, X. (2020). The occurrence and distribution characteristics of microplastics in the agricultural soils of Shaanxi Province, in north-western China. *Science of the Total Environment, 720*, Article 137525. https://doi.org/10.1016/j.scitotenv.2020.137525

Dissanayake, P. D., Kim, S., Sarkar, B., Oleszczuk, P., Sang, M. K., Haque, M. N., Ahn, J. H., Bank, M. S., & Ok, Y. S. (2022). Effects of microplastics on the terrestrial environment: a critical review. *Environmental Research, 209*, 112734. https://doi.org/10.1016/j.envres.2022.112734

Duis, K., & Coors, A. (2016). Microplastics in the aquatic and terrestrial environment: Sources (with a specific focus on personal care products), fate and effects. *Environmental Sciences Europe, 28*(1), 2. https://doi.org/10.1186/s12302-015-0069-y

FICCI. (2014). Potential of plastics industry in northern India with special focus on plasticulture and food processing, http://ficci.in/spdocument/20396/Knowledge-Paper-ps.pdf

Guo, J. J., Huang, X. P., Xiang, L., Wang, Y. Z., Li, Y. W., Li, H., Cai, Q. Y., Mo, C. H., & Wong, M. H. (2020). Source, migration and toxicology of microplastics in soil. *Environment international, 137*, 105263. https://doi.org/10.1016/j.envint.2019.105263

Han, N., Zhao, Q., Ao, H., Hu, H., & Wu, C. (2022). Horizontal transport of macro-and microplastics on soil surface by rainfall induced surface runoff as affected by vegetations. *Science of the Total Environment, 831*, Article 154989. https://doi.org/10.1016/j.scitotenv.2022.154989

Hartline, N. L., Bruce, N. J., Karba, S. N., Ruff, E. O., Sonar, S. U., & Holden, P. A. (2016). Microfiber masses recovered from conventional machine washing of new or aged garments. *Environmental Science & Technology, 50*(21), 11532–11538. https://doi.org/10.1021/acs.est.6b03045

He, L., Gielen, G., Bolan, N. S., Zhang, X., Qin, H., Huang, H., & Wang, H. (2015). Contamination and remediation of phthalic acid esters in agricultural soils in China: A review. *Agronomy for Sustainable Development, 35*, 519–534. https://doi.org/10.1007/s13593-014-0270-1

He, D., Luo, Y., Lu, S., Liu, M., Song, Y., & Lei, L. (2018). Microplastics in soils: Analytical methods, pollution characteristics and ecological risks. *TrAC Trends in Analytical Chemistry, 109*, 163–172. https://doi.org/10.1016/j.trac.2018.10.006

Huang, Y., Liu, Q., Jia, W., Yan, C., & Wang, J. (2020). Agricultural plastic mulching as a source of microplastics in the terrestrial environment. *Environmental Pollution, 260*, Article 114096. https://doi.org/10.1016/j.envpol.2020.114096

Iqbal, S., Xu, J., Khan, S., Arif, M. S., Yasmeen, T., Nadir, S., & Schaefer, D. A. (2021). Deciphering microplastic ecotoxicology: impacts on crops and soil ecosystem functions. *Circular Agricultural Systems, 1*(1), 1–7. https://doi.org/10.48130/CAS-2021-0008

Issac, M. N., & Kandasubramanian, B. (2021). Effect of microplastics in water and aquatic systems. *Environmental Science and Pollution Research, 28*, 19544–19562.

Kapur-Bakshi, S., Kaur, M., & Gautam, S. (2021). Circular economy for plastics in India: a roadmap. 85. The Energy and Resources Institute, New Delhi, https://www.teriin.org/sites/default/files/2021-12/Circular-Economy-Plastics-India-Roadmap.pdf

Karthik, R., Robin, R. S., Purvaja, R., Ganguly, D., Anandavelu, I., Raghuraman, R., Hariharan, G., Ramakrishna, A., & Ramesh, R. (2018). Microplastics along the beaches of southeast coast of India. *Science of the Total Environment, 645*, 1388–1399. https://doi.org/10.1016/j.scitotenv.2018.07.242

Kataria, N., Yadav, S., Garg, V. K., Rene, E. R., Jiang, J. J., Rose, P. K., Kumar, M., & Khoo, K. S. (2024). Occurrence, transport, and toxicity of microplastics in tropical food chains: perspectives view and way forward. *Environmental Geochemistry and Health, 46*(3), 98. https://doi.org/10.1007/s10653-024-01862-2

Kole, P. J., Löhr, A. J., Van Belleghem, F. G., & Ragas, A. M. (2017). Wear and tear of tyres: A stealthy source of microplastics in the environment. *International Journal of Environmental Research and Public Health, 14*(10), 1265. https://doi.org/10.3390/ijerph14101265

Kukkola, A., Chetwynd, A. J., Krause, S., & Lynch, I. (2024). Beyond microbeads: Examining the role of cosmetics in microplastic pollution and spotlighting unanswered questions. *Journal of Hazardous Materials*, 135053. https://doi.org/10.1016/j.jhazmat.2024.135053

Lee, M., Kim, H., Ryu, H. S., Moon, J., Khant, N. A., Yu, C., & Yu, J. H. (2022). Review on invasion of microplastic in our ecosystem and implications. *Science Progress, 105*(4), 00368504221140766. https://doi.org/10.1177/00368504221140766

Li, X., Chen, L., Mei, Q., Dong, B., Dai, X., Ding, G., & Zeng, E. Y. (2018). Microplastics in sewage sludge from the wastewater treatment plants in China. *Water Research, 142*, 75–85. https://doi.org/10.1016/j.watres.2018.05.034

Li, H., Lu, X., Wang, S., Zheng, B., & Xu, Y. (2021). Vertical migration of microplastics along soil profile under different crop root systems. *Environmental Pollution, 278*, Article 116833. https://doi.org/10.1016/j.envpol.2021.116833

Li, S., Ding, F., Flury, M., Wang, Z., Xu, L., Li, S., Jones, D. L., & Wang, J. (2022). Macro- and microplastic accumulation in soil after 32 years of plastic film mulching. *Environmental Pollution, 300*, 118945. https://doi.org/10.1016/j.envpol.2022.118945

Liu, K., Wang, X., Fang, T., Xu, P., Zhu, L., & Li, D. (2019). Source and potential risk assessment of suspended atmospheric microplastics in Shanghai. *Science of the Total Environment, 675*, 462–471. https://doi.org/10.1016/j.scitotenv.2019.04.110

Lozano, Y. M., Lehnert, T., Linck, L. T., Lehmann, A., & Rillig, M. C. (2021). Microplastic shape, polymer type, and concentration affect soil properties and plant biomass. *Frontiers in Plant Science, 12*, Article 616645. https://doi.org/10.3389/fpls.2021.616645

Lwanga, E. H., Gertsen, H., Gooren, H., Peters, P., Salánki, T., van der Ploeg, M., Besseling, E., Koelmans, A. A & Geissen, V. (2017). Incorporation of microplastics from litter into burrows of Lumbricus terrestris. *Environmental Pollution, 220*, 523-531. https://doi.org/10.1016/j.envpol.2016.09.096

Madhav, N. V., Gopinath, K. P., Krishnan, A., Rajendran, N., & Krishnan, A. (2020). A critical review on various trophic transfer routes of microplastics in the context of the Indian coastal ecosystem. *Watershed Ecology and the Environment, 2*, 25–41. https://doi.org/10.1016/j.wsee.2020.08.001

Mason, S. A., Garneau, D., Sutton, R., Chu, Y., Ehmann, K., Barnes, J., Fink, P., Papazissimos, D., & Rogers, D. L. (2016). Microplastic pollution is widely detected in US municipal wastewater treatment plant effluent. *Environmental pollution, 218*, 1045–1054. https://doi.org/10.1016/j.envpol.2016.08.056

Mészáros, E., Bodor, A., Szierer, Á., Kovács, E., Perei, K., Tölgyesi, C., Batori, Z., & Feigl, G. (2022). Indirect effects of COVID-19 on the environment: How plastic contamination from disposable surgical masks affect early development of plants. *Journal of Hazardous Materials, 436*, 129255. https://doi.org/10.1016/j.jhazmat.2022.129255

Miraj, S. S., Parveen, N., & Zedan, H. S. (2019). Plastic microbeads: Small yet mighty concerning. *International Journal of Environmental Health Research, 31*(7), 788–804. https://doi.org/10.1080/09603123.2019.1689233

NITI Aayog. (2021). SDG India Index & Dashboard 2020–21, Partnerships in the Decade of Action, https://sdgindiaindex.niti.gov.in/assets/Files/SDG3.0_Final_04.03.2021_Web_Spreads.pdf. Accessed 28 Apr 2023.

O'Connor, D., Pan, S., Shen, Z., Song, Y., Jin, Y., Wu, W. M., & Hou, D. (2019). Microplastics undergo accelerated vertical migration in sand soil due to small size and wet-dry cycles. *Environmental Pollution, 249*, 527–534. https://doi.org/10.1016/j.envpol.2019.03.092

OECD. (2019). Environment-Global Plastics Outlook [WWW Document], https://stats.oecd.org/

Patchaiyappan, A., Dowarah, K., Ahmed, S. Z., Prabakaran, M., Jayakumar, S., Thirunavukkarasu, C., & Devipriya, S. P. (2020). Prevalence and characteristics of microplastics present in the street dust collected from Chennai metropolitan city India. *Chemosphere, 269*, Article 128757. https://doi.org/10.1016/j.chemosphere.2020.128757

Pathak, G., & Nichter, M. (2021). Ecocommunicability, citizenship, and discourses on plastic control in India. *Geoforum, 125*, 132–139. https://doi.org/10.1016/j.geoforum.2021.04.027

Peixoto, D., Pinheiro, C., Amorim, J., Oliva-Teles, L., Guilhermino, L., & Vieira, M. N. (2019). Microplastic pollution in commercial salt for human consumption: A review. *Estuarine, Coastal and Shelf Science, 219*, 161–168. https://doi.org/10.1016/j.ecss.2019.02.018

Pérez-Reverón, R., Álvarez-Méndez, S. J., Kropp, R. M., Perdomo-González, A., Hernández-Borges, J., & Díaz-Peña, F. J. (2022). Microplastics in agricultural systems: Analytical methodologies and effects on soil quality and crop yield. *Agriculture, 12*(8), 1162. https://doi.org/10.3390/agriculture12081162

Plastics Europe. (2022). Plastics—the facts 2022 an analysis of European plastics production, demand, conversion and end-of-life management [WWW Document],https://plasticseurope.org/knowledge-hub/plastics-the-facts-2022/

Prokić, M. D., Radovanović, T. B., Gavrić, J. P., & Faggio, C. (2019). Ecotoxicological effects of microplastics: Examination of biomarkers, current state and future perspectives. *TrAC Trends in Analytical Chemistry, 111*, 37–46. https://doi.org/10.1016/j.trac.2018.12.001

Ragusa, A., Svelato, A., Santacroce, C., Catalano, P., Notarstefano, V., Carnevali, O., Papa, F., Rongioletti, M. C., Baiocco, F., Draghi, S., D'Amore, E.,Rinaldo, D., Matta, M., & Giorgini, E. (2021). Plasticenta: First evidence of microplastics in human placenta. *Environment international, 146*, 106274. https://doi.org/10.1016/j.envint.2020.106274

Rai, M., Pant, G., Pant, K., Aloo, B. N., Kumar, G., Singh, H. B., & Tripathi, V. (2023). Microplastic pollution in terrestrial ecosystems and its interaction with other soil pollutants: A potential threat to soil ecosystem sustainability. *Resources, 12*(6), 67. https://doi.org/10.3390/resources 12060067

Ren, Z., Gui, X., Xu, X., Zhao, L., Qiu, H., & Cao, X. (2021). Microplastics in the soil-groundwater environment: Aging, migration, and co-transport of contaminants–a critical review. *Journal of Hazardous Materials, 419*, Article 126455. https://doi.org/10.1016/j.jhazmat.2021.126455

Rezaei, M., Abbasi, S., Pourmahmood, H., Oleszczuk, P., Ritsema, C., & Turner, A. (2022). Microplastics in agricultural soils from a semi-arid region and their transport by wind erosion. *Environmental Research, 212*, Article 113213. https://doi.org/10.1016/j.envres.2022.113213

Roy, P., Mohanty, A. K., & Misra, M. (2022). Microplastics in ecosystems: their implications and mitigation pathways. Environmental Science: Advances, *1*(1), 9-29

Sahasa, R. G. K., Dhevagi, P., Poornima, R., Alagirisamy, B., Moorthy, P. S., & Karthikeyan, S. (2022). Influence of microplastics on seedling growth of blackgram under different soil types. *International Journal of Environment and Climate Change, 12*(11), 642–649. https://doi.org/ 10.9734/ijecc/2022/v12i1131016

Sarkar, A., Deb, S., Ghosh, S., Mandal, S., Quazi, S. A., Kushwaha, A., Hoque, A., & Choudhury, A. (2022). Impact of anthropogenic pollution on soil properties in and around a town in Eastern India. *Geoderma Regional, 28*, e00462. https://doi.org/10.1016/j.geodrs.2021.e00462

Shahul Hamid, F., Bhatti, M. S., Anuar, N., Anuar, N., Mohan, P., & Periathamby, A. (2018). Worldwide distribution and abundance of microplastic: How dire is the situation? *Waste Management & Research, 36*(10), 873–897. https://doi.org/10.1177/0734242X18785730

Shi, M., Sun, Y., Wang, Z., He, G., Quan, H., & He, H. (2019). Plastic film mulching increased the accumulation and human health risks of phthalate esters in wheat grains. *Environmental Pollution, 250*, 1–7. https://doi.org/10.1016/j.envpol.2019.03.064

Shi, J., Wang, J., Lv, J., Wang, Z., Peng, Y., Shang, J., & Wang, X. (2022). Microplastic additions alter soil organic matter stability and bacterial community under varying temperature in two contrasting soils. *Science of the Total Environment, 838*, Article 156471. https://doi.org/10. 1016/j.scitotenv.2022.156471

Singh, S., Chakma, S., Alawa, B., Kalyanasundaram, M., & Diwan, V. (2023). Identification, characterization, and implications of microplastics in soil–A case study of Bhopal, central India. *Journal of Hazardous Materials Advances, 9*, Article 100225. https://doi.org/10.1016/j.hazadv. 2022.100225

De Souza Machado, A. A., Lau, C. W., Kloas, W., Bergmann, J., Bachelier, J. B., Faltin, E., Becker, R., Golrich, A. S., & Rillig, M. C. (2019). Microplastics can change soil properties and affect plant performance. *Environmental science & technology, 53*(10), 6044–6052. https://doi.org/ 10.1021/acs.est.9b01339

Steinmetz, Z., Wollmann, C., Schaefer, M., Buchmann, C., David, J., Tröger, J., Munoz, K., Fror, O., & Schaumann, G. E. (2016). Plastic mulching in agriculture. Trading short-term agronomic benefits for long-term soil degradation?. *Science of the Total Environment, 550*, 690–705. https:// doi.org/10.1016/j.scitotenv.2016.01.153

Surendran, U., Jayakumar, M., Raja, P., Gopinath, G., & Chellam, P. V. (2023). Microplastics in terrestrial ecosystem: Sources and migration in soil environment. *Chemosphere, 318*, Article 137946. https://doi.org/10.1016/j.chemosphere.2023.137946

Tian, L., Jinjin, C., Ji, R., Ma, Y., & Yu, X. (2022). Microplastics in agricultural soils: Sources, effects, and their fate. *Current Opinion in Environmental Science & Health, 25*, Article 100311. https://doi.org/10.1016/j.coesh.2021.100311

Tun, T. Z., Kunisue, T., Tanabe, S., Prudente, M., Subramanian, A., Sudaryanto, A., Viet, P. H., & Nakata, H. (2022). Microplastics in dumping site soils from six Asian countries as a source of plastic additives. *Science of the Total Environment, 806*, 150912. https://doi.org/10.1016/j.sci totenv.2021.150912

Vaid, M., Mehra, K., & Gupta, A. (2021). Microplastics as contaminants in Indian environment: a review. *Environmental Science and Pollution Research*, 1–28. https://doi.org/10.1007/s11356-021-16827-6

Wang, F., Wang, Q., Adams, C. A., Sun, Y., & Zhang, S. (2022a). Effects of microplastics on soil properties: Current knowledge and future perspectives. *Journal of Hazardous Materials, 424*, Article 127531. https://doi.org/10.1016/j.jhazmat.2021.127531

Wang, K., Chen, W., Tian, J., Niu, F., Xing, Y., Wu, Y., Zhang, R., Zheng, J., & Xu, L. (2022). Accumulation of microplastics in greenhouse soil after long-term plastic film mulching in Beijing, China. *Science of the Total Environment, 828*, 154544. https://doi.org/10.1016/j.scitotenv.2022.154544

Wong, J. K. H., Lee, K. K., Tang, K. H. D., & Yap, P. S. (2020). Microplastics in the freshwater and terrestrial environments: Prevalence, fates, impacts and sustainable solutions. *Science of the Total Environment, 719*, Article 137512. https://doi.org/10.1016/j.scitotenv.2020.137512

Wright, S. L., Thompson, R. C., & Galloway, T. S. (2013). The physical impacts of microplastics on marine organisms: A review. *Environmental Pollution, 178*, 483–492. https://doi.org/10.1016/j.envpol.2013.02.031

Xu, B., Liu, F., Cryder, Z., Huang, D., Lu, Z., He, Y., Wang, H., Lu, Z., Brookes, P. C., Tang, C., Gan, J., & Xu, J. (2020). Microplastics in the soil environment: occurrence, risks, interactions and fate–a review. *Critical Reviews in Environmental Science and Technology, 50*(21), 2175–2222. https://doi.org/10.1080/10643389.2019.1694822

Yadav, S., Kataria, N., Khyalia, P., Rose, P. K., Mukherjee, S., Sabherwal, H., Chai, W. S., Rajendran, S., Jiang, J. J, & Khoo, K. S. (2023). Recent analytical techniques, and potential ecotoxicological impacts of textile fibrous microplastics (FMPs) and associated contaminates: A review. *Chemosphere, 326*, 138495. https://doi.org/10.1016/j.chemosphere.2023.138495

Yan, Z., Xuegeng, C. H. E. N., Haojun, W. E. N., Xuan, Z. H. E. N. G., Qi, N. I. U., & Jianming, K. A. N. G. (2017). Research status and prospect of control technology for residual plastic film pollution in farmland. *Nongye Jixie Xuebao/Transactions of the Chinese Society of Agricultural Machinery, 48*(6).

Yoon, J. H., Kim, B. H., & Kim, K. H. (2024). Distribution of microplastics in soil by types of land use in metropolitan area of Seoul. *Applied Biological Chemistry, 67*(1), 15. https://doi.org/10.1186/s13765-024-00869-8

You, X., Wang, S., Li, G., Du, L., & Dong, X. (2022). Microplastics in the soil: A review of distribution, anthropogenic impact, and interaction with soil microorganisms based on meta-analysis. *Science of the Total Environment, 832*, Article 154975. https://doi.org/10.1016/j.scitotenv.2022.154975

Yuan, Z., Nag, R., & Cummins, E. (2022). Human health concerns regarding microplastics in the aquatic environment-From marine to food systems. *Science of the Total Environment, 823*, Article 153730. https://doi.org/10.1016/j.scitotenv.2022.153730

Zhang, J., Wang, L., & Kannan, K. (2020). Microplastics in house dust from 12 countries and associated human exposure. *Environment International, 134*, Article 105314. https://doi.org/10.1016/j.envint.2019.105314

Zhang, S., Wang, W., Yan, P., Wang, J., Yan, S., Liu, X., & Aurangzeib, M. (2023). Microplastic migration and distribution in the terrestrial and aquatic environments: A threat to biotic safety. *Journal of Environmental Management, 333*, Article 117412. https://doi.org/10.1016/j.jenvman.2023.117412

Zhou, Y., Liu, X., & Wang, J. (2020). Ecotoxicological effects of microplastics and cadmium on the earthworm Eisenia foetida. *Journal of Hazardous Materials, 392*, Article 122273. https://doi.org/10.1016/j.jhazmat.2020.122273

Zhu, F., Zhu, C., Wang, C., & Gu, C. (2019). Occurrence and ecological impacts of microplastics in soil systems: A review. *Bulletin of Environmental Contamination and Toxicology, 102*, 741–749. https://doi.org/10.1007/s00128-019-02623-z

Microplastic in the Inland Water Ecosystem of India

Alka Rao, Shikha Kumari, Jitender Singh Laura, and Geeta Dhania

Abstract Microplastics have been described as widely distributed and new emerging contaminants. Microplastics with land-based sources including domestic sewage, industrial runoff, and littering, were carried by rivers and streams mainly responsible for polluting the inland water system. Microplastics also have adverse effects on the aquatic biota and might even be dangerous to human health. Since microplastics may end up in the marine environment, being one of their last stops via inland water systems as their primary transporting route, their movement and transformation into the inland water systems appear to be more instructive than in the marine environment. This chapter overviews the various sources of microplastics, their migration, and their fate in inland water systems. The distribution of microplastic in Indian inland water systems is also discussed in detail and its risks and effects are also summarized.

Keywords Microplastics · Distribution · Sources · Impacts · Inland water system

1 Introduction

The distribution and impact of plastic debris, notably microplastics, on natural ecosystems have recently received much attention. As microplastics are dangerous to the biosphere, they have been recognized as one of the key markers of the Anthropocene (Waters et al., 2016) and are listed with other chemical pollutants as "Contaminants of Emerging Concern" (CEC). Microplastics are plastic fragments (pieces,

A. Rao · S. Kumari · J. S. Laura · G. Dhania (✉)
Department of Environmental Science, Maharshi Dayanand University, Rohtak, Haryana, India
e-mail: Geeta.env.sc@mdurohtak.ac.in

A. Rao
e-mail: alka.rs.envsc@mdurohtak.ac.in

S. Kumari
e-mail: shikha.rs.envsc@mdurohtak.ac.in

P. Khyalia and S. Yadav (eds.), *Microplastic Pollution in India and Its Environmental Impacts*, Earth and Environmental Sciences Library,
https://doi.org/10.1007/978-3-031-92530-6_4

fibers, or beads) with a diameter of less than 5 mm (Sarkar et al., 2020). The breakdown of huge plastic trash, fabrics and polymers, and small manufactured items like the tiny plastic beads used as scrubbers in cosmetics are the sources of these small plastic particles (Cesa et al., 2017). The microplastics migrate from land-based sources to lakes by rivers and streams (Yadav et al., 2024). Inland water bodies such as lakes and estuaries work as the microplastics' last stop before reaching the sea (Elagami et al., 2023; Yang et al., 2022). Microplastic s spend a longer period of time in the lake's water because of their small size and densities that are similar to those of water. Since microplastics spend so much time in the water column, they are exposed to various lake biota, which increases the likelihood that lake creatures will consume microplastics (Elagami et al., 2022). Through entanglement, plastics have an impact on terrestrial birds and mammals. They have significant negative effects on organism health when consumed. Reproductive toxicity, carcinogenicity, and mutagenicity are also caused by plastic waste and its related pollutants. Furthermore, various chronic and acute impacts can be imposed on ecosystems that might result in irreversible consequences for a longer period, depending on the intensity and volume of the contamination (Patil et al., 2021).

According to CPCB (2017), India generates over 25,940 tons of plastic trash per day, of which 40% is mismanaged and accumulates in landfills (FICCI, 2018), rendering land infertile, clogs drains, litters the ecosystem, and is consumed by organisms. Gujarat, Goa, Lakshadweep, Karnataka, and the Andaman Islands are most plagued by plastic trash (Kaladharan et al., 2017). Land-based sources, coastal populations, recreational activities, domestic activities, and beachgoers are the main input channels (Dowarah & Devipriya, 2019).

2 Sources of Microplastics

Identifying sources of microplastics plays an essential role regulating microplastic pollution. Land litter, sewage treatment plants, shipping industries, ports and harbors, fishing activities, tourism, recreation, and other anthropogenic activities are the main sources of microplastic pollution (Jeyasanta et al., 2020). On the basis of source, these are categorized into primary and secondary microplastics. Both primary and secondary microplastics, which have different sources are eventually released into the water bodies, accompanied by various chemical, physical, and biological processes. Some of the primary microplastics are released into the aquatic systems directly while other large-sized microplastics get fragmented during or after their transfer to the water systems. On the other hand, secondary microplastics are marine-based litter, released from the breakdown of marine litter. It has been determined that the main conduits for microplastic pollution in aquatic environments are rivers and small streams.

2.1 *Primary Microplastics*

Primary microplastics are commonly utilized in many aspects of daily life such as lotions, toothpaste, cosmetics (Carr et al., 2016), medicinal sciences, and industrial manufacturing. Man-made fibers like rayon, nylon, and acrylic used in textile industries are also considered sources of primary microplastics (Browne et al., 2011). These primary microplastics then reach the wastewater treatment plants via the sewage system through washing processes. However, this systematic process of reaching wastewater treatment plants only happens in some developed urban regions; otherwise, the wastewater is directly released into rivers and other water streams with little or no treatment. As a result, microplastics reach the water system without any obstacles (Zhang et al., 2021). In freshwater bodies, small pieces of microplastics are present in minute quantities, showing that microplastics in the inland water system do not originate from primary sources (Zhang et al., 2018). Wastewater resulting from household cleaning and laundry is directly discharged into the stormwater systems and water bodies. Despite of removal of microplastic from sewage water, microplastic still remains in the sewage sludge. When the sludge is transferred to the landfill system, the microplastics pollute the soil and finally reach the waterbodies (Van Den Berg et al., 2020; Guo et al., 2020).

2.2 *Secondary Microplastics*

Microbial and environmental degradation processes of larger plastic particles generate secondary microplastics. Their sources of origin are fishing nets, fibers, films, landfill leachate, beads, industrial resin pellets, foams, and some other degradable plastic wastes, etc. mainly in the inland water system (Silva et al., 2021).

3 Migration of Microplastics into Inland Water Systems

Microplastics generated by humans directly or indirectly reach the inland water systems. Waves and tides, stormwater runoff, beach littering, sewage, and industrial wastewater discharge also add microplastics to the inland water systems (Vanapalli et al., 2021). Treated effluents from the wastewater treatment plants contain microplastics released into the river upstream. Certain surrounding environmental conditions lead certain microplastics to stay in rivers longer (Ding et al., 2019) and others might keep flowing due to certain hydrodynamic conditions (Ballent et al., 2016). Microplastics naturally flow from upstream, and migrate along with rivers and streams. Before they finally reach the oceans, they may face particular possibilities like colonists and predators, who live or feed upon microplastics (Kataria et al., 2024). During all these processes of migration and feeding, microplastics

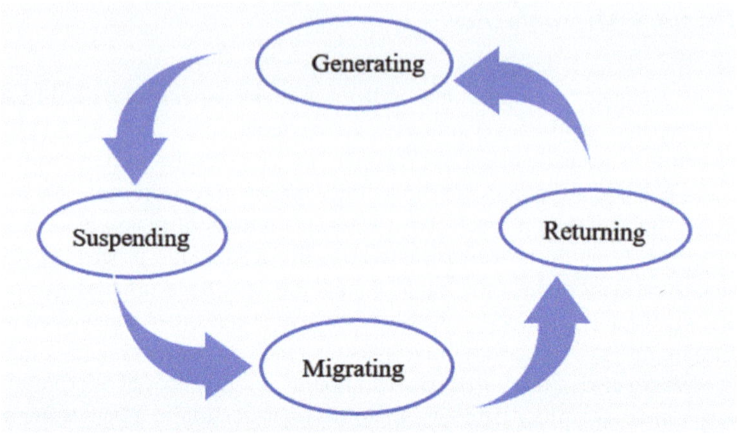

Fig. 1 The process of migration of microplastics in the inland water system

reach and end up in inland water systems and the remaining particles continue to travel until they reach the drinking water treatment plants, which is the final destination for the majority of the microplastics (Barchiesi et al., 2021). These inland water systems may contain microplastics that seep into the groundwater or return to the urban water supply systems through groundwater exploitation (Panno et al., 2019). When humans consume water, microplastics finally reach or accumulate in the human body. Figure 1 shows the steps involved in the migration of microplastic, as human societies generate them, discharged and migrated to the water bodies, contaminate the inland water system, and return to human societies. Microplastics present in an open, dynamic freshwater system may eventually find their way to the ocean (Li et al., 2018). However, this could occur in rivers or estuaries that flow into the oceans. Estuaries play a significant role in the migration of microplastics because they are areas where human settlements, rivers, and oceans meet (Suteja et al., 2021).

4 Behaviour of Microplastics in the Inland Water System

Microplastics originating from terrestrial systems are responsible for microplastic pollution of the marine environment, reaching through inland water systems. The aggregation, settlement, and transportation process of microplastics in the inland water system describes the transportation path of microplastics in the entire water system. The aggregation of microplastics depends on the pH, particle size, aging, degree of microplastics, surface modification, natural organic matter, metal ions surfactant, etc. pH and metal ions may affect the aggregation of polystyrene microplastics through steric effects like altered surface charges and steric repulsion. The stability and aggregation of microplastics with bigger particle sizes is greater than that of those with smaller ones in the same aquatic environment. The aging process prevented the aggregation of polystyrene nanoparticles by increasing the negative charge on the surface of particle and the amount of organic matter in the NaCl solution. Additionally, ageing can cause carbonyl groups to develop on the surface of the microplastics, improving their water stability. Water conditions, organism species, and microplastic characteristics impact the colonization of organisms on microplastics. Hydraulic conditions, precipitation, river morphology, dams, vegetation, etc. all impact the long-distance horizontal transportation of microplastics in water (Yan et al., 2021). Since lakes have longer water retention times and lower flow rates than rivers, microplastics can be deposited considerably more readily in lakes. This longer water retention time and lower flow rates are also conducive to the growth of organisms, particularly microalgae, which leads to significantly increased microplastic aggregation and settlement in lakes (Hu et al., 2020). Most microplastics may be kept permanently in some lakes since they are relatively closed systems and have extended recycling cycles.

5 The Fate of Microplastics in Inland Water Bodies

The transformation and fate of microplastics in inland water is not yet understood properly. Microplastic debris of different shapes and sizes goes in different directions in different water systems. In river systems, debris can reach deep in water, precipitate in soils or sediments, or be mistakenly ingested by aquatic animals. This mistakenly ingested microplastic is a significant issue for aquatic life as well as the people who rely on it for food. The size, colour, and density of microplastic, as well as species, habitat, life cycle, dietary preferences, and environmental contamination with microplastic, all affect the bioaccumulation of microplastic in the aquatic biota. Microplastics remaining in the water were going to find their way into the ocean via wide, flowing rivers or would stay and build up in solitary, silent rivers. Additionally, they could settle in the bottom of the riverbed (Xiong et al., 2019). Decreased flow velocity in lakes and reservoirs encourages the precipitation of wastes that are heavier than water (Zhang et al., 2018). Higher amounts of microplastic were found near the dam (Di & Wang, 2018). The microplastics first undergo morphological alteration by hydrologic function to more appropriate size for deposition and floatation. The microplastics in suspension run the risk of being ingested by aquatic creatures unintentionally or being colonized by microbes, which would subsequently fall to the riverbed due to gravity. Microplastics that have been dumped will be resuspended from the bottom due to certain hydrological effects (Kumari et al., 2025). On the other hand, they will remain deposited at the lake's or river's bottom forever. The resuspended microplastics will also follow a similar suspending, depositing, or burying trend. In a freshwater environment, this causes microplastics to float, suspend, deposit, and eventually bury themselves, as the steps are shown in Fig. 2 (Cai et al., 2021).

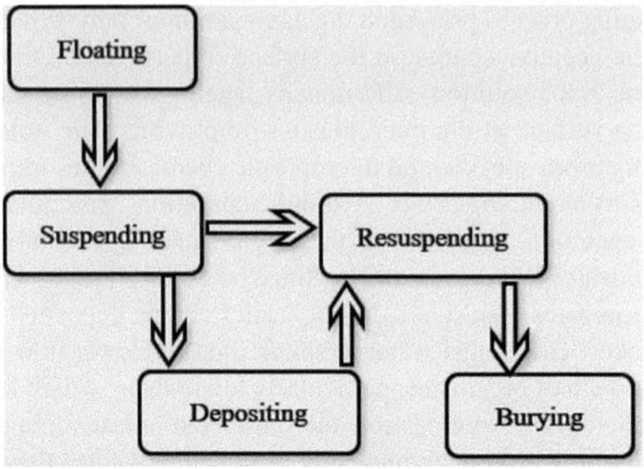

Fig. 2 The transformation process of microplastics in the water column

6 Distribution of Microplastics in the Indian Inland Water System

India's microplastic pollution research is still in its infancy. Many factors affect the distribution of microplastics in the inland water system. A study on microplastic pollution, conducted on the sandy beaches sideways the Tamil Nadu coast of India, discovered that beaches near rivers had relatively greater microplastic abundances than beaches affected by tourism and fishing. It is observed that plastic pieces made up the majority of the trash (47 to 50%), followed by line/fibers (24 to 27%) and foams (10 to 19%) (Karthik et al., 2018). In another study by conducted Sathish et al. (2019), the distribution of microplastics was investigated in the high-tidal and low-tidal five sedimentation areas along the coasts of Tamil Nadu, India. The concentration of microplastics varies from 119 ± 72 to 439 ± 172 (in high tidal regions) and 33 ± 30 to 179 ± 68 (in low tidal regions) items kg^{-1} of sediments. The concentration of microplastics in the high tide region was found to be almost double, that of the concentration of microplastics in the low tide region. Microplastic concentration solely depends on the level of plastic pollution along the coastline. The tendency for microplastics to accumulate there is directly related to anthropogenic and recreational activity levels in the chosen research regions (Sathish et al., 2019). A study was conducted on the distribution of microplastic in lake ecosystems of rural and urban areas of Tamil Nadu. The findings show that microplastic concentration is higher in urban lakes (areas having more population density and larger sewage discharge), with an average microplastic concentration of 88.06 items/L in water and 115.24 items/kg in sediments as compared to rural lakes, showing 42.98 items/L in water and 53.29 items/kg in sediment (Laju et al., 2023). Table 1 compiles the distribution of microplastic in Indian inland water systems, including average concentration, types, and adopted detection tetechniques.

Physical, chemical, and biological factors can contribute to the negative consequences of microplastics in inland water systems.

Table 1 Distribution and abundance of microplastic in Indian Inland water systems

S. no	Location	Sample media	Average microplastic concentration	Microplastic types	Detection methods	References
1	Vembanad lake, Kerala	Subsurface water	26.79 ± 3.74 items per L	PA, PEVA, PP, PET, PVC, LDPE, HDPE	SEM, Raman spectrometer	Anagha et al. (2023)
		Bottom water	52.70 ± 5.43 items per L			
2	Ganga River, Patna, Bihar	Sediment	144–250 particles per 25 mg		Microscope	Singh et al. (2022)
3	Kaveri River, South India	Sediment	386.07items per kg	PP, PS, PA, PET, PVC	FTIR, SEM–EDS	Maheswaran et al. (2022)
4	Coastal waters of Mumbai	Water	372 ± 143 items/liter	PP, PE, PS, PVC, PET, Nylon, PC, PAM, PES, PMMA	Raman spectroscopy	Gurjar et al. (2022)
		Sediment	9630 ± 2947 items/kg			
5	Vembanad Lake, Kochi, Kerala	Water	460 to 1880 particles per m^3	CPVC, PP, PES, PE	Raman spectrometer	Nikki et al. (2021)
6	Sal Estuary, Goa	Water	48 ± 19 MP particles/L	PAM, PVC, Nylon, EVOH	m-FTIR spectroscopy	Saha et al. (2021)
		Sediment	3950 ± 930 MP particles/ kg			
7	Ganga River, Uttar Pradesh and Bihar	Water	380–684 particles per 1000 m^3	PP, PEST, PS, PE, PVC	ATR-FTIR, Stereomicroscope,	Singh et al. (2021)
		Sediment	17–36 particles per kg			
8	Ganga River, India and Bangladesh	Water	0–0.125 particle per L	PVC, Rayon, PEST, PET, Nylon, Acrylic,	FTIR, Microscopy	Napper et al. (2021)
9	Sabarmati River, Ahmadabad, Gujarat	Sediment	134.53 mg per kg and 581.706 mg/ kg dry		SEM	Patel et al. (2020)
10	Netravathi River Karnataka	Water	288 particles per m^3	PP, PVC, PET, PE	ATR-FTIR, Stereozoom Microscope	Amrutha and Warrier (2020)
		Soil sediment	96 particles per kg			

(continued)

Table 1 (continued)

S. no	Location	Sample media	Average microplastic concentration	Microplastic types	Detection methods	References
11	Red Hills Lake, Chennai City, Tamil Nadu	Water	5900 items per m^3	PE, PP, EPS	Stereomicroscope, ATR-FTIR, SEM, EDX	Gopinath et al. (2020)
		Sediment	27 items per kg			
12	Veeranam Lake, Cuddalore, Tamil Nadu	Water,	13–54 items per km^2	PP, PE, PS, PVC, Nylon	ATR-FTIR spectroscopy Stereomicroscope	Manikanda et al. (2021)
		Sediment	92–604 items per kg			
13	Ganga river, eastern India	Sediment	99.27–409.86 items/kg	PET, PE	FTIR spectroscopy	Sarkar et al. (2019)
14	Vembanad lake, Kochi, Kerala,	Sediment	252.8 particles per m^2	PP, HDPE, PS, LDPE,	Micro Raman spectrometer	Sruthy and Ramasamy (2017)

6.1 Physical Impacts

In water systems, entanglement, and ingestion are the two main categories of physical impacts. Entanglement is frequently reported and could be a serious threat to aquatic organisms. These aquatic organisms are more likely to starve, drown, suffocate, or strangle once they are entangled. They will surely perish if their predators approach while they are entangled. Aquatic creatures can continuously absorb microplastics, which can build throughout trophic levels and eventually reach human bodies (Tang et al., 2021).

6.2 Chemical Impacts

The chemical effects predominate significantly. Through ingestion, microplastics may eventually lead to harmful effects on both human bodies and aquatic organisms. These occurrences are brought on by the fact that heavy metals and hazardous compounds may be transported by microplastics (Iri et al., 2021). The disintegration of the microplastics causes their release into the water systems, where they enter the food chains and eventually build up in creatures through a process known as bioaccumulation (Driscoll et al., 2021).

6.3 *Biological Impacts*

The biological impacts claim that it's conceivable for microbes to cling and colonize to the desired microplastics. According to Thiel and Gutow (2004), bacteria can easily cling to the microplastic surface and travel to more sturdy and harder ones. The other biological impact is a physical characteristic that can be changed as a result of the pervasive presence of biofilms as coverings that occupies the surface of microplastics and are typically seen to totally enclose the microplastics (Wang et al., 2021).

7 Conclusion

Although microplastics are found everywhere in the world, most of the research on their distribution has been done in marine habitats. Studies on the prevalence of microplastics and their negative impact in inland open seas are still in their early stages. The study concluded that microfibers from washing and disposing of clothing along riverbanks and packaging waste from tourist and pilgrim destinations were some of the main sources of microplastic pollution in the Indian context, in addition to the minimal land- and marine-based litter transported by the extensive network of rivers and streams. Along with other regional topographical, environmental, and anthropogenic factors, the monsoon season significantly impacted on the amount and distribution of microplastic pollution. Different microplastic features, such as their size, shape, kind of polymer, and colour, are all significant factors that aid in tracing the sources of pollution and identifying the main polymers of concern. Microplastics' movement through inland open waters and their conversion to smaller particles have a terrible impact on aquatic life and, human health.

8 Recommendations

Future research should focus on comprehensive monitoring programs to assess microplastic pollution in major inland water bodies across different seasons. Identifying key sources, such as industrial discharge, urban runoff, agricultural activities, and household waste, is crucial for effective mitigation. Investigating the ecotoxicological impact of microplastics on freshwater organisms, biodiversity, and ecosystem functions will provide insights into long-term environmental consequences. Additionally, studies on the potential health risks associated with microplastic-contaminated drinking water and food chains are necessary to safeguard public health. Research on eco-friendly degradation methods, including microbial degradation and bio-based adsorbents, can aid in developing sustainable remediation strategies. Finally, policy-oriented studies evaluating the effectiveness of current regulations and proposing science-based solutions will support decision-makers in implementing targeted interventions. This could aid in having a greater understanding of the scope of the issue and, consequently, help to focus on more appropriate mitigation solutions. Officials need to take quick action to prevent plastic pollution in Indian inland water systems. Monitoring of the pollution of plastic in Indian inland eater systems needs to be prioritized.

References

Amrutha, K., & Warrier, A. K. (2020). The first report on the source-to-sink characterization of microplastic pollution from a riverine environment in tropical India. *Science of the Total Environment, 739*, Article 140377.

Anagha, P. L., Viji, N. V., Devika, D., & Ramasamy, E. V. (2023). Distribution and abundance of microplastics in the water column of Vembanad Lake–A Ramsar site in Kerala, India. *Marine Pollution Bulletin, 194*, Article 115433.

Ballent, A., Corcoran, P. L., Madden, O., Helm, P. A., & Longstaffe, F. J. (2016). Sources and sinks of microplastics in Canadian Lake Ontario nearshore, tributary and beach sediments. *Marine Pollution Bulletin, 110*(1), 383–395.

Barchiesi, M., Chiavola, A., Marcantonio, C. D., & Boni, M. R. (2021). Presence and fate of microplastics in the water sources: Focus on the role of wastewater and drinking water treatment plants. *Journal of Water Process Engineering, 40*, Article 101787.

Browne, M. A., Crump, P., Niven, S. J., Teuten, E., Tonkin, A., Galloway, T., & Thompson, R. (2011). Accumulation of microplastic on shorelines woldwide: Sources and sinks. *Environmental Science & Technology, 45*(21), 9175–9179.

Cai, Y., Li, C., & Zhao, Y. (2021). A review of the migration and transformation of microplastics in inland water systems. *International Journal of Environmental Research and Public Health, 19*(1), 148.

Carr, S. A., Liu, J., & Tesoro, A. G. (2016). Transport and fate of microplastic particles in wastewater treatment plants. *Water Research, 91*, 174–182.

Cesa, F. S., Turra, A., & Baruque-Ramos, J. (2017). Synthetic fibers as microplastics in the marine environment: A review from textile perspective with a focus on domestic washings. *Science of the Total Environment, 598*, 1116–1129.

CPCB. (2017). Consolidated guidelines for segregation, collection and disposal of plastic waste. https://cpcb.nic.in/uploads/plasticwaste/Consolidate_Guidelines_for_disposal_of_PW.pdf.

Di, M., & Wang, J. (2018). Microplastics in surface waters and sediments of the Three Gorges Reservoir, China. *Science of the Total Environment, 616*, 1620–1627.

Ding, L., fan Mao, R., Guo, X., Yang, X., Zhang, Q., & Yang, C. (2019). Microplastics in surface waters and sediments of the Wei River, in the northwest of China. *Science of the Total Environment, 667*, 427–434.

Dowarah, K., & Devipriya, S. P. (2019). Microplastic prevalence in the beaches of Puducherry, India and its correlation with fishing and tourism/recreational activities. *Marine Pollution Bulletin, 148*, 123–133.

Driscoll, S. C., Glassic, H. C., Guy, C. S., & Koel, T. M. (2021). Presence of microplastics in the food web of the largest high-elevation lake in North America. *Water, 13*(3), 264.

Elagami, H., Ahmadi, P., Fleckenstein, J. H., Frei, S., Obst, M., Agarwal, S., & Gilfedder, B. S. (2022). Measurement of microplastic settling velocities and implications for residence times in thermally stratified lakes. *Limnology and Oceanography, 67*(4), 934–945.

Elagami, H., Frei, S., Boos, J. P., Trommer, G., & Gilfedder, B. S. (2023). Quantifying microplastic residence times in lakes using mesocosm experiments and transport modelling. *Water Research, 229*, Article 119463.

FICCI. (2018). Accelerating India's circular economy shift. A half-trillion USD opportunity future-proofing growth in a resource-scarce world. http://www.ficcices.in/pdf/FICCI-Accenture_Circular%20Economy%20Report_OptVer.pdf.

Gopinath, K., Seshachalam, S., Neelavannan, K., Anburaj, V., Rachel, M., Ravi, S., & Achyuthan, H. (2020). Quantification of microplastic in red hills lake of Chennai city, Tamil Nadu, India. *Environmental Science and Pollution Research, 27*, 33297–33306.

Guo, J. J., Huang, X. P., Xiang, L., Wang, Y. Z., Li, Y. W., Li, H., & Wong, M. H. (2020). Source, migration and toxicology of microplastics in soil. *Environment International, 137*, Article 105263.

Gurjar, U. R., Xavier, K. M., Shukla, S. P., Jaiswar, A. K., Deshmukhe, G., & Nayak, B. B. (2022). Microplastic pollution in coastal ecosystem off Mumbai coast, India. *Chemosphere, 288*, Article 132484.

Hu, D., Zhang, Y., & Shen, M. (2020). Investigation on microplastic pollution of Dongting Lake and its affiliated rivers. *Marine Pollution Bulletin, 160*, 111555.

Iri, A. H., Shahrah, M. H., Ali, A. M., Qadri, S. A., Erdem, T., Ozdur, I. T., & Icoz, K. (2021). Optical detection of microplastics in water. *Environmental Science and Pollution Research, 28*, 63860–63866.

Jeyasanta, K. I., Sathish, N., Patterson, J., & Edward, J. P. (2020). Macro-, meso-and microplastic debris in the beaches of Tuticorin district, Southeast coast of India. *Marine Pollution Bulletin, 154*, Article 111055.

Kaladharan, P., Vijayakumaran, K., Singh, V. V., Prema, D., Asha, P. S., Sulochanan, B., & Bhint, H. M. (2017). Prevalence of marine litter along the Indian beaches: A preliminary account on its status and composition. *Journal of the Marine Biological Association of India, 59*(1), 19–24.

Karthik, R., Robin, R. S., Purvaja, R., Ganguly, D., Anandavelu, I., Raghuraman, R., & Ramesh, R. (2018). Microplastics along the beaches of southeast coast of India. *Science of the Total Environment, 645*, 1388–1399.

Kataria, N., Yadav, S., Garg, V. K., Rene, E. R., Jiang, J. J., Rose, P. K., & Khoo, K. S. (2024). Occurrence, transport, and toxicity of microplastics in tropical food chains: Perspectives view and way forward. *Environmental Geochemistry and Health, 46*(3), 98.

Kumari, S., Yadav, D., Yadav, S., Selvaraj, M., Sharma, G., Karnwal, A., & Yadav, S. (2025). From macro to micro: The key parameters influencing the degradation mechanism and the toxicity of microplastics in the environment. *Polymer Degradation and Stability*, 111174.

Laju, R. L., Jayanthi, M., Jeyasanta, K. I., Patterson, J., Bilgi, D. S., Sathish, N., & Edward, J. P. (2023). Microplastic contamination in Indian rural and urban lacustrine ecosystems. *Science of the Total Environment, 895*, Article 165146.

Li, J., Liu, H., & Chen, J. P. (2018). Microplastics in freshwater systems: A review on occurrence, environmental effects, and methods for microplastics detection. *Water Research, 137*, 362–374.

Maheswaran, B., Karmegam, N., Al-Ansari, M., Subbaiya, R., Al-Humaid, L., Raj, J. S., & Govarthanan, M. (2022). Assessment, characterization, and quantification of microplastics from river sediments. *Chemosphere, 298*, Article 134268.

Manikanda, B. K., Srinivasalu, S., Natesan, U., Ayyamperumal, R., Kalam, N., Anbalagan, S., Sujatha, K., & Alagarasan, C. (2021). Microplastics as an emerging threat to the freshwater ecosystems of Veeranam lake in south India: A multidimensional approach. *Chemosphere, 264*, Article 128502.

Napper, I. E., Baroth, A., Barrett, A. C., Bhola, S., Chowdhury, G. W., Davies, B. F., & Koldewey, H. (2021). The abundance and characteristics of microplastics in surface water in the transboundary Ganges River. *Environmental Pollution, 274*, Article 116348.

Nikki, R., Jaleel, K. A., Ragesh, S., Shini, S., Saha, M., & Kumar, P. D. (2021). Abundance and characteristics of microplastics in commercially important bottom dwelling finfishes and shellfish of the Vembanad Lake, India. *Marine Pollution Bulletin, 172*, Article 112803.

Panno, S. V., Kelly, W. R., Scott, J., Zheng, W., McNeish, R. E., Holm, N., & Baranski, E. L. (2019). Microplastic contamination in karst groundwater systems. *Groundwater, 57*(2), 189–196.

Patel, A. K., Bhagat, C., Taki, K., & Kumar, M. (2020). Microplastic vulnerability in the sediments of the Sabarmati River of India. *Resilience, Response, and Risk in Water Systems: Shifting Management and Natural Forcings Paradigms*, 127–138.

Patil, S., Bafana, A., Naoghare, P. K., Krishnamurthi, K., & Sivanesan, S. (2021). Environmental prevalence, fate, impacts, and mitigation of microplastics—A critical review on present understanding and future research scope. *Environmental Science and Pollution Research, 28*(5), 4951–4974.

Saha, M., Naik, A., Desai, A., Nanajkar, M., Rathore, C., Kumar, M., & Gupta, P. (2021). Microplastics in seafood as an emerging threat to marine environment: A case study in Goa, west coast of India. *Chemosphere, 270*, Article 129359.

Sarkar, D. J., Sarkar, S. D., Das, B. K., Manna, R. K., Behera, B. K., & Samanta, S. (2019). Spatial distribution of meso and microplastics in the sediments of river Ganga at eastern India. *Science of the Total Environment, 694*, Article 133712.

Sarkar, D. J., Sarkar, S. D., Manna, R. K., Samanta, S., & Das, B. K. (2020). Microplastics pollution: An emerging threat to freshwater aquatic ecosystem of India. *Journal of Inland Fisheries Society of India, 52*(1), 05–15.

Sathish, N., Jeyasanta, K. I., & Patterson, J. (2019). Abundance, characteristics and surface degradation features of microplastics in beach sediments of five coastal areas in Tamil Nadu, India. *Marine Pollution Bulletin, 142*, 112–118.

Silva, A. L., Prata, J. C., Duarte, A. C., Soares, A. M., Barceló, D., & Rocha-Santos, T. (2021). Microplastics in landfill leachates: The need for reconnaissance studies and remediation technologies. *Case Studies in Chemical and Environmental Engineering, 3*, Article 100072.

Singh, N., Mondal, A., Bagri, A., Tiwari, E., Khandelwal, N., Monikh, F. A., & Darbha, G. K. (2021). Characteristics and spatial distribution of microplastics in the lower Ganga River water and sediment. *Marine Pollution Bulletin, 163*, Article 111960.

Singh, R., Kumar, R., & Sharma, P. (2022). Microplastic in the subsurface system: Extraction and characterization from sediments of River Ganga near Patna, Bihar. In *Advances in remediation techniques for polluted soils and groundwater* (pp. 191–217). Elsevier.

Sruthy, S., & Ramasamy, E. V. (2017). Microplastic pollution in Vembanad Lake, Kerala, India: The first report of microplastics in lake and estuarine sediments in India. *Environmental Pollution, 222*, 315–322.

Suteja, Y., Atmadipoera, A. S., Riani, E., Nurjaya, I. W., Nugroho, D., & Cordova, M. R. (2021). Spatial and temporal distribution of microplastic in surface water of tropical estuary: Case study in Benoa Bay, Bali, Indonesia. *Marine Pollution Bulletin, 163*, Article 111979.

Tang, Y., Liu, Y., Chen, Y., Zhang, W., Zhao, J., He, S., & Yang, Z. (2021). A review: Research progress on microplastic pollutants in aquatic environments. *Science of the Total Environment, 766*, Article 142572.

Thiel, M., & Gutow, L. (2004). The ecology of rafting in the marine environment. I. The floating substrata. *Oceanography and Marine Biology: An Annual Review, 42*, 181–264.

Van den Berg, P., Huerta-Lwanga, E., Corradini, F., & Geissen, V. (2020). Sewage sludge application as a vehicle for microplastics in eastern Spanish agricultural soils. *Environmental Pollution, 261*, Article 114198.

Vanapalli, K. R., Dubey, B. K., Sarmah, A. K., & Bhattacharya, J. (2021). Assessment of microplastic pollution in the aquatic ecosystems–An indian perspective. *Case Studies in Chemical and Environmental Engineering, 3*, Article 100071.

Wang, G., Lu, J., Li, W., Ning, J., Zhou, L., Tong, Y., Liu, Z., Zhou, H., & Xiayihazi, N. (2021). Seasonal variation and risk assessment of microplastics in surface water of the Manas River Basin, China. *Ecotoxicology and Environmental Safety, 208*, Article 111477.

Waters, C. N., Zalasiewicz, J., Summerhayes, C., Barnosky, A. D., Poirier, C., Gałuszka, A., & Wolfe, A. P. (2016). The Anthropocene is functionally and stratigraphically distinct from the Holocene. *Science, 351*(6269), aad2622.

Xiong, X., Wu, C., Elser, J. J., Mei, Z., & Hao, Y. (2019). Occurrence and fate of microplastic debris in middle and lower reaches of the Yangtze River–from inland to the sea. *Science of the Total Environment, 659*, 66–73.

Yadav, P., Dahiya, S., Yadav, S., Dahiya, D., Rani, M., & Chaudhary, S. (2024). Examining the ecological footprint of microplastics: a holistic exploration from genesis to demise. *Oriental Journal of Chemistry, 40*(3).

Yan, M., Wang, L., Dai, Y., Sun, H., & Liu, C. (2021). Behavior of microplastics in inland waters: aggregation, settlement, and transport. *Bulletin of environmental contamination and toxicology*, 1–10.

Yang, R., Xu, X., Xue, Y., Zhang, L., Guo, J., Wang, L., Peng, M., Zhang, Q., & Zhu, Y. (2022). Horizontal and vertical distribution of microplastics in Gehu Lake, China. *Water Supply, 22*, 8669–8681.

Zhang, K., Shi, H., Peng, J., Wang, Y., Xiong, X., Wu, C., & Lam, P. K. (2018). Microplastic pollution in China's inland water systems: A review of findings, methods, characteristics, effects, and management. *Science of the Total Environment, 630*, 1641–1653.

Zhang, L., Liu, J., Xie, Y., Zhong, S., & Gao, P. (2021). Occurrence and removal of microplastics from wastewater treatment plants in a typical tourist city in China. *Journal of Cleaner Production, 291*, Article 125968.

Microplastic in the Sediment Ecosystem of India

Divya Yadav, Sweta Kumari, Santosh Bhukal, Sumistha Das, and Sangita Yadav

Abstract There is widespread concern about the accumulation of microplastics, which are microscopic plastic particles less than 1 mm in marine ecosystems. While their presence in many marine settings is well documented, it is uncertain if they are found in deep-sea sediments. Recent research, however, has shown that microplastic pollution exists everywhere on Earth, including in remote and unexplored deep-sea strata. These minute plastic bits, which result from the widespread use of plastic, pose a risk because marine organisms may swallow them and suffer unfavourable health consequences. This chapter goes into detail of the genesis, distribution, and environmental impacts of microplastics. It investigates the channels by which microplastics enter aquatic environments, identifying the mechanisms involved in their transport and breakdown. According to the findings, the majority of microplastics eventually degrade into smaller particles that fall to the ocean floor. The chapter also investigates microplastics' role as transporters of toxins and bacteria, as well as their impact on aquatic life when combined with other pollutants. To better understand how microplastics influence higher-level organisms, the transit of microplastics through the food chain is studied. Although their implications are yet unknown because to their diverse physical and chemical properties, microplastics have been seen as signs of a new era known as the "Plasticene" due to their ubiquitous distribution. These particles, which serve as transporters for poisonous substances, also include a cocktail of hazardous chemicals that were purposely added during manufacturing. The purpose of this chapter is to identify and grasp these hazardous compounds, their fate after disposal, and the possible impacts on human health when associated with micro- and nanoplastics. The objective of this chapter is to analyse available data on exposure routes and potential risks to human health in order to guide future research and address knowledge gaps in this critical field.

D. Yadav · S. Kumari · S. Bhukal (✉) · S. Yadav
Department of Environmental Science and Engineering, Guru Jambheswar University of Science & Technology, Hisar, Haryana, India
e-mail: santoshesegju@gmail.com

S. Das
Amity Institute of Biotechnology, Amity University Haryana, Gurugram, India
e-mail: sdas1@ggn.amity.edu

P. Khyalia and S. Yadav (eds.), *Microplastic Pollution in India and Its Environmental Impacts*, Earth and Environmental Sciences Library,
https://doi.org/10.1007/978-3-031-92530-6_5

1 Introduction

The word 'microplastics' was first introduced by Richard Thompson, a highly distinguished professor of Marine Science and Engineering at Plymouth University, located in England, in 2004 (Thompson et al., 2004). The ubiquitous presence of microplastics (MPs), a term used for particles in 1–5000 μm size range is well documented in the literature (Dris et al., 2018; Lambert & Wagner, 2018). The growing use of plastic in packaging and single-use items has led to a very rapid turnaround of plastics into the waste streams. Studies have indicated that packaging is typically discarded within a year of manufacture (Thompson et al., 2009; Woodall et al., 2014). The indiscriminate disposal into the environment has predominantly led to the widespread accumulation of plastic debris in the marine environment (Aragaw, 2020). This plastic debris in the long term may disintegrate into smaller particles and may ultimately become MPs. Several studies have investigated the pathways of MPs reaching the marine environments (Vivekanand et al., 2021; Waldschläger et al., 2020; Priya et al., 2022). Microplastic pollution in western India, specifically in Gujrat, Maharastra, and Kerala, is a significant issue. In Maharastra, 41.85% of plastic content is microplastics, with the highest concentration found on Juhu beach. In Kerala, the highest concentration of microplastics is found in Central Kerala, followed by Southern Kerala, and the least from Northern Kerala. Tamil Nadu, a Coromandel state, recorded the highest amount of microplastic pollution. The Southern parts of the states, particularly along the Gulf of Mannar and Palk Bay, recorded the highest microplastic content. Microplastics in the regions of Thoothukudi and Dhanushkodi Beach at Pamban Island in Tamil Nadu were made up of 51% fibers, 40% granules, and 9% films. Microplastic pollution in Tamil Nadu was not limited to the coast, with concentrations higher in downstream than upstream. The study also recorded species from the region of Tamil Nadu, focusing on the marine ecosystems, freshwater lakes, brackish water setups, and marine ecosystems. The impact of microplastic pollution was not uniform on the surveyed beaches, with Marina beach having a lot of visible microplastic agglomerates. The Palikaranai marshland, one of the largest wetlands in the state, was declared a reserve forest but still suffers from high pollution due to ongoing dumping. The study also surveyed both brackish and freshwater lakes, and the Important Birding Areas (IBAs), Mudhuliyarkuppam and Pulicat Lake. The study also surveyed Kelambakkam backwaters, Sirthavur Lake, and Thaiyur Lake, which are less disturbed but threatened by illegal land grabbing and poaching. The study recorded microplastic pollution from all beaches in Chennai, with some areas taken as control sets to show that protected areas have lesser threats from microplastic pollution (Amrutha et al., 2021; Madhav et al., 2020).

Worldwide, this has led to significant progress in understanding the origin of microplastics, their destiny, and the impacts of microplastics and the chemicals they carry. These tiny plastic particles have been found in marine species across various trophic levels, spanning from the Polar Regions to the equator and from the ocean surface to the seafloor. However, researchers have recently initiated the process of documenting and analyzing microplastics in freshwater and terrestrial sediment environments (Campanale et al., 2020). This development is considered favorable, as research indicates that approximately 80% of microplastic pollution in the ocean originates from terrestrial sources (Dissanayake et al., 2022). Moreover, rivers are recognized as prominent conduits through which microplastics are transported to the oceans from sediments (Yuan et al., 2022). The extent of research on microplastics in sediment environments is currently quite limited.

1.1 Microplastic Pollution in Deep-Sea Sediments

Microplastics have been reported in the water column and marine sediments worldwide (Moore et al., 2001), from low, back-ground concentrations of 3 particles/m^3 in water and 8 particles/kg in sediment, to very high, hot-spot concentrations of 10.2 $\times$ 10^4 particles/m^3 in water and 6.2 $\times$ 10^4 particles/kg in sediments. The observed sediment concentrations all originate from sites located on the continental shelf. However, accumulation zones of floating plastic debris and associated microplastics, the so-called garbage patches, are located far from any continental margin. Hence, the question arises whether the degradation products of larger marine debris are present in deep-sea sediments as well since surface particulate material can be rapidly exported to abyssal depths (Liebezeit & Dubaish, 2012; Van Cauwenberghe et al., 2013).

River, rain, flood, and storm events are the key drivers of MP transport and contamination in estuarine, coastal, and marine environments (Veerasingam et al., 2016; Hurley et al., 2018). Storm surge is another major contributor to the transportation of microplastics between ocean and land (Ockelford et al., 2020). The frequency and severity of storm surges that occur along the East Coast of India (ECI) is higher than those on the West Coast of India (WCI). In the Bay of Bengal and Arabian Sea, the storm surges generated by the tropical cyclones normally occur during premonsoon (March–May) and post-monsoon (October–November) seasons. Therefore, tropical cyclones and associated storm surges not only facilitate the adjustment or migration of shorelines but also transport MPs from offshore to the beach (Van Der Mheen et al., 2020).

1.2 Plastic Degradation

Plastics can break down into micro- and nanoplastics in various ways which can be defined as either via biodegradation or non-biodegradation process (Fig. 1). Biodegradation is a process that involves the use of microbial agents to break down materials without the need for heat. It occurs at a moderate pace and is considered environmentally friendly and cost-effective. For example, it has also been demonstrated that bacteria in the guts of waxworms, or Indian mealmoths (the larvae of *Plodia interpunctella*) can eat polyethylene films. Processes such as thermal degradation, physical degradation, photodegradation, thermo-oxidative degradation, and hydrolysis are all examples of non-biodegradation (Ali et al., 2021). Thermal, or heat degradation is a non-natural, commercial process whilst physical degradation, through weathering, causes larger plastics to be fragmented into smaller pieces. On the other hand, hydrolysis and photodegradation are naturally occurring chemical processes that use water molecules, and UV–visible light, respectively, to break down the chemical bonds in plastics, converting them into monomeric forms (Ga ko & Sajdak, 2022). Plastic non-biodegradation processes decompose polymeric structures, altering their mechanical properties, and increasing their specific surface area, resulting in enhanced physical–chemical reactions and interactions with microorganisms.

Once plastics are broken down into microplastics, they can be transported through the water column and eventually settle in sediments. The mixing and ultimate settling of microplastics (MPs) in sediments are influenced by several factors, including their density, biofouling, and aggregation with organic matter and other particles. Here's an explanation of why and how these factors affect plastic mixing in sediments:

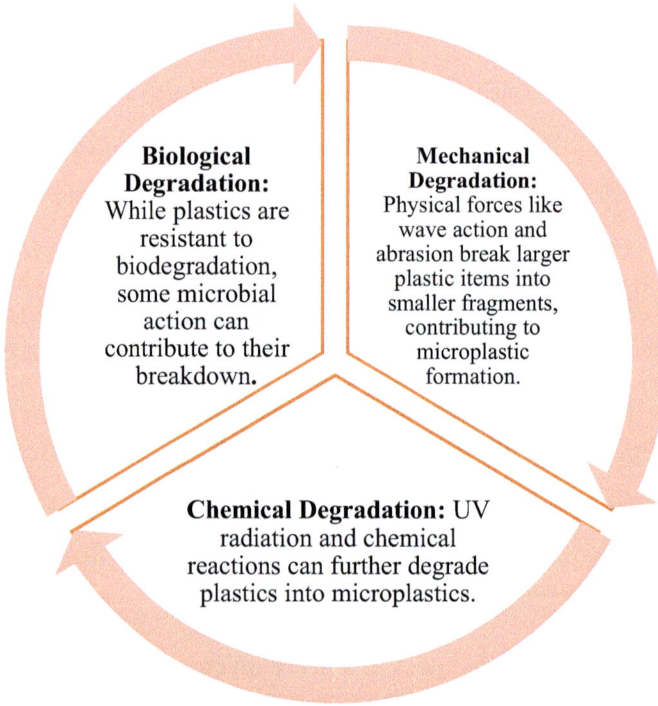

Fig. 1 The three stages of plastic degradation

1. Density

The density of microplastics plays a crucial role in their settling behaviour and their ability to mix within sediments. Microplastics can have a range of densities depending on their polymer type and additives. Polymers like polyvinyl chloride (PVC) have densities greater than that of seawater, causing them to sink rapidly and mix with bottom sediments. On the other hand, Plastics like polyethylene (PE) and polypropylene (PP) have lower densities and tend to float. However, over time, processes such as weathering and biofouling can alter their buoyancy, causing them to sink. The diverse range of densities and shapes means that microplastics can be widely distributed throughout the water column and sediments, contributing to heterogeneous mixing in sediment layers (Karkanorachaki et al., 2021).

2. Biofouling

Biofouling refers to the colonization of microplastics by microorganisms, algae, and other marine life. This process significantly impacts the settling and mixing of microplastics in sediments. As biofilms of bacteria, algae, and other organisms form on the surface of microplastics, their overall density increases. This added weight can cause initially buoyant plastics to sink. Biofouling changes microplastics' surface texture and chemistry, which can affect their interactions with water and sediment particles. Biofouled microplastics tend to settle more quickly than their clean counterparts. The process can lead to a substantial accumulation of microplastics in sediment layers, especially in nutrient-rich areas where biofouling is more intense (Amaral-Zettler et al., 2021).

3. Aggregation

Microplastics can aggregate with organic matter (such as detritus and plankton) and inorganic particles (such as silt and clay). When microplastics aggregate with other particles, they form larger, denser particles that settle more readily at the bottom of water bodies. Also, aggregated microplastics are less likely to be resuspended by water currents and more likely to remain within sediment layers. The aggregation process is facilitated by extracellular polymeric substances (EPS) produced by microorganisms acting as natural adhesives, binding microplastics and other particles together. The charges on the surfaces of microplastics and sediment particles can lead to electrostatic attraction and aggregation. The combined effect of biofouling and aggregation means that microplastics can be efficiently transported from the water column to the sediment, where they become mixed and embedded within sediment layers. This interaction not only aids in settling microplastics but also impacts the vertical distribution and long-term fate of these particles in aquatic environments (Alimi et al., 2018).

Understanding the factors of density, biofouling, and aggregation is essential for predicting the behavior and impact of microplastics in aquatic sediment systems. Each factor plays a unique role in determining how microplastics settle, mix, and become part of the sediment matrix. This knowledge is crucial for assessing the environmental impact of microplastics and developing strategies to mitigate their presence in marine and freshwater ecosystems.

2 Routes of Plastic to Sediments in Aquatic Systems

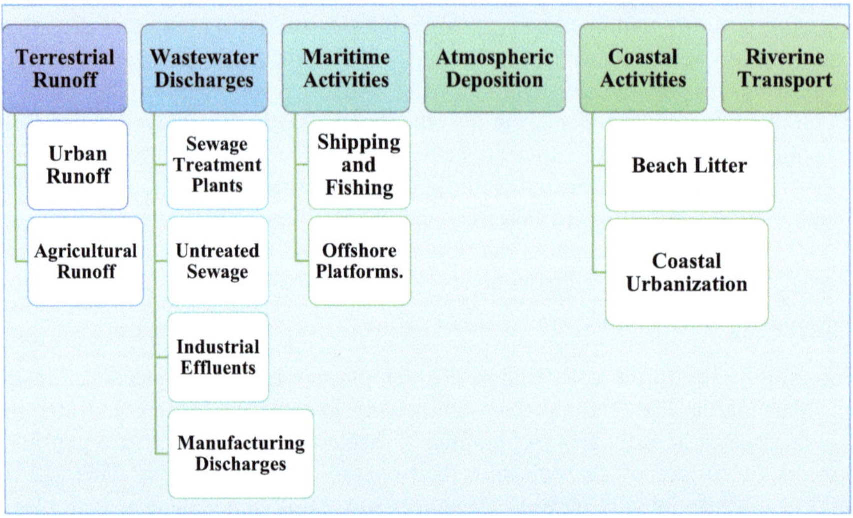

Plastics enter aquatic systems through a variety of pathways, eventually breaking down into microplastics and accumulating in sediments. By understanding these routes and mechanisms, we can better address the sources of plastic pollution and implement effective measures to reduce their impact on aquatic ecosystems.

2.1 Quantification and Identification Methods of Microplastics in Sediments

- Raman Spectroscopy: Commonly used for identifying microplastic components.
- Fourier Transform Infrared Spectroscopy (FTIR): Used for identifying microplastics.
- Thermal Desorption Pyrolysis Gas Chromatography/Mass Spectrometry (TD-Pyr -GC/MS): Identifies microplastics at minimum particle sizes of 1, 20, and 100 μm.

Microplastics in environmental samples can be modified by mechanical crushing, chemical and biological weathering, and additives. Current methods for analyzing microplastics in sediments vary across studies. Inconsistencies were noted in size range, recovery rate, and quantification measures. Density separation is the most common method to extract microplastics in sediments. Zhu et al. (2018) improved the density separation method by including a continuous flow-air floatation separation device. Hurley et al. (2018) recommended performing each separation step twice to ensure a recovery. New Analytical approaches the electrostatic separation method takes advantage of differences in electrostatic behaviors between microplastics and sediment matrix. Grbic et al. (2019) used magnets for the separation of microplastics and the recovery rate of PP, PVC, polyurethane (PU), PS, and PE were 49% ± 12%, 85% ± 27%, 79% ± 21%, 87% ± 26%, and 90% ± 47%, respectively. Digestion can further improve the separation of microplastics from sediments by removing natural organic matter and biofilms. The digestion reagents include acids, alkalis, oxidants, and enzymes. The temperature of digestion should be less than 60 °C due to the loss of microplastics at higher temperatures. Organic Solvent Separation Method extracts microplastics based on their compatibility with organic solvents. Fuller and Gautam 2016) used methanol and dichloromethane as solvents to extract microplastics using the pressurized fluid extraction method. However, this method extracts not only plastics but also other organic matter (e.g., biofilms) and thus has to rely on infrared spectroscopy or other method to distinguish microplastics.

Study area	Main polymer	Particle size (mm)	Shape	References
Northeast Atlantic	Polyester, Nylon, Rayon	0.2–43.2	Fiber, Fragment, Pellet, Foam	(La Daana et al., 2017)
Atlantic	Polyester, polyamide, Rayon	N/A	Fiber, others	(Lusher et al., 2015)
Arctic	Polyester, Polyamide, Polyethylene, Acrylic, Polyvinyl chloride, cellulose	2.50–7.71	Fiber, Fragment, Film	(Lusher et al., 2014)
Ross Sea	Polyester, Polytetrafluoroethylene, Polymethyl methacrylate, Polyamide	N/A	Fiber, Fragment, Others	(Cincinelli et al., 2017)
Southern waters of Sri Lanka	Polyethylene, Polypropylene, polystyrene	1.5–2.5	Fragment, Fiber, Pellet, Film	(Koongolla et al., 2018)
North Eastern Indian Ocean (Southwestern Sumatra)	NA	20 μm–500 μm	Granules and fibers	(Thathsarani et al., 2023)
East Indian Ocean	Polyethylene terephthalate polypropylene	58.59 μm–2251.54 μm	Fibrous and fragments	(Wang et al., 2020)
South Eastern Indian Ocean (Malaysia)	Low-density polyethylene and polypropylene	3 μm–17 μm		(Praveena et al., 2018)
North Eastern Indian Ocean (Southern coastal belt of Sri Lanka)	High-density polyethylene, polypropylene, and polystyrene	1.01 mm–5 mm	Filaments and films	(Athapaththu et al., 2020)

(continued)

(continued)

Study area	Main polymer	Particle size (mm)	Shape	References
Northern Indian Ocean (Marine protected areas in Southern Sri Lanka)	Polyethylene (PE), followed by polypropylene (PP) and Polystyrene (PS). Polyvinyl chloride (PVC) and Nylon (PA)	N/A	Fragments, filaments, foam, pellets and films	(Dharmadasa et al., 2021)
Noth Western Indian Ocean (Kerala Coast, India	Polyethylen, Polypropylene, Polypropylene, Polystyrene and Polycarbonate-Urethane N/A	1–5 mm	Fragments, fibers, films, beads, pellets	(SK & Varghese, 2020)

3 Case Studies

3.1 *Spatial Distribution of Meso and Microplastics in River Ganga Sediments*

Microplastic pollution in river sediments is a significant environmental concern due to its potential impacts on aquatic ecosystems and human health. This case study focuses on the spatial distribution of meso (5–10 mm) and microplastics (less than 5 mm) in the sediments of the Ganga River in Eastern India. The study covers seven different locations along the lower and estuarine stretches of the river, highlighting the sources, distribution, and environmental correlations of plastic particles in this major waterway. The Ganga River, flowing through densely populated and industrialized regions, was divided into seven sampling sites spanning a longitudinal distance of 1092 km. These sites included areas with varying hydrodynamic disturbances, population densities, and industrial activities, extending from Buxar in the upstream to Fraserganj in the estuarine region.

Sediment Sampling and Plastic Particle Extraction: Sediment samples were collected from the river shoreline during the post-monsoon season (October to December 2018). The samples were dried, sieved, and then processed using saturated $ZnCl_2$ solution to separate plastic particles. Microplastics were categorized into two size fractions: 63–850 μm and 850 μm–5 mm, while mesoplastics were analyzed in the 5–10 mm range. The extracted plastic particles were analyzed using microscopic and infrared techniques. FT-IR was employed to identify the polymer types, while the physical characteristics of the sediments and water (e.g., pH, specific conductivity, organic carbon, and available phosphate) were measured to understand their correlation with plastic pollution.

All seven sites were contaminated with mesoplastics and microplastics. The average mass concentration of mesoplastics ranged from 0.68 to 148.31 ng/g, while microplastics ranged from 11.48 to 63.79 ng/g. The number fraction of plastic particles was highest in Godakhali (215.43 items/kg) and lowest in Bhagalpur (4.37 items/kg). Polyethylene terephthalate (PET) (39%) and polyethylene (PE) (30%) were the predominant types of plastic debris. The most common morphotypes were fibers, filaments, films, foams/beads, and fragmented plastics. Fibers were most abundant in Patna, whereas monofilaments were found in Fraserganj and Godakhali.

Correlation with Environmental Parameters: Statistical analysis revealed a strong correlation between microplastic abundance and pollution indicators such as Biological Oxygen Demand (BOD) and available phosphate in both water and sediments. Sites with higher pollution levels showed greater concentrations of microplastics, indicating that poor water and sediment quality contribute to higher plastic pollution. The microplastic abundance in the Ganga was lower compared to rivers like the Antuã River in Portugal but higher than in the Bloukrans River in South Africa. The plastic debris levels in the Ganga were also lower than those reported in the Rhine, Beijiang, and Thames rivers. The study on the spatial distribution of meso and microplastics in the Ganga River sediments underscores the significant pollution levels and the complex interplay between plastic particles and environmental parameters. The findings emphasize the need for long-term monitoring and mitigation strategies to address plastic pollution in this vital river system. The strong correlations with BOD and phosphate levels suggest that improving water and sediment quality could help reduce the prevalence of microplastics (Sarkar et al., 2019).

3.2 *Ecological Risk Posed by MPs in Coastal Sediments Along the Indian Coastline*

India has a coastline of approximately 7500 km, with an exclusive economic zone (EEZ) of 2 million km^2. The climate is dominated by monsoon systems, with 11 major rivers, 47 medium rivers, and 162 small rivers. The primary objectives were to evaluate the presence, distribution, and potential ecological risk of MPs in Indian coastal sediments using meta-data analysis. Key indices like the Polymer Hazard Index (PHI), Pollution Load Index (PLI), and Potential Ecological Risk Index (PERI) were employed for comprehensive risk assessment.

Data collection involved systematic literature searches across environmental, oceanographic, and ecological disciplines using reputable databases. MP concentrations in sediments were assessed using established indices, considering polymer types and abundance as key variables. The study incorporated wind data from monsoon seasons to model MP transport patterns.

The study calculates the average percentage of polymer types of microplastics (MPs) found in terrestrial and marine sediments. The most abundant polymer types include high-density/low-density polyethylene (HD/LD-PE), polypropylene (PP), polystyrene (PS), polyethylene terephthalate (PET), and polyamide (PA). PE (47.7%) and PP (18.8%) are the most abundant polymer types found in sediment samples. These polymers are widely used in food packaging, agricultural film, plastic bags, and plastic bottles. Coastal regions of Tamil Nadu, Maharashtra, and Kerala have high PHI values due to the presence of MPs with high hazard scores, such as PA and PS. The chemical composition of MPs has a minor effect on PLI, as it is calculated based on the ratio between the abundance of measured MPs and the background value. The potential ecological risk index (PERI) values of terrestrial and marine sediments along the Indian coast show high ecological risk from combined MP polymers in sediments from Tuticorin, Kerala coast, Tamil Nadu coast, Vembanad Lake, Goa, Maharashtra, and Karnataka. No noticeable correlation between the abundance of MPs (also, PLI) and PHI was observed, but the higher abundance of MPs may cause potential ecological risk. This study provides the preliminary ecological risk assessment caused by MP contamination in terrestrial and marine sediments along the Indian coast.

Most studies reported that fibers and/or particles <300 μm are the types of MPs ingested by terrestrial and marine organisms. However, most existing data on the physiological and ecological effects of MPs is based on particle sizes that most seawater sampling studies are not quantified. Accurate quantification of MPs and their polymer types is important for planning relevant MP toxicology-related investigations and linking the gap between field and laboratory analyses of potential ecological risk. In conclusion, the abundance of different polymer compositions of MPs along the Indian coast is closely related to different types of human activities on land and sea. Policymakers and environmental managers should consider the ecological risk of MPs in sediments along the Indian coast and establish strict regulations for the discharge of plastic waste to conserve the environment and human health (Ranjani et al., 2021).

3.3 Kanyakumari

The rise in urbanization and human activities over the past several decades has resulted in substantial quantities of plastic waste accumulating in coastal and marine environments. This study aimed to determine the quantity of microplastics (particles less than 5 mm) in the beach sediments of Kanyakumari, the southernmost region of India. Sediment samples were collected from harbors, coastal fishing villages/ residential beaches, tourist beaches, and undisturbed coastal regions along the Indian Ocean. We obtained a total of 343 articles from eight distinct stations by analyzing 50 g of dry sediments from each location. Of these particles, 67% were identified as fiber and 33% as fragments. All of the particles were classified as secondary microplastics, with the majority of them being transparent (specifically, 331 out of 343 particles). The concentration of particles at tourist beaches was the highest, with 150 particles per 50 g of dry sediment. Harbors had the second highest concentration, with 99 particles per 50 g of dry sediment. The pristine beach had the lowest concentration of microplastics, accounting for just 4.3% of all the particles that were collected. Our study examines the correlation between the prevalence of microplastics and varying levels of human-induced activities. Microplastics were found along the southern coast of Kanyakumari, India, with tourist beaches having the highest abundance (44%). The majority of microplastics were found in residential beaches near fishing villages, with minimal amounts in undisturbed areas due to less anthropogenic activities. The presence of microplastics in coastal sediments is primarily due to human activities such as tourism, beach nourishment, and litter disposal. However, comparing the microplastic abundance at Kanyakumari coast with other studies in the southern coast of India is challenging due to differences in extraction procedures and units of expression. The study used the same extraction method, but differences between the microplastics of previous studies could be due to different extraction methods. The vertical profile of microplastic distribution is crucial in determining temporal variations in environmental conditions. The study's findings highlight the importance of understanding the impact of human activities on microplastics in coastal sediments. Additional investigation is necessary to determine the process that impacted the transportation and deposition of these particles in coastal sediments. It is also important to assess the potential interaction between microplastic particles and marine ecosystems (Sundar et al., 2021).

3.4 *Karnataka*

An assessment was conducted to determine the prevalence of microplastic particles at five distinct beach areas along the Arabian Sea coast in the Indian state of Karnataka. Five beaches have been chosen for sediment sampling in the Mangalore region of DK and Udupi. The selected beaches are Tannirubavi (S1) (12∘53′ 12.8″N 74∘48′ 54.6″E), Panambur (S2) (12∘56′ 16.7″N 74∘48′ 16.7″E), Surathkal (S3) (13∘00′ 11.0″N 74∘47′ 23.9″E), Kapu (S4) (13∘13′ 31.5″N 74∘44′ 13.8″E), and Malpe (S5) (13∘21′ 43.2″N 74∘41′ 51.0″E).

The sieving and density separation technique was adapted to optimize the output of microplastics. Isolated microplastics were characterized using three distinct techniques: optical fluorescence microscopy, Raman spectroscopy, and FESEM-EDX. The concentrations of microplastics in beach sediments ranged from 264 ± 62 n/kg to 1002 ± 174 n/kg of dry sand. The average abundance of microplastics across the five beaches was 664 ± 114 n/kg. The bulk of the microplastics were fragmented, measuring less than 1 mm in size, and appeared white and clear. Field emission scanning electron microscopy (FESEM) pictures of microplastics reveal the presence of surface roughness, fissures, and signs of mechanical and oxidative weathering. These observations suggest that microplastics have been consistently exposed to the environment. EDX spectra indicate the presence of different elements on the surface of microplastics. The ranking of microplastic abundance in beach sediment was as follows: Malpe (1002 ± 174 MP kg $^{-1}$) > Panambur (931 ± 156 MP kg $^{-1}$) > Tannirubavi (781 ± 121 MP kg $^{-1}$) > Surathkal (350 ± 108 # MP kg^{-1}) > Kapu

(264 ± 62 MP kg $^{-1}$). The Malpe beach (S-5) had the greatest concentration of microplastics, while the Kapu beach (S-4) had the lowest concentration (Yaranal et al., 2021).

3.5 *Odisha*

Odisha has a long coastline of around 485 km, forming a major coastline of India. The sampling was carried out during January 2020, in Golden Beach, Swargadwara Beach, Gopalpur, Sunapur, Chandipur, Beleswar, Abhayachandpur, Chandrabhaga, and Paradeep which host different activities like tourism, fishing, and harbor (Fig. 1). The sampling was carried out along the selected locations' high-tide line by laying six quadrats (0.25 m $\times$ 0.25 m) at each location and maintaining a gap of at least 200 m between each quadrat. Subsequently, the top sediment layer was scooped off, collected in paper bags, and finally transferred to the laboratory (Dowarah & Devipriya, 2019). The density of microplastics in the Odisha coast was in the order: Swargadwara > Chandipur > Chandrabhaga > Golden Beach > Beleswar > Gopalpur > Paradeep > Sunapur > Abhayachandpur. Sites such as Swargadwara, Chandipur, Chandrabhaga, and Golden Puri beaches had higher microplastic abundance, which may be attributed to the prominence of beaches with the huge inflow of human activities for aesthetics, spiritual attraction, etc. Increased human activities and improper solid waste management resulted in a larger concentration of microplastic in the beach sediments. Further, sites like Beleswar, Gopalpur, Paradeep, Sunapur, and Abhayachandpur recorded lesser microplastic abundance as they are less prominent with limited activities and access. Of the 81 representative particles, 76 particles are confirmed as polymers from the Raman spectroscopy analysis. From the spectral analysis within the database, 12 polymers were identified; these include High-Density Poly Ethylene, Poly- butadiene, Ethylene/Vinyl acetate copolymer, Poly(N Vinyl carbazole), Polyester urethane, Polyvinyl benzoate, Polyvinyl Chloride, P(tetrafluoroethylene), Poly Methacrylamide, Styrene/Methyl methacrylate copolymer, Acrylonitrile/Butadiene/Styrene copolymer, and Polypropylene (Patchaiyappan et al., 2021).

3.6 *Kerela*

India, a major plastic consumer, generates 5.6 million tons of plastic waste annually. Kerala, a densely populated state, generates large quantities of plastic waste, with 4–6% of municipal solid waste being plastic. Kerala lacks sufficient waste management facilities, leading to waste pile accumulation. The prolonged monsoons in Kerala wash plastic debris into rivers, streams, lakes, and the Arabian Sea. Vembanad Lake, the largest lake in Kerala, is part of the Vembanad wetland system, recognized as a Ramsar site. The lake receives discharge from seven major rivers and many streams and canals, and is a sink for contaminants, including heavy metals. Due to its importance for local livelihoods, Vembanad Lake has been selected as a study area for pollution studies. Despite 80% of marine litter originating from land sources and riverine systems directing plastic litter into the marine environment, no study has addressed the presence of metals in Indian freshwater/estuarine water or sediment.

Microplastic particles were found in all sediment samples, with a mean abundance of 252.80 ± 25.76 items per m^2, indicating their widespread presence in lake sediment. The abundance varied significantly among ten sites, with site 2 having the highest number ($496 \pm 2.0\ m^{-2}$). The marine-influenced sites had a higher abundance of MPs, possibly due to higher salinity (4.5) in the north of the TM Bund, compared to the riverine-influenced sites (south of TM Bund). This finding aligns with a study in Mobile Bay estuary, USA, which reported higher MP abundance in sites with higher salinity. Additionally, MPs discharged from Kochi City and Eloor may contribute to the higher MPs abundance (Sruthy & Ramasamy, 2017).

4 Impacts of Microplastic in Sediments on Aquatic Life

Microplastics, despite their various sources, are discharged into oceans, posing a significant threat to marine life. The accumulation of microplastics along shorelines and sandy beaches poses a threat to the species that inhabit these ecosystems.

The Bay of Bengal is a risky location for microplastic pollution, with the east coast of India in a red zone. Studies related to the Bay of Bengal have shown that microplastics in the gut lining of *Daphnia magna* can prevent or inhibit their food intake. Studies have shown that *D. magna* ingests and egests microplastics without significant changes in survival or reproduction. However, the number of ingested particles increases with increased exposure time, suggesting that the organism would have a higher intake of plastics in its natural lifespan than in the lab. Plastic particles in natural ecosystems exist in different shapes, sizes, and types, with spiky particles such as fragments and fibers not being egested as easily. Slow or no ingestion of microplastic particles can lead to a decreased ingestion rate, resulting in reduced food intake, which may lead to starving and adverse effects. *Scleractinia corals* can ingest microplastics, potentially damaging the health of coral reefs. Some species of corals mistake microplastics for their prey, consuming up to 50 μg of plastic at a rate similar to their consumption of plankton. Different responses to microplastic exposure include ingestion of microplastic polymers, mucus production on reef-building corals, overgrowth, and attachment of microplastic to tentacles or mesenterial filaments. The most detrimental effects on corals are coral bleaching and tissue necrosis. Acute exposure of *Pocillopora damicornis* to microplastics activates the stress response, repressing detoxification and the immune system through JNK and ERK signal pathways. Microplastics disturb coral polyps through direct or indirect interaction or photosynthesis perturbation. Hard corals show nematocyst discharge and ingestion of different microplastics, suggesting the presence of phagostimulants, potentially related to toxic compounds.

Microplastics in sediment pose a threat to microfauna like plankton, oysters, and higher-order organisms. Oysters can ingest and accumulate microplastic debris, leading to physical effects.• Polystyrene beads found in oysters' digestive glands post-ingestion can transfer microparticles to higher-order organisms through bioaccumulation. Pacific mole crabs in California have been found with microplastic in their diet, increasing mortality rates and reducing egg-carrying capacity. In Tamil Nadu, *Emerita emeritus*, a species similar to Pacific mole crabs, has been recorded with microplastic on all beaches. The accumulation of microplastics in benthic species is a concern as they constitute the diet of several higher-order organisms. The accumulation of microplastics in marine habitats could change the quality of marine habitats and potentially affect humans.• Many waders, birds that feed on benthic species, may also be affected by microplastics. More studies are needed to confirm the impact on waders. E.J. Carpenter and K. L. Smith were the first to identify microplastics in fish in the United States in 1972. Microplastics were found in the gastrointestinal tracts of fish species like *Aplodactylus punctatus* and *Basilichthys australis* in central Chile. *Amberstripe scads* in Chile also ingested microplastic content. Microplastics can be transferred into the fish's system without conscious ingestion. The extent of microplastics' effect depends on the size distribution, concentration, and toxic nature of the microplastic. Fish ingestion can occur during suction feeding on biofilms or externally, where microplastics are transferred into the fish's system. Microplastics can accumulate in the gills, affecting the larval stages of fish. Microplastics have been linked to increased mercury bioconcentration in gills and liver bioaccumulation. Microplastics can also be found in the stomach and intestines of various species, including Goby species and Indian Mackerel. The presence of microplastics in fish species like Goby, Common goby, and Indian Mackerel is a cause of concern.

Due to their persistence in nature, (micro)plastics exhibit a longer half-life than other marine substrates, creating an interesting habitat for microorganisms. For fouling, microbial biofilm formation is the initial step, while in the consecutive steps, so-called epiplastic organisms like diatoms, ciliates, and a wide variety of other organisms will attach to the formed biofilm. The formation of these biofilms on microplastics is of concern as they might contain human or animal pathogens that could potentially endanger animal and human health, and impact economic activities. Additionally, the nutritional value of these biofilms might encourage grazing and ingestion of the covered microplastics. Associated impacts hence 544 include food web effects (Van Cauwenberghe et al., 2015).

Microplastics, including polystyrene particles, have been found to have toxic effects on inflammation, oxidative stress, apoptosis, and metabolic homeostasis. These particles can induce upregulation of pro-inflammatory cytokines in lung cells and human malignancies, alter scavenger receptors, and increase IL-10 production in certain cell types. They also disrupt cell viability, disrupt ATP levels, and increase ROS concentrations. Microplastics can also affect metabolic homeostasis, leading to gut microbiota dysbiosis and intestinal barrier dysfunction. The impact of microplastics on organisms can be categorized into physical and chemical factors. Microplastics can exert a direct mechanical influence on aquatic species through entanglement and ingestion, resulting in altered feeding behavior, internal obstruction, and harm to the digestive system. The size of microplastic particles affects their accumulation within organisms, affecting growth and reproductive processes. Microplastics can facilitate the transportation of chemicals, potentially inducing carcinogenic, teratogenic, and mutagenic effects on species. Microplastics in aquatic environments may contribute to bioaccumulating contaminants inside aquatic species. The transmission of microplastics via trophic levels is a concern within aquatic ecosystems, impacting around 700 species. Microplastics can be observed in various nutrition levels and ingested directly or indirectly. However, some scholars argue that microplastics can undergo quick purification inside organisms, resulting in minimal influence on species with high nutritional levels.

5 Conclusion

In conclusion, the abundance of different polymer compositions of MPs along the Indian coast is closely related to different types of human activities on land and sea. Policymakers and environmental managers should consider the ecological risk of MPs in sediments along the Indian coast and establish strict regulations for the discharge of plastic waste to conserve the environment and human health. A study was conducted in Karnataka, Odisha, and Kerala to assess the prevalence of microplastic particles in beach sediments. The concentrations of microplastics in Karnataka's five beaches ranged from 264 ± 62 n/kg to 1002 ± 174 n/kg of dry sand. The average abundance of microplastics across the five beaches was 664 ± 114 n/kg. The majority of microplastics were fragmented and appeared white and clear. Odisha's coastline, which spans around 485 km, had a density of microplastics in the order of Swargadwara, Chandipur, Chandrabhaga, Golden Beach, Beleswar, Gopalpur, Paradeep, Sunapur, and Abhayachandpur. Of the 81 representative particles, 76 were confirmed as polymers from Raman spectroscopy analysis. Kerela, India's largest state, generates 5.6 million tons of plastic waste annually. The state generates large quantities of plastic waste, with 4–6% of municipal solid waste being plastic. The prolonged monsoons in Kerala wash plastic debris into rivers, streams, lakes, and the Arabian Sea. Vembanad Lake, the largest lake in Kerala, is a sink for contaminants, including heavy metals. Microplastic particles were found in all sediment samples, with a mean abundance of 252.80 ± 25.76 items per m^2, indicating their widespread presence in lake sediment. Microplastics, despite their various sources, are discharged into oceans, posing a significant threat to marine life. The accumulation of microplastics along shorelines and sandy beaches poses a threat to the species that inhabit these ecosystems. The Bay of Bengal is a risky location for microplastic pollution, with the east coast of India in a red zone. Studies related to the Bay of Bengal have shown that microplastics in the gut lining of Daphnia magna can prevent or inhibit their food intake. However, the number of ingested particles increases with increased exposure time, suggesting that the organism would have a higher intake of plastics in its natural lifespan than in the lab. Microplastics in sediment pose a threat to microfauna like plankton, oysters, and higher-order organisms. Oysters can ingest and accumulate microplastic debris, leading to physical effects. Polystyrene beads found in oysters' digestive glands post-ingestion can transfer microparticles to higher-order organisms through bioaccumulation. The accumulation of microplastics in marine habitats could change the quality of marine habitats and potentially affect humans. Microplastics, including polystyrene particles, have toxic effects on inflammation, oxidative stress, apoptosis, and metabolic homeostasis. They can induce upregulation of pro-inflammatory cytokines, alter scavenger receptors, and increase IL-10 production in certain cell types. The transmission of microplastics via trophic levels is a concern within aquatic ecosystems, impacting around 700 species.

References

Ali, S. S., Elsamahy, T., Al-Tohamy, R., Zhu, D., Mahmoud, Y.A.-G., Koutra, E., Metwally, M. A., Kornaros, M., & Sun, J. (2021). Plastic wastes biodegradation: Mechanisms, challenges and future prospects. *Science of the Total Environment, 780*, Article 146590.

Alimi, O. S., Farner Budarz, J., Hernandez, L. M., & Tufenkji, N. (2018). Microplastics and nanoplastics in aquatic environments: Aggregation, deposition, and enhanced contaminant transport. *Environmental Science & Technology, 52*(4), 1704–1724.

Amaral-Zettler, L. A., Zettler, E. R., Mincer, T. J., Klaassen, M. A., & Gallager, S. M. (2021). Biofouling impacts on polyethylene density and sinking in coastal waters: A macro/micro tipping point? *Water Research, 201*, Article 117289.

Amrutha, K., Unnikrishnan, V., Shajikumar, S., & Warrier, A. K. (2021). Current state of microplastics research in SAARC countries—a review. *Microplastic Pollution*, 27–63.

Aragaw, T. A. (2020). Surgical face masks as a potential source for microplastic pollution in the COVID-19 scenario. *Marine Pollution Bulletin, 159*, Article 111517.

Athapaththu, A., Thushari, G. G. N., Dias, P. C. B., Abeygunawardena, A. P., Egodauyana, K., Liyanage, N. P. P., Pitawala, H., & Senevirathna, J. D. M. (2020). Plastics in surface water of southern coastal belt of Sri Lanka (Northern Indian Ocean): Distribution and characterization by FTIR. *Marine Pollution Bulletin, 161*, Article 111750.

Campanale, C., Stock, F., Massarelli, C., Kochleus, C., Bagnuolo, G., Reifferscheid, G., & Uricchio, V. F. (2020). Microplastics and their possible sources: The example of Ofanto river in southeast Italy. *Environmental Pollution, 258*, Article 113284.

Cincinelli, A., Scopetani, C., Chelazzi, D., Lombardini, E., Martellini, T., Katsoyiannis, A., Fossi, M. C., & Corsolini, S. (2017). Microplastic in the surface waters of the Ross Sea (Antarctica): Occurrence, distribution and characterization by FTIR. *Chemosphere, 175*, 391–400.

Dharmadasa, W. L. S. S., Andrady, A. L., Kumara, P. B. T. P., Maes, T., & Gangabadage, C. S. (2021). Microplastic pollution in marine protected areas of Southern Sri Lanka. *Marine Pollution Bulletin, 168*, Article 112462.

Dissanayake, P. D., Kim, S., Sarkar, B., Oleszczuk, P., Sang, M. K., Haque, M. N., Ahn, J. H., Bank, M. S., & Ok, Y. S. (2022). Effects of microplastics on the terrestrial environment: A critical review. *Environmental Research, 209*, Article 112734.

Dris, R., Gasperi, J., & Tassin, B. (2018). Sources and fate of microplastics in urban areas: A focus on Paris megacity. *Freshwater Microplastics: Emerging Environmental Contaminants?*, 69–83.

Fuller, S., & Gautam, A. (2016). A procedure for measuring microplastics using pressurized fluid extraction. *Environmental Science & Technology, 50*(11), 5774–5780.

Ga ko, G., & Sajdak, M. (2022). Trends for the thermal degradation of polymeric materials: Analysis of available techniques, issues, and opportunities. *Applied Sciences, 12*(18), 9138.

Grbic, J., Nguyen, B., Guo, E., You, J. B., Sinton, D., & Rochman, C. M. (2019). Magnetic extraction of microplastics from environmental samples. *Environmental Science & Technology Letters, 6*(2), 68–72.

Hurley, R., Woodward, J., & Rothwell, J. J. (2018). Microplastic contamination of river beds significantly reduced by catchment-wide flooding. *Nature Geoscience, 11*(4), 251–257.

Karkanorachaki, K., Syranidou, E., & Kalogerakis, N. (2021). Sinking characteristics of microplastics in the marine environment. *Science of the Total Environment, 793*, Article 148526.

Koongolla, J. B., Andrady, A. L., Kumara, P. B. T. P., & Gangabadage, C. S. (2018). Evidence of microplastics pollution in coastal beaches and waters in southern Sri Lanka. *Marine Pollution Bulletin, 137*, 277–284.

La Daana, K. K., Officer, R., Lyashevska, O., Thompson, R. C., & O'Connor, I. (2017). Microplastic abundance, distribution and composition along a latitudinal gradient in the Atlantic Ocean. *Marine Pollution Bulletin, 115*(1–2), 307–314.

Lambert, S., & Wagner, M. (2018). *Microplastics are contaminants of emerging concern in freshwater environments: an overview*. Springer International Publishing.

Liebezeit, G., & Dubaish, F. (2012). Microplastics in beaches of the East Frisian islands Spiekeroog and Kachelotplate. *Bulletin of Environmental Contamination and Toxicology, 89*, 213–217.

Lusher, A. L., Burke, A., O'Connor, I., & Officer, R. (2014). Microplastic pollution in the Northeast Atlantic Ocean: Validated and opportunistic sampling. *Marine Pollution Bulletin, 88*(1–2), 325–333.

Lusher, A. L., Tirelli, V., O'Connor, I., & Officer, R. (2015). Microplastics in Arctic polar waters: The first reported values of particles in surface and sub-surface samples. *Scientific Reports, 5*(1), 14947

Madhav, N. V., Gopinath, K. P., Krishnan, A., Rajendran, N., & Krishnan, A. (2020). A critical review on various trophic transfer routes of microplastics in the context of the Indian coastal ecosystem. *Watershed Ecology and the Environment, 2*, 25–41.

Moore, C. J., Moore, S. L., Leecaster, M. K., & Weisberg, S. B. (2001). A comparison of plastic and plankton in the North Pacific central gyre. *Marine Pollution Bulletin, 42*(12), 1297–1300.

Ockelford, A., Cundy, A., & Ebdon, J. E. (2020). Storm response of fluvial sedimentary microplastics. *Scientific Reports, 10*(1), 1865.

Patchaiyappan, A., ZakiAhmed, S., Dowarah, K., Khadanga, S. S., Singh, T., Jayakumar, S., Thirunavukkarasu, C., & Devipriya, S. P. (2021). Prevalence of microplastics in the sediments of Odisha beaches, southeastern coast of India. *Marine Pollution Bulletin, 167*(March), Article 112265. https://doi.org/10.1016/j.marpolbul.2021.112265

Praveena, S. M., Shaifuddin, S. N. M., & Akizuki, S. (2018). Exploration of microplastics from personal care and cosmetic products and its estimated emissions to marine environment: An evidence from Malaysia. *Marine Pollution Bulletin, 136*, 135–140.

Priya, K. L., Renjith, K. R., Joseph, C. J., Indu, M. S., Srinivas, R., & Haddout, S. (2022). Fate, transport and degradation pathway of microplastics in aquatic environment—A critical review. *Regional Studies in Marine Science, 56*, Article 102647.

Ranjani, M., Veerasingam, S., Venkatachalapathy, R., Mugilarasan, M., Bagaev, A., Mukhanov, V., & Vethamony, P. (2021). Assessment of potential ecological risk of microplastics in the coastal sediments of India: A meta-analysis. *Marine Pollution Bulletin, 163*(January), Article 111969. https://doi.org/10.1016/j.marpolbul.2021.111969

Sarkar, D. J., Das Sarkar, S., Das, B. K., Manna, R. K., Behera, B. K., & Samanta, S. (2019). Spatial distribution of meso and microplastics in the sediments of river Ganga at eastern India. *Science of the Total Environment, 694*, Article 133712. https://doi.org/10.1016/j.scitotenv.2019.133712

SK, A., & Varghese, G. K. (2020). Environmental forensic analysis of the microplastic pollution at "Nattika" Beach, Kerala Coast, India. *Environmental Forensics, 21*(1), 21–36.

Sruthy, S., & Ramasamy, E. V. (2017). Microplastic pollution in Vembanad Lake, Kerala, India: The first report of microplastics in lake and estuarine sediments in India. *Environmental Pollution, 222*, 315–322. https://doi.org/10.1016/j.envpol.2016.12.038

Sundar, S., Chokkalingam, L., Roy, P. D., & Usha, T. (2021). Estimation of microplastics in sediments at the southernmost coast of India (Kanyakumari). *Environmental Science and Pollution Research, 28*(15), 18495–18500. https://doi.org/10.1007/s11356-020-10333-x

Thathsarani, N., Rajakaruna, R., & Jayawardena, U. (2023). Microplastic pollution in the Indian Ocean: Passageways and the contamination of the open ocean and its marine environment. *Toxicology Advances, 5*(2), 9. https://doi.org/10.53388/ta202305009

Thompson, R. C., Olsen, Y., Mitchell, R. P., Davis, A., Rowland, S. J., John, A. W. G., McGonigle, D., & Russell, A. E. (2004). Lost at sea: Where is all the plastic? *Science, 304*(5672), 838.

Thompson, R. C., Swan, S. H., Moore, C. J., & Vom Saal, F. S. (2009). Our plastic age. In *Philosophical Transactions of the Royal Society B: Biological Sciences* (Vol. 364, Issue 1526, pp. 1973–1976). The Royal Society Publishing.

Van Cauwenberghe, L., Vanreusel, A., Mees, J., & Janssen, C. R. (2013). Microplastic pollution in deep-sea sediments. *Environmental Pollution, 182*, 495–499. https://doi.org/10.1016/j.envpol.2013.08.013

Van Cauwenberghe, L., Devriese, L., Galgani, F., Robbens, J., & Janssen, C. R. (2015). Microplastics in sediments: A review of techniques, occurrence and effects. *Marine Environmental Research, 111*, 5–17. https://doi.org/10.1016/j.marenvres.2015.06.007

Van Der Mheen, M., Van Sebille, E., & Pattiaratchi, C. (2020). Beaching patterns of plastic debris along the Indian Ocean rim. *Ocean Science, 16*(5), 1317–1336.

Veerasingam, S., Saha, M., Suneel, V., Vethamony, P., Rodrigues, A. C., Bhattacharyya, S., & Naik, B. G. (2016). Characteristics, seasonal distribution and surface degradation features of microplastic pellets along the Goa coast, India. *Chemosphere, 159*, 496–505.

Vivekanand, A. C., Mohapatra, S., & Tyagi, V. K. (2021). Microplastics in aquatic environment: challenges and perspectives. *Chemosphere, 282*, Article 131151.

Waldschläger, K., Lechthaler, S., Stauch, G., & Schüttrumpf, H. (2020). The way of microplastic through the environment—Application of the source-pathway-receptor model. *Science of the Total Environment, 713*, 136584

Wang, X., Li, C., Liu, K., Zhu, L., Song, Z., & Li, D. (2020). Atmospheric microplastic over the South China Sea and East Indian Ocean: Abundance, distribution and source. *Journal of Hazardous Materials, 389*, Article 121846.

Woodall, L. C., Sanchez-Vidal, A., Canals, M., Paterson, G. L. J., Coppock, R., Sleight, V., Calafat, A., Rogers, A. D., Narayanaswamy, B. E., & Thompson, R. C. (2014). The deep sea is a major sink for microplastic debris. *Royal Society Open Science, 1*(4), Article 140317.

Yaranal, N. A., Subbiah, S., & Mohanty, K. (2021). Distribution and characterization of microplastics in beach sediments from Karnataka (India) coastal environments. *Marine Pollution Bulletin, 169*(June), 112550. https://doi.org/10.1016/j.marpolbul.2021.112550

Yuan, W., Christie-Oleza, J. A., Xu, E. G., Li, J., Zhang, H., Wang, W., Lin, L., Zhang, W., & Yang, Y. (2022). Environmental fate of microplastics in the world's third-largest river: Basin-wide investigation and microplastic community analysis. *Water Research, 210*, Article 118002.

Zhu, L., Bai, H., Chen, B., Sun, X., Qu, K., & Xia, B. (2018). Microplastic pollution in North Yellow Sea, China: Observations on occurrence, distribution and identification. *Science of the Total Environment, 636*, 20–29.

Microplastics in the Marine Ecosystem of India

Soniya Kaushik, Manisha Saini, Jitender Singh Laura, Meenakshi Nandal, and Pradeep Khyalia

Abstract There are multiple types of plastic polymers in the marine environment. Understanding the composition of polymers offers scientists valuable insights into its physicochemical properties and degradation process. The environmental deterioration of plastics plays an important role in the aquatic ecosystem in the creation, dispersion, and buildup of microplastics. Therefore, this chapter provides a concise overview of the processes by which synthetic polymers break down in the environment, several procedures for sampling water and sediment, as well as sample treatments. These treatments include separating plastic particles, which is essential before identifying microplastics. Furthermore, this chapter presents the methodologies employed to determine the chemical composition of plastic particles. The approach outlined in this chapter successfully identified microplastics in soil and water samples taken from estuaries, beaches, and the sea in India.

Keywords Marine pollution · Microplastic · Degradation · Fate · Indian ecosystem impacts

1 Introduction

Plastic pollution is a significant and urgent problem for humanity in the present day. The widespread production and use of plastics, and inadequate waste management, have led to the careless disposal of billions of tons of plastic waste into the environment. Roughly 80% of all plastics produced since 1950 have been found in either aquatic or terrestrial ecosystems (Kataria et al., 2024). The contemporary era is often called plasticine (age of plastic), distinguished by the widespread use of synthetic polymers (Rose et al., 2023). These polymers have been found in biodiversity hotspots such as the Bue Marino cave in Italy, in sediments at concentrations ranging from 10 to 27 items kg^{-1}, and in water at concentrations ranging from 18

S. Kaushik · M. Saini · J. S. Laura (✉) · M. Nandal · P. Khyalia
Department of Environmental Science, Maharshi Dayanand University, Rohtak, Haryana, India
e-mail: Jslmduu@gmail.com; Jitender.env.sc@mdurohtak.ac.in

P. Khyalia and S. Yadav (eds.), *Microplastic Pollution in India and Its Environmental Impacts*, Earth and Environmental Sciences Library,
https://doi.org/10.1007/978-3-031-92530-6_6

to 22 items l^{-1}. In the Argentarola cave, in the shell of the agglutinated foraminifer Lagenammina difflugiformis, the polymers are found at concentrations ranging from 5.4 to 11.9 items kg^{-1}(Romano et al., 2024). Yu and Singh (2023) analysis indicated that the levels of marine microplastics are predominantly within the moderate range (0.005–1 pieces/m^3) across all five continents, with comparatively higher magnitude in North America and the northern coast of Asia. Once plastic waste is introduced into the environment, several mechanisms initiate the degradation of the material and convert large pieces of plastic into microplastics. The physiochemical characteristics of the material entirely determine the magnitude of these effects and the duration of their occurrence. Primary microplastics are produced in the form of tiny particles measuring below 5 mm, such as resin pellets and micro beads. Secondary microplastics are derived from larger plastic fragments, such as the mechanical fragmentation of plastic. MPs are made up of various types of polymers, including polyethylene (PE), polyvinyl chloride (PVC), polyethylene terephthalate (PET), poly (methyl methacrylate) (PMMA), polypropylene (PP), polystyrene (PS), polyamide (PA), polylactide (PLA), and polytetrafluoroethylene (PTFE). The shapes comprise fragments, irregular particles, films, spheres and fibers, with dimensions typically ranging from above 1 micron to below 5 mm (Yu & Singh, 2023). Both natural and synthetic polymers are believed to be widespread in the maritime environment, where they can enter through untreated sewage sludge, textile wastewater discharge and agricultural runoff (Yadav et al., 2023). Colored particles pose significant threats to marine organisms as they consume plastic particles (Qi et al., 2022), which impedes feeding efficiency and leads to malnutrition and mortality. Sometimes, particles are sharp or abrasive, which can cause damage to the digestive tract, resulting in tears, ulcers, and inflammation. Microplastics have the ability to accumulate and absorb environmental pollutants. The transfer of microplastics across the food web has an impact on multiple trophic levels by accumulation in the tissues of smaller organisms when they are ingested and get biomagnified and exacerbating their ecological impacts (Bhusare et al., 2024; Nandal et al., 2022).

Research on microplastics in marine environments is still in progress, and advancing quickly. Although there is not a single, universally accepted method for studying microplastics, several new techniques for their isolation and characterization have been developed recently. There is still an incomplete understanding of microplastics' sources, transport, distribution and microplastics fate across various environmental compartments, which hinders the development of effective strategies to tackle this issue. Given the increasing focus on this issue, the main objective of this chapter is to provide a current overview of microplastics in marine ecosystems, particularly within the Indian subcontinent. This chapter aims to enhance understanding of recent progress, pinpoint challenges, and predict future trends of MPs in marine environments. It covers methods for sampling, isolating, detecting, and characterizing microplastics, along with their sources, degradation processes, occurrence, and eventual fate. The present chapter offers significant insights and tools for future research, proving advantageous to specialists and novices.

2 Sources of MPs in the Marine Ecosystem

The presence of microplastics in the oceans can be attributed to a range of activities occurring at sea, on land, or in the air. The improper disposal of personal care items, cosmetic goods, and plastic packaging material into sewage or wastewater treatment facilities and subsequently into rivers results in the release of substantial amounts of freshwater saturated with microplastics into the ocean (Bhuyan et al., 2021). Personal care products like toothpaste Microbeads of PE, PP, PS, PU, PET, PMMA, and PTFE are found in hand wash, face wash, shampoos, shower gel, bubble bath, face scrubs, hair coloring products, shaving cream, sunscreens, nail paints, eye shadows, and other cosmetics as shown through Fig. 1 (Bashir et al., 2021; Sharma et al., 2024). There are many different causes of marine microplastic contamination, however they can be broadly categorized as air, sea, and inland-based (Auta et al., 2017). The most significant routes for the movement of microplastics from inland regions to the ocean are considered rivers. Approximately 80% of the plastic debris found in the water comes from land-based environments (Lebreton et al., 2017) Plastic garbage from beach-related tourism is dumped into the environment directly, and plastic debris from sewage effluents and municipal drainage systems, as well as from poorly managed inland areas, is blown into the sea through rivers and sea-based sources originate from fishing, shipping, and offshore industries (Bell et al., 2017). Large shipping is considered as a significant source of microplastics due to its emissions and leaks. Plastic particles can be readily introduced into the water by losing or damaging fishing and aquaculture equipment. The main offshore sources are the world's fishing fleet, illegally thrown trash from ships or offshore platforms, marine aquaculture, and a significant percentage of things from lost containers. Furthermore, MPs in the air are also significant sources (Yang et al., 2021). Furthermore, the direct discharge of MPs into oceanic basins is attributed to plastic debris that originates from marine sources and activities. Examples encompass several sectors, such as offshore oil or gas companies, aquaculture operations and equipment, shipping, and recreational boating. Discarded or lost fishing gear, which is a form of plastic garbage, frequently poses a problem for marine organisms, leading to what is known as ghost fishing (Bhuyan et al., 2021).

Bashir et al. (2021) emphasized that there were between 259–1743 micro beads of personal care products, mostly smaller than 1 mm and made of PE/l of coastal sediments in Taipa and Coloane islands in Macao, China. Fibrous microplastics from domestic washing and textile microfibres from manufacturing processes have been noted as a major source of microplastic in wastewater treatment facilities (WWTPs) (Pothiraj et al., 2023). Zhang et al. (2021) have calculated that between 50.6 and 1180 kg of fibrous microplastics are released annually per lakh people from washing synthetic textiles. The researchers emphasized that the total emissions amount to around 2.53–59 kg of fibrous MPs annually in the water that is received from WWT plants operating with a 95% effectiveness in removing these microplastics. Microplastic residues are also lost during the recycling process of plastics. Microplastics infiltrate the environment and can ultimately find their way

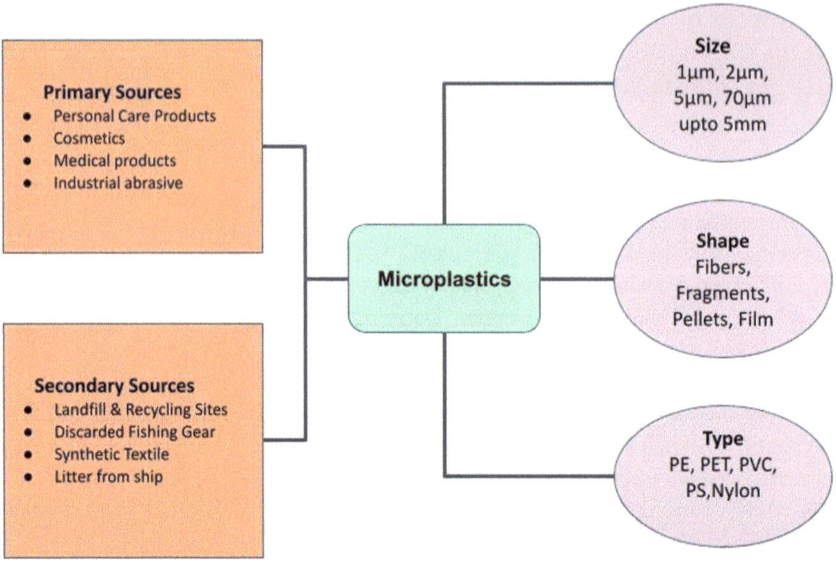

Fig. 1 Classification of MPs based on their origins and its morphological characteristics

into the oceans through many pathways. Categorizing the origin of the microplastics found in the input of the WWTP is difficult due to the intricate composition of the microplastics (Pothiraj et al., 2023; Sharma et al., 2024).

Rubber tires have been recognized as significant contributors to the presence of MPs contamination in the environment. Tire microplastics can be released through tire abrasion, recycling, and repair processes, taking the form of tire wear particles, recycled tire crumb, and debris from tire repair polishing (Luo et al., 2021). Goßmann et al. (2021) have documented that road dust contains 0.5 g of tire wear particles/kg (dry weight). In addition, substantial quantities of microplastic resins in the form of pellets, powder or flakes are utilized in the production of plastic goods, which might result in microplastic pollution when they are lost during transportation or accidents. A study conducted by Aragaw (2020) has suggested that facemasks may contribute to the presence of MPs. Research has shown that used masks emit much greater levels of microplastics (1246 particles/piece) rather than new masks (183 particles/piece) (Chen et al., 2021). Additional notable sources are drilling fluids employed in the discovery of oil and gas, disposable utensils and single-use plastics, medical applications (such as medication delivery devices and imaging), and urban particulate matter. A study about Chennai Metropolitan City, India, showed an average of 227.94 MP per 100 g of street dust (Patchaiyappan et al., 2021a). Paints that are used to cover and protect boats, urban structures, and ships and that contain materials such as PE, PU, PS, polyacrylates, and epoxies are a significant contributor to the presence of microplastics in the waters (Gaylarde et al., 2021; Turner, 2021). Spills, uncontrolled paint transportation, air blasting removal, aging-induced abrasive and erosive release, and release from abandoned materials and structures lead MP to enter wastewater or runoff. Turner (2021) claims that the majority of synthetic micro-debris input into the oceans, around 35%, comes from paint particle emissions.

3 Degradation of Plastics in Marine Ecosystem

Plastics in the maritime environment undergo degradation by physical, chemical, and biological mechanisms (see Fig. 2). UV radiation in the photic zone of seawater can photodegrade plastics, causing weight loss, appearance and texture changes, mechanical degradation, oxidation, and physicochemical changes (Zhang et al., 2021). The deterioration rate is further reduced in the deep ocean due to lower temperatures and the absence of UV light (Coyle et al., 2020). Beaches are the primary sites where plastic breaks down in the marine environment due to favorable conditions that promote rapid photodegradation. Brandon et al. (2016) showed polypropylene, low-density PE, and high-density PE pre-production pellets were weathered for three years. Experimental treatments comprised dry environments with sunlight, seawater with sunlight, and seawater in darkness. Weathering altered hydroxyl, carbonyl, and carbon-oxygen bonds, which were assessed using Fourier Transform Infrared (FTIR) spectroscopy. There are some major degradation mechanisms discussed below:

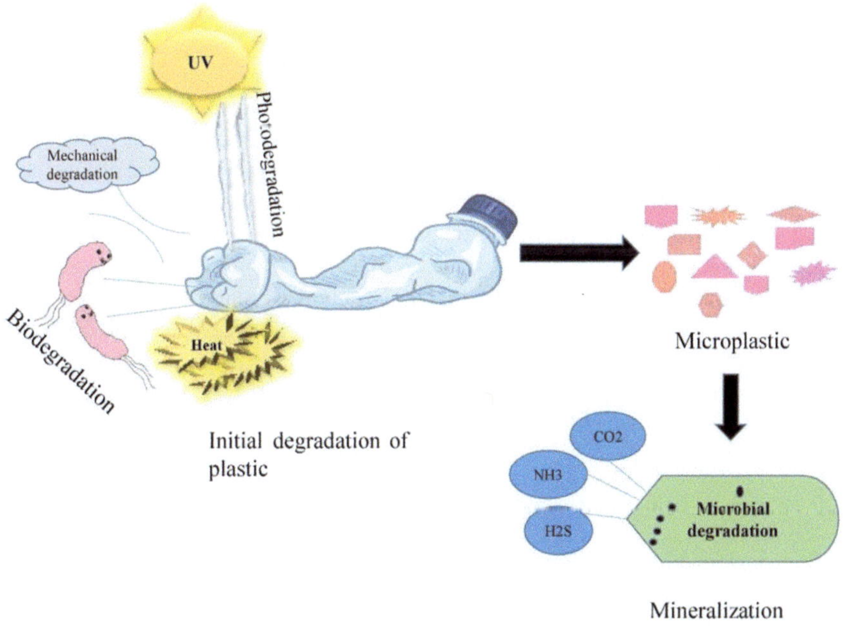

Fig. 2 Degradation mechanism of plastic by various environmental factors (Mechanical, thermal, and biological)

3.1 Photodegradation

Photodegradation is widely acknowledged as the primary mechanism in the environment that initiates the deterioration of plastic. The process of photodegradation of plastics typically occurs through sun irradiation, which triggers free radical-mediated processes. High-energy ultraviolet (UV) irradiation, specifically UV-B (290–315 nm) and medium-energy UV-A (315–400 nm), are primarily accountable (Zhang et al., 2021). The photodegradation of MPs is both cost-effective and environmentally benign. However, it is essential to note that photodegradation efficiency is often poor under natural conditions. Several studies have shown that the inclusion of certain catalysts, specifically TiO_2, can enhance the effectiveness of photodegradation. According to a study by Ariza-Tarazona et al. (2020), the weight loss effectiveness of MPs at 0 °C was nearly six times higher than at 40 °C when using C, N–TiO_2 powder as a catalyst at pH 3. The findings imply that the addition of the photocatalyst C, N–TiO_2 may improve the photodegradation of MP at low temperatures and acidic conditions. In addition to the catalyst, temperature is a significant factor in determining the photodegradation of plastics. Raising the temperature of the photoreaction can directly or indirectly influence the degradation of MP. This alter the activation energy of the chemical reaction and the quantum yield of the photochemical reaction (which indicates the free volume of the polymer), consequently affecting the efficiency of photodegradation. Additionally, it can impact the rate at which free radicals diffuse and alter the efficiency of plastic degradation by modifying the movement of polymer chains at a molecular level (Lin et al., 2022). Hebner and Maurer-Jones (2020) evaluated the influence of the combined processes of photodegradation and water movement on the formation of microplastic particles resulting from alterations in the chemical and physical characteristics of the materials. They exposed polypropylene, polyethylene, and polyethylene terephthalate thin films and polypropylene injection-molded sheets to 254 nm UV light and placed them in aquatic microcosms and observed that extended irradiation resulted in increased formation of microplastics (MPs) in the polyethylene films. A study by Ding et al. (2021) investigated the photodegradation characteristics of polyvinyl chloride (PVC) and poly (ethylene terephthalate) (PET), two widely used plastics when exposed to clay minerals (kaolinite and montmorillonite). Their findings revealed that due to the generation of •OH species by clay minerals, mainly kaolinite, greatly enhanced the photo-degradation of MPs (Ding et al., 2021).

3.2 Thermal Degradation

Thermal degradation is the process by which polymers break down as a result of increased temperature. Plastics can experience thermo-oxidative reactions when exposed to high temperatures. When sufficient heat absorbed by the polymers to surpass the energy barrier, it can break the long chains of the polymer, resulting in the production of radicals (Pirsaheb et al., 2020). These radicals have the ability to undergo a reaction with oxygen, resulting in the production of hydroperoxide. This process is similar to the one that occurs during the deterioration of plastics through exposure to light, where the polymers break down to form hydroxyl free radicals and alkoxy radicals. The reaction can continue to replicate itself until the energy supply is stopped or two radicals collide and generate inert products. This leads to both the decrease and expansion of molecules due to chain scission and cross-linking (Crawford & Quinn, 2016). In addition, temperature, UV radiation, and humidity can have a synergic effect on the degradation of plastics. The primary chemical parameters affecting plastic breakdown in aquatic environments are pH value and salinity. Elevated levels of H+ (acidic) or OH− (basic) in a water-based setting have the potential to accelerate the breakdown of plastics that are prone to hydrolysis, such as polyamides (PA). These two elements can additionally modify the surface of various plastics and microplastics, affecting their behavior in a water environment and their interaction with other components and contaminants present in the water (Zhang et al., 2021). The rate and amount of degradation, occurring via radical, ionic, or mixed routes, can be assessed through variations in the sample's mass, molecular weight, detection and measurement of reaction enthalpy changes, and qualitative examination of volatile by-products (Pielichowski et al., 2023). Material qualities generally determine the rate of thermal deterioration. The thickness of the thermally oxidized layer in low-density polyethylene (LDPE) is contingent upon the temperature (Vahabi et al., 2014). A hydrothermal coupled Fenton system has been established to degrade ultrahigh-molecular-weight polyethylene, resulting in a 95.9% weight reduction in 16 h and a 75.6% mineralization efficiency in 12 h. The significant efficacy is ascribed to the combination of hydrothermal hydrolysis, a proton-rich environment, and the extensive generation of hydroxyl radicals (Hu et al., 2021).

3.3 Mechanical Degradation

The mechanical degradation of microplastic waste occurs as a result of contact with both natural and human-made objects in aquatic ecosystems. Microplastics can undergo mechanical weathering in the water column due to exposure to shear stress pressures. Mechanical degradation of polymers can occur as a consequence of the freezing and thawing of plastics in aquatic environments. The impact of external forces is influenced by the mechanical characteristics of the plastics, such as the ability of a plastic product to withstand changes in shape without the development of cracks, known as Elongation at break, which ranges from 1% to around 900% for various types of plastics (Crawford & Quinn, 2016; Zhang et al., 2021). Plastics with a reduced elongation at break value are more prone to fracturing when subjected to external tensile stresses. The findings by Enfrin et al. (2020) indicate that microplastics from face cleansers undergo fragmentation into nanoplastics when subjected to modest shear stress, resulting in an increased quantity of plastic particles being introduced into the environment. Crushing minimizes the specific surface area and maximizes the contact area for reactions, whereas abrasion greatly accelerates the degradation rate of polymers, and persistent stress leads to the disruption of carbon-carbon bonds on the polymer backbone, inducing fracture. Furthermore, the effectiveness of plastic degradation could be influenced by temperatures, qnd mechanical combination (Lin et al., 2022). Mechanical degradation plays a crucial role in the deterioration of synthetic fibers like polyester, polyolefin, acrylic, and PA. Other mechanical degradation processes encompass temperature fluctuations and alternating wet and dry conditions (Corcoran, 2022). Kalogerakis et al. (2017) showed that exposure to sunlight and mechanical stress increases the amount of weathering. In the latter, plastic strips were inserted into sand-filled bottles, and the bottles were rotated for 24 h at a constant speed. Mechanical abrasion alone can cause some degree of polymer degradation, as evidenced by weight drop of plastic by 14% and the appearance of microplastics that were imperceptible to the unaided eye. Song et al. (2017) examined the impact of mechanical abrasion and ultraviolet exposure on MPs in a simulated coastal environment and demonstrated that the extent of mechanical deterioration differed according to the kind of polymer. Polypropylene (PP) and polyethylene (PE) particles exhibit a minimal tendency for degradation through mechanical weathering alone, whereas expanded polystyrene (EPS) disintegrates into multiple smaller fragments only due to frictional forces.

3.4 Biotic Degradation

Plastic biodegradation can occur through aerobic and anaerobic processes. The former process is commonly seen in natural environments and produces carbon dioxide and water. On the other hand, the latter process takes place in sediments and landfills and produces carbon dioxide, water, and methane (Lin et al., 2022). Fungal and bacterial communities frequently release organic compounds, including acids, as a component of their decomposition processes. Consequently, this can establish a micro-environment that promotes enhanced breakdown of specific plastic materials, such as polyamides (Nylon).

Du et al. (2021) reviewed that MPs are utilized as substrates for the formation of biofilms. As the biofilm grows, pits and cracks form in the structure of the MPs, which weakens them. Othman et al. (2021) reviewed various types of plastic group degradation by microbial enzymes, which selectively target the polymer structure of microplastic, breaking it down into its constituent monomers. These monomers are then utilized as a carbon source in the microorganism's energy generation cycle. Numerous properties of the plastic material, including its molecular weight, crystallinity, functional groups, and additives, affect how quickly it breaks down (Corcoran, 2022). Streptomyces sp., B. pumilus, B. pumilus M27, B. subtilis H1584, K. palustris M16, P. simplicissimum, and B. sphericus are some microorganisms that break down plastics and MPs in rivers and the ocean (Lin et al., 2022).

In addition to bacteria, MPs are also utilized and broken down by fungi. Novel fungi like Aspergillus flavus (Zhang et al., 2020) with enhanced capacity for decomposing microplastics (MPs) have recently been discovered. Nevertheless, most plastic materials resist biotic breakdown since fungi and bacteria have not yet developed sufficiently efficient enzymes can break down these artificially produced materials. Auta et al. (2021) assess the growth response and degradation mechanism of polypropylene (PP) by Bacillus sp. strain 27 and Rhodococcus sp. strain 36, which were obtained from mangrove sediments. Both bacterial strains were found to be capable of using PP microplastics for their growth, as evidenced by a decrease in polymer mass. Specifically, Rhodococcus sp. strain 36 exhibited a weight loss of 6.4%, while Bacillus sp. strain 27 measured 4.0% after 40 days of incubation. Harshvardhan and Jha (2013) examined bacteria obtained from pelagic waters to determine their capacity to break down low-density polyethylene as the only carbon source for growth. Following a 30-day incubation period, the isolates of Kocuria palustris M16, Bacillus pumilus M27, and Bacillus subtilis H1584 exhibited weight loss of polyethylene of 1%, 1.5%, and 1.75% respectively.

4 Distribution Processes of MPs in the Marine Ecosystem

Microplastics are dispersed horizontally and vertically throughout ocean waters through various physical and biological processes. The quantities of microplastics vary according to the environmental variables and the specific sampling location. Microplastic concentrations are greater in areas with high industrialization rates or those close to river mouths as documented by (Pothiraj et al., 2023). Qi et al. (2022) stated the declining dispersion of microplastics from coastal areas to the open ocean as a result of contributions from terrestrial sources, rivers, and human activities. After input from different sources, the MPs in the water remain scattered in the surface layer, making them more accessible for ingestion by marine creatures. Multiple factors lead to microplastic buildup in specific regions, amplifying the ecological risks to marine ecosystems. In horizontal distribution, the dispersion of microplastics from their origins occurs due to hydrodynamic mechanisms, such as coastal currents, drift, and river outflow. Larger basin-scale and regional currents transport microplastics to distant areas, while nearshore currents transport them to beaches, where they may subsequently re-enter the open ocean. In addition, rotational ocean currents facilitate the transportation of surface plastics to convergence zones inside oceanic gyres, forming concentrated areas where these plastics accumulate (Coyle et al., 2020). There are multiple possibilities for the vertical distribution of microplastics Polymer density is a determining factor in the distribution of polymers in the aqueous column. Lower-density polymers will float, whereas those with denser densities than seawater would sink. High-density microplastics tend to accumulate in sediments, whereas some are found in intermediate depths. PE-MP and PP-MP, with a density less than seawater (1.03 g/cm^3), have a tendency to remain on the surface (Shamskhany et al., 2021; Gokul et al., 2023). Vertical dispersal is significantly impacted by the oceanic conditions and the features of the microplastics (Li et al., 2020). The combination of MP with organic matter and minerals in the ocean increases particle density, resulting in their sinking. Turbidity currents, characterized by significant sediment content, significantly impact the dispersion and deposition of microplastics in water. Their morphology also influences the rate of descent of microplastics. For instance, fibers with a higher surface area to volume ratio exhibit slower sinking rates in comparison to fragments. The irregularly shaped fragments are distributed at various depths in the water column of ocean. Dimension is another crucial component. The particle size significantly influences long-distance movement. It influences both the speed of motion and the process of descending. Biofouling is the accumulation of microorganisms, plants, algae, or animals on surfaces in aquatic environments. Research has emphasized the correlation between biofouling and the dimensions and surface area-to-volume ratio of MPs. Microplastics may experience vertical motion as suspended sediments settle on them, forming flocs and enhancing the settling velocity (Laursen et al., 2022). Marine organisms, particularly zooplankton, consume and excrete microplastics, accumulating these particles in sediment. Bioturbation refers to the process by which organisms, such as animals or plants, disturb and mix sediment or soil through their activities. Bioturbation also modifies the distribution of microplastics in sediments and can potentially increase sedimentation. The vertical distribution of marine microplastics is influenced by the rate at which they settle and resuspend, as well as the settling and resuspension of their aggregates. The distribution status of various polymer types of plastic has been shown on the basis of density in the ocean through Table 1.

5 Occurrence and Fate of MP in the Marine Environment

An understanding of microplastics' occurrence and fate is essential to estimate the magnitude of microplastic pollution. Invariably, any debris present in a water body will eventually find its way into the ocean. The ecosystems most severely contaminated by microplastics are marine circulation, estuaries, and other coastal locations frequently inhabited by humans (Yang et al., 2021). Microplastics are transported by water and wind power forces, progressively spreading and dispersing across the ocean. Over time, they have become widespread, including in the vast ocean gyres. Small and colored microplastics can rapidly infiltrate the bodies of aquatic organisms. This indicates that larger plastics in the lake and its adjacent sources (including domestic and industrial waste) are fragmenting into smaller particles. This suggests that the dispersion of microplastics is likely to increase during periods of rainfall. The Kumaraswamy Lake included microplastics of various sizes, primarily categorized as 5–3 mm, 3–1 mm, 1–0.5 mm, and 0.5–1 μm. The majority of microplastics in samples had a size range of 3–1 mm. Microplastic particles measuring 5–3 mm and 3–1 mm in size were found to be more prevalent during the monsoon season, accounting for 34.89% and 30.20% of the total, respectively, in the outlet area (Ephsy & Raja, 2023).

Table 1 Classification, characteristics and applications of commonly used plastics (Bhuyan et al., 2021; Coyle et al., 2020; Ranjani et al., 2021)

Resin type	Monomer unit	Characteristics	Application	Specific gravity (g/cm^3)	Float/sink
Polyethylene (PE)	Ethylene	Chemically stable and extremely resistant soft plastics	Bags, microbeads, bottles, pipes	0.4–0.5	Float
Polypropylene (PP)	Propylene	Stable, chemical and high-temperature-resistant	Rope, food containers, fishing gear	0.90–0.92	Float
Polystyrene (PS)	Styrene	Shiny, heat insulating	Jewel boxes, electronic appliance enclosures	0.28–1.04	Float or sink
Polyethylene terephthalate (PET)	Ethanediol	Clear, lightweight, tear-resistant, water repellent	Water bottles, home appliances, medical implants	1.37–1.38	Sink
Polyvinyl chloride (PVC)	vinyl chloride	Robust, high fire resistance	Pipes, window profiles	1.16–1.30	Sink
Polyamide (PA)	Hexamethylenediamine	Fibre, high molecular weight polyamides	Tents, clothes, tires, fishing nets	1.14–1.15	Sink
Polyester (Poly)	Terephthalic acid and ethylene glycol	High strength, resistant to chemicals, durable	Drilling fluids, air blasting media, textiles	1.4	Sink

A study by Qiu et al. (2015) examined the number and composition of microplastics at five sediment sites in China for the first time. The quantity of microplastics ranged from 251 to 436 items/50 g of sediment. Polyesters, polyethylene terephthalate, high-density polyesters, and polystyrene were characterized by u-FTIR. The movement of microplastics is a constantly changing process, which can be spread throughout the entire marine environment not only through physical actions like crushing and deposition along the coast but also through chemical reactions like oxidation or hydrolysis. Additionally, microplastics can be transported to all parts of the ocean through biological processes such as absorption, digestion, and excretion (Yeng et al., 2021).

Transportation from these shores and beaches to deeper columns pertains to the ultimate repositories where pollutants are accumulated and retained. Many of these micro-fragments can aggregate in the central oceanic areas known as gyres. These ocean currents patterns facilitate the transport of plastic trash while simultaneously contributing to its accumulation. Estimates indicate that ocean gyres have resulted in the Pacific Ocean retaining approximately 35% of the global total of oceanic plastic (Cózar et al., 2014; Tursi et al., 2022). Various factors like microbial interactions, biofilm formation, residence time in the photic zone, particle aggregation in the water column, organism uptake, and the vertical and horizontal movement of pollutants, plastic particles, microorganisms, and intermediate degradation products in marine environments all affect the fate of microplastics. Additionally, sedimentation and diagenetic processes may also affect the fate of microplastics (Fig. 3). Microplastics have been discovered on the seafloor at depths ranging from 2200 to 10,000 m. These microplastics consist of both high (Courtene et al., 2017) and low (Peng et al., 2018) density particles in comparison to seawater. Microplastics collect in the global ocean circulation because of their lower density than seawater, causing them to float on the sea surface. The converging sea currents then concentrate and retain this trash for extended periods. Environmental elements such as salinity, temperature, oxygen levels, and high nutrient concentrations can also potentially affect the growth of microorganisms and can potentially change the density and movement of microplastics (Rogers et al., 2020). A meta-analysis of worldwide marine microplastic occurrence features was integrated with a Monte Carlo simulation to evaluate worldwide marine microplastic research hotspots and their occurrence features and examination of the drivers and source-sink dynamics of marine microplastics (Tang et al., 2023).

Liu et al. (2022) compiled a MP dataset comprising 649 samples from 24 bays to elucidate the geographical variations, drivers, transfer mechanisms, compositional profiles, and environmental risks associated with sedimentary microplastics. The abundance of microplastics ranged from 0.72 to 1963.96 items/kg (d.w), with elevated concentrations predominantly found in East Asian bays. The distribution of abundance was influenced by river plastic pollution, aquaculture productivity, and hydrodynamic conditions. A strong positive correlation was observed between MP abundance in water and sediment, with microplastic sedimentation linked to polymer density, hydrodynamic circumstances, and sediment characteristics. The predominant morphology and polymer of sedimentary MPs were fibers and PP, respectively,

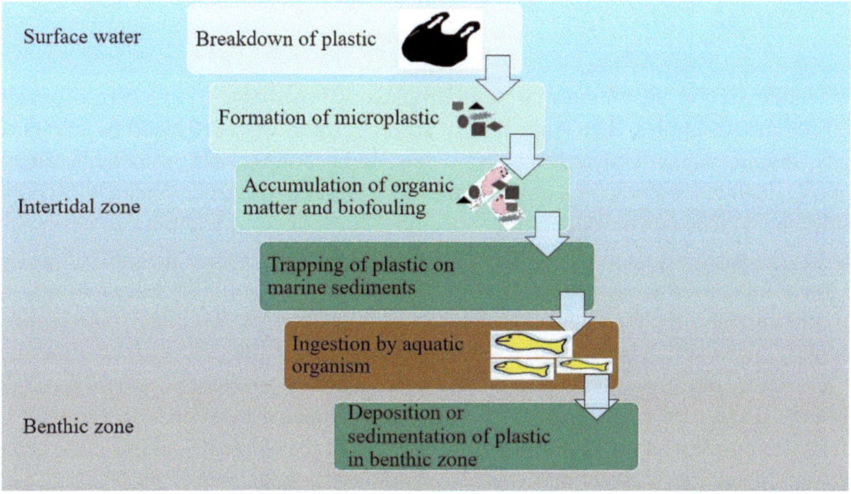

Fig. 3 Distribution process of MPs in different zones, including surface water zone, intertidal zone, and benthic zone, comprising their breakdown and deposition at the bottom of the marine ecosystem

and the compositional similarity of microplastics reduced with increasing geographical distance. Alarif et al. (2023) reviewed the studies where the concentration of MPs in the Mediterranean Sea was found to be 2.60 $\times$ 10^5 MPs/km^2 and in the Arabian Gulf found up to 4.6 $\times$ 10^4 particles/km^2, which has been reported to be higher than the water samples from the other Middle East region.

6 Detection and Analysis Methods of MPs in the Marine Environment

The identification and characterization of microplastics involves a series of processes, including sample collection, preservation, processing, and subsequent analysis utilizing various techniques. The methodologies for each of these procedures vary depending on the specific samples. Various techniques and devices are employed to gather water and sediment samples, as an illustration. Diverse techniques are employed for the processing of both biological and non-biological substances. The selection of an analytical approach is contingent upon the sample types, the properties of microplastics (MPs) inside the sample, their attributes, and their connection to other organic or mineral substances.

6.1 Sample Collection

The initial step in analyzing marine microplastics involves the gathering of water and sediment samples. The collection of surface water was performed by manta tr awls, andmid-water depths, using bongo nets, and benthic trawls in sea beds. There are two distinct methods for sampling marine water to detect microplastics: (a) bulk sampling, which involves collecting a specific volume of water without decreasing its quantity during the process, and (b) water-reduced sampling, where the sample quantity is gradually reduced throughout the process, retaining only the relevant portion for subsequent analysis. These procedures are effective for obtaining microplastic samples by extracting marine microplastics that have been filtered from nets (Sharma et al., 2023). Improved hydraulic pressure separates low-density polymers through flotation (Bhuyan et al., 2021). MPs in sediment samples depend on winds and tides. Marine microplastics settle in sediments. Forceps, grab samplers, and box corers can be used to sample. Results may vary by sampling location and depth (Prata et al., 2019). Beach sediments can be grabbed with hand spades or stainless steel spoons. Grabs and corers are used to acquire samples from deeper depths (Razeghi et al., 2021). A large tube is used to remove a cylindrical sediment sample by corers. Large box corers may collect sediment samples from any depth with minimal disturbance. Van Veen Grab samplers work with several sediment types. The designs reduce blockage and sample loss, according to Brander et al. (2020). Sathish et al. (2019) collected five random sediment samples from a 50 × 50 cm quadrat at a depth of roughly 5 cm using pre-cleaned stainless-steel tweezers. A total of fifty samples were collected from 5 sites and subsequently air-dried for further examination. Dutta et al. (2022) in his study, they collected samples from each sampling location across all beaches, resulting in a total of three sediments and three saltwater samples from each beach during every sampling event. A 1 m × 1 m square quadrat was positioned at the sampling site, and the upper 1–2 cm of silt was removed with a stainless-steel spoon and collected in 1 L glass containers. Seawater samples were obtained concurrently with sediment samples from a depth of approximately 50 cm in the water column at the same location (Dutta et al., 2022). Alternative methods such as beachcombing, marine observer surveys, and biological samples have been developed to measure ocean plastics with low logistical preparation and ease of management and are cost-effective (OSPAR, 2007; Ryan et al., 2009).

6.2 Separation of Microplastic from the Matrixes

In order to maintain the integrity of the samples, they can be either refrigerated or preserved using ethanol. Microplastics must be extracted from the water and sediment samples before quantification and characterization. seiving, filteration and density separation (mixing the sample with a saturated salt solution and using the resulting filtrate for further filtration), elutriation (injecting liquid at the bottom of a column and allowing the microplastics to float), and the use of a specific separator, such as the Korona-Walzen-Scheider electrostatic bell separator (Prata et al., 2019), have been performed for sample segregation. Removal of debris is a necessary step to obtain microplastics from the natural environment frequently as matrices possess biological or organic substances sticking to their surface, necessitating the process of digestion. Digestion is essential because it helps differentiate between biological dead matter and microplastics, preventing an overestimation of their concentration. An approach that can decrease the amount of organic debris without affecting the integrity of the microplastics should be utilized. Various techniques might be employed to carry out the digestion of the material. Acid digestion is commonly used to break down organic materials from microplastics, but optimizing time, acid concentration, and temperature is crucial to effectively remove organic matter while preserving the quality of the plastics. Nitric acid is frequently employed; however, it can cause melting of low-density polystyrene (PS) and the yellowing of other polymers, necessitating careful use. An alternative, alkali digestion, often utilizes potassium hydroxide (KOH) at a 10% concentration and 60 °C for 24 h, although it can leave oily or tissue residues that complicate subsequent vibrational spectroscopic analyses. Sodium hydroxide (NaOH) is also effective but may stain PVC and PET. Oxidizing agents, particularly a 30–35% hydrogen peroxide solution, show greater efficiency than acid and alkali methods without significantly deteriorating the polymers. Additionally, enzymatic digestion using enzymes like papain, trypsin, and collagenase demonstrates high activity (72–88%) without harming the polymers, presenting a viable alternative for digestion. During the processing of biological samples, acid or base digestion is used to isolate microplastics. These acids and bases efficiently break down the tissue, leaving behind stubborn remnants. The assessment of plastic ingestion in the gastrointestinal tract of animals can be readily determined through the dissection of bigger marine creatures, such as whales and sharks. Dissection is a simple and cost-effective technique for identifying plastic particles larger than 500 μm.

6.3 *Analytical Methods for Characterization of Microplastics*

Sorting and visually identifying large-sized microplastics can be easily accomplished using forceps and the naked eye. However, there is a risk of overlooking microplastics that are approximately 1 mm in size and have the same color as other particles present in the sample. Many techniques can be employed to identify these particles, including optical microscopy, stereoscopy, fluorescence microscopy, and hyperspectral imaging (HIS) (Baruah et al., 2021). The scanning electron microscope (SEM) combined with energy-dispersive X-ray analysis is a valuable tool for accurately determining the structure and composition of MPs (Luo et al., 2021). Fourier transform infrared spectroscopy (FTIR) is the highly recommended technique for characterizing microplastics. While this procedure is dependable, the duration of the operation lengthens as the quantity of particles grows (Veerasingam et al., 2021). ATR-FTIR is a more precise method for characterizing the opaque and thick samples than FTIR. The micro-FTIR technique is applicable to samples with a maximum size of 20 μm. Micro-FTIR, while a dependable method, is somewhat costly and has the potential to contribute to pollution (Meyers et al., 2022). Other than FTIR, Raman spectroscopy is a functional method for identifying and analysing microplastics. The molecular structure can be determined using this vibrational spectroscopic approach, which relies on inelastic light scattering and has the ability to measure particles as small as 1 micronmeter. Advanced techniques have been employed to improve the effectiveness of this procedure. The chemical picture of microplastics is obtained using a combination of Raman spectroscopy and an optical microscope called micro-Raman spectroscopy (μRS). SERS (surface-enhanced Raman scattering) can be employed to detect more efficient signals in cases where low-energy vibrational domains are occasionally overlooked (Chakraborty et al., 2022). In SERS, the sample is absorbed onto a plasmonic surface consisting of a nano-substrate composed of gold and silver. The Raman scattering is enhanced by the plasmonic oscillation of the surface electrons (Zhou et al., 2021). A portable Raman device equipped with an additional tree algorithm is utilized to automatically identify microplastics in seawater. This device can categorize various types of microplastics (Yang et al., 2020). Stimulated Raman scattering (SRS) has also been used to identify microplastics. To fully analyze the molecular properties of microplastics, regardless of their size restrictions, Pyr-GC/MS is used (Bouzid et al., 2022; Rødland et al., 2022). The detection process involves subjecting the microplastics to thermal decomposition in the absence of oxygen to produce gas. This gas is then trapped at low temperatures on a chromatographic column for the purpose of separating and identifying its components using mass spectrophotometry. This approach does not provide any information about the size and shape of the molecule being identified, and it also causes damage to the molecule (Prata et al., 2019). The use of thermal extraction and desorption (TED) in combination with gas chromatography (GC) and mass spectrometry (MS) approach simplifies the manipulation of samples and is significantly more effective in comparison to Pyr-GC/MS (Rødland et al., 2022; Sorolla-Rosario et al., 2022). Currently, additional techniques such as liquid chromatography, portable XRF (X-ray fluorescence spectrometry), and atomic force microscopy combined with infrared microscopy are being employed successfully to identify microplastics smaller than 500 μm with high accuracy (Luo et al., 2021). There has been growing interest in biosensors and microfluidic chips in recent years due to their multiple advantages, such as high sensitivity, mobility, and ease of detection (Elsayed et al., 2021).

7 Microplastics in the Indian Marine Ecosystem

India is currently grappling with a significant issue of plastic pollution. Extreme events and pandemics exacerbate the situation by increasing the amount of plastic garbage produced. The pronounced impact of the monsoon-induced seasonality has significantly changed the presence of plastics in the surface water, as evidenced by research conducted in the north Indian Ocean and other areas (Alimi et al., 2021; Gupta et al., 2021). This indicates that seasonal variations in the environment, such as the higher volume of water flowing during the monsoon season, are important in carrying significant quantities of litter and plastics from sources on land into the coastal ecosystem. As evident in 2015, the flood had a significant effect on the presence of MP pellets along the Chennai coast in India, as well as the presence of MP resin pellets in sediments near Agatti Island and the beach areas of the southeast coast of India (Pavithran, 2021; Veerasingam et al., 2016).

Significant variations in the distribution of MPs have been observed due to spatial areas like the Indian Ocean and the Bay of Bengal. Li et al. (2021) findings indicate that the levels of microplastics pollution in the Eastern Indian Ocean (EIO), whether in the epeiric sea or the open ocean, are among the highest compared to other oceans worldwide. Their field observations at thirty-six stations in the EIO, utilizing a standard manta trawl with a mesh size of 330 μm to collect samples from the surface water. The abundance of MPs in the EIO ranged from 0.01 items/m^2 to 4.53 items/m^2, with an average concentration of 0.34 ± 0.80 items/m^2. The average concentration of microplastics in the Bay of Bengal was 2.04 ± 2.26 items per square meter, while in the open ocean of the Eastern Indian Ocean (EIO), it was 0.16 ± 0.17 items per sq m. The polymer composition study using Micro-FTIR revealed that the bulk of MPs were composed of polypropylene (PP, 51.11%) and polyethylene (PE, 20.07%) (Li et al., 2021).

7.1 *Microplastic in Beaches*

The polymers that were identified on beaches include plastic bags, styrofoam cups, beverage bottles, plastic rope, net pieces, bottle caps, and food wrappers. The waters in South Juhu Creek, Mumbai, India, contain a concentration of plastic waste at a rate of 2 g/kg (Manickavasagam et al., 2020). Gopinath et al. (2020) investigated the levels of microplastic (MP) concentrations in water and sediments. They found that the concentration of MP was 6 particles per liter of water and 27 items per kilogram of sediment. The particle size in water ranges from 0.33 to 2 mm, whereas in the sediments of the Red Hills Lake in India, it is 2 mm. The polymers, namely HDPE, LDPE, PP, and PS, were identified utilizing SEM and ATR-FTIR techniques.

Sathish et al. (2020) investigated the levels of microplastic (MP) concentrations in sediments and water. They found that the average MP concentration in sediments was 134.29 items per kilogram, while in water, it was 19.87 items per liter. The distribution of MPs in sediments and water was 44% and 58%, respectively. Karthikeyan et al. (2023) studied the particle size in the sediment and water near the coast of Tuticorin, Gulf of Mannar (GoM), India, ranging from 1 to 5 mm and from 500 microns to 1 mm, respectively. The polymer types PE, PP, PP–PE, PA, PET, PEST, PVC, PS, and PVA were identified using FTIR-ATR, SEM, and EDAX techniques.

7.2 Microplastic in Sea Salt

Various types of plastic polymers have been identified in edible salt from different salt-producing regions in India. Selvam et al. (2020) investigated the occurrence of MPs in sea salt and found that MP accounted for 60% of the overall distribution. A particle size of 100 microns was identified at the Tuticorin coastal salt pan stations in the Gulf of Mannar. The polymer types polyethylene (PE), polypropylene (PP), cellulose (CL), and nylon (NY) were detected using Fourier Transform Infrared Spectroscopy (FTIR) and Atomic Force Microscopy (AFM). Sathish et al. (2020) discovered that the sea salt of Tuticorin included a concentration of 72 items/kg of MPs. These MPs were distributed in sizes ranging from 100 to 500 microns, making up 100% of the total distribution. The polymer kinds, namely PE, PP, PA, and PEST were identified using SEM-EDAX.

Samples of crystal and powdered salt from Tamil Nadu and Gujarat were collected and examined for MPs contamination. The microplastic content in these salts varied between 46 to 115 particles/200 g in the Gujarat samples and 23 to 101 particles/200 g in the Tamil Nadu samples. The predominant microplastics observed were red and blue fibrous materials. The most frequently found microplastics in these edible salts were PE, polyester, and PVC, originating from marine environments and salt processing facilities (Vidyasagar et al., 2021).

7.3 Microplastics in Marine Biota

Microplastics (MPs) are manmade contaminants capable of adsorbing hazardous compounds from surrounding water and being absorbed into the bodies of fish. In a study conducted by Gurjar et al. (2022), on the coastal waters of Mumbai, the gastrointestinal tracts of marine organisms were sampled. They found the average abundance of microplastics in pelagic fish species ranged from 6.74 ± 2.74 to 9.12 ± 3.57 items/ individual, whereas in demersal species, ranged from 5.62 ± 2.27 to 6.6 ± 2.98 items/ individual. Seven distinct colors of MPs (red, blue, black, transparent, brown, green, and yellow) were identified from the analyzed samples. MPs of less than 250 μm constituted the predominant size category in the biota samples, with the exception of Bombay duck and Malabar sole fish (Gurjar et al., 2022). The degree of microplastic present in zooplankton, shellfish, and finfish and samples collected from Port Blair Bay, A&N Islands, was quantified, revealing average concentrations of 0.12 ± 0.07 pieces/ zooplankter and 10.65 ± 7.83 particles/ specimen, respectively. Peak MP consumption was recorded in adult *Carangoides malabaricus*. Fibre was the most abundant, followed by fragments and pellets. Nevertheless, pieces were prominent in zooplankton (Goswami et al., 2020). The study conducted by Saha et al. (2021) reveals the occurrence of MPs in commercially significant shellfish and finfish specimens from the Sal estuary in Goa, with average concentrations of 4 ± 2 MPs/g body weight (bw) for Crassostrea sp., 3.2 ± 1.8 MPs/g bw for Perna viridis, and

0.7 ± 0.3 MPs/g bw for Paphia malbarica, respectively. The highest concentrations were seen in finfish (gastro-intestinal tract) (*Mugil cephalus*) at 7.8 ± 4, followed by (*Gerres filamentosus*) at 5.3 ± 4.9, (*Arius jella*) at 4.6 ± 2.6, and (*Etroplus suratensis*) at 1.4 ± 0.3 MP/g bw. The most abundant polymer types were polyacrylamide (PAM), polyacetylene, polyvinyl chloride (PVC), ethylene vinyl alcohol (EVOH), and polyamide (nylon) (Table 2).

8 Conclusion

This book chapter highlights the severity of marine microplastic contamination, which presents substantial risks to marine ecosystems. We have analyzed the routes of plastic entering the oceans and the process of plastic accumulation, specifically focusing on its creation and subsequent destiny. By conducting a thorough analysis of microplastic levels in marine environments throughout the Indian sub-continent, we have uncovered the specific distribution patterns of microplastics in this region. In addition, we have examined the analytical approach and detection technologies currently employed in microplastic investigations. The identification and analysis of microplastics (MPs) in various samples is crucial for comprehending several elements related to marine pollution caused by MPs. Moreover, researchers are currently investigating traditional techniques to assess their efficacy in removing microplastics from various samples.

9 Recommendations

It is necessary to enhance the current procedures to improve their efficiency, viability, and cost-effectiveness. This information highlights the immediate necessity for further scientific research and policy actions to tackle the significant risks posed by microplastic contamination to both worldwide marine ecosystems and human health. We advocate for more stringent regulatory measures to decrease plastic consumption, foster sustainable manufacturing and consumption practices, provide substitute materials, and augment public consciousness regarding environmental preservation. Future studies should prioritize investigating unexplored coastal regions, the diverse marine biota, the patterns of microplastic (MP) weathering, the impacts of biofouling, and the ultimate fate of microplastics in the marine environment. Furthermore, it is essential to establish standardized methodologies for the separation and analysis of microplastics to facilitate meaningful comparisons across various research endeavors.

Table 2 Characteristics of various types of microplastics found in marine ecosystems of India

Location	Sample type	Polymer identified	Polymer quantity	Particle size	Color	Shape	Reference
Eastern Indian ocean	Deep basin sediment	Rayon, polyester, PE	170.5 ± 140.2 particles/kg	44–5000 μm	Black, white, and blue, yellow, transparent	Fragment, fibers	Qi et al. (2022)
Marina beach	Beach sediment	Polyethylene	459 particles/kg	< 2 mm	Transparent, white, red	Unidentified particles, fragments, fibers, foam, pellets	Venkatramanan et al. (2022)
Kayamkulam estuary	Sediment	Polyester, PP, PE	433 particles/kg	< 1000 μm	White & coloured	Fibers, film	Radhakrishnan et al. (2021)
Odisha coast	Beach sediment	HDPE, PS, PVC, Acrylonitrile	258.7 ± 90.0 particles/kg	500–1000 μm	White, transparent, blue	Fragments, fibers, spherules	Patchaiyappan et al. (2021b)
Kanyakumari	Beach sediment	N.M	343 particles/ 400 g	< 5 mm	Transparent, blue, green	Fibers and fragments	Sundar et al. (2020)
Silver beach	Beach sediment	PVC, PE, nylon	204 particles/kg	> 5 mm, < 5 mm, < 1micronmeter	Blue green	Fragments, pellets, fibers	Vidyasakar et al. (2020)
Tamil Nadu	Sediment	PE, PP, nylon	439 ± 172 to 119 ± 72 particles/kg	0.5 to 3 mm	White, green, red, blue	Fibre, fragment, foam	Sathish et al. (2019)

References

Alarif, W. M., Budiyanto, F., Bawakid, N. O., Alshehri, N. D., Al-Farawati, R. K., Shaban, Y. A., & Ali, A. M. (2023). The occurrence of microplastic in marine ecosystems in the Middle East: A review. *Regional Studies in Marine Science, 67*, Article 103208. https://doi.org/10.1016/j.rsma.2023.103208

Alimi, O. S., Fadare, O. O., & Okoffo, E. D. (2021). Microplastics in African ecosystems: Current knowledge, abundance, associated contaminants, techniques, and research needs. *Science of the Total Environment, 755*, Article 142422.

Aragaw, T. A. (2020). Surgical face masks as a potential source for microplastic pollution in the COVID-19 scenario. *Marine Pollution Bulletin, 159*, Article 111517.

Ariza-Tarazona, M. C., Villarreal-Chiu, J. F., Hernández-López, J. M., De la Rosa, J. R., Barbieri, V., Siligardi, C., & Cedillo-González, E. I. (2020). Microplastic pollution reduction by a carbon and nitrogen-doped TiO2: Effect of pH and temperature in the photocatalytic degradation process. *Journal of Hazardous Materials, 395*, Article 122632.

Auta, H., Abioye, O., Aransiola, S., Bala, J., Chukwuemeka, V., Hassan, A., Aziz, A., & Fauziah, S. (2021). Enhanced microbial degradation of PET and PS microplastics under natural conditions in mangrove environment. *Journal of Environmental Management, 304*, 114273. https://doi.org/10.1016/j.jenvman.2021.114273

Auta, H. S., Emenike, C. U., & Fauziah, S. H. (2017). Distribution and importance of microplastics in the marine environment: A review of the sources, fate, effects, and potential solutions. *Environment International, 102*, 165–176.

Auta, H., Emenike, C., Jayanthi, B., & Fauziah, S. (2018). Growth kinetics and biodeterioration of polypropylene microplastics by Bacillus sp. And Rhodococcus sp. Isolated from mangrove sediment. *Marine Pollution Bulletin, 127*, 15–21. https://doi.org/10.1016/j.marpolbul.2017.11.036

Baruah, A., Sharma, A., Sharma, S., & Nagraik, R. (2021). An insight into different microplastic detection methods. *International Journal of Environmental Science and Technology, 19*(6), 5721–5730. https://doi.org/10.1007/s13762-021-03384-1

Bashir, S. M., Kimiko, S., Mak, C. W., Fang, J. K. H., & Gonçalves, D. (2021). Personal care and cosmetic products as a potential source of environmental contamination by microplastics in a densely populated Asian city. *Frontiers in Marine Science, 8*, Article 683482.

Bell, J. D., Watson, R. A., & Ye, Y. (2017). Global fishing capacity and fishing effort from 1950 to 2012. *Fish and Fisheries, 18*(3), 489–505.

Bhusare, S., Satkar, S. G., Sahu, A., Savaliya, B., Karale, T., & Gautam, R. (2024). Microplastic (MP) pollution in aquatic ecosystems and environmental impact on aquatic animals. *Uttar Pradesh Journal of Zoology, 45*(5), 59–68. https://doi.org/10.56557/upjoz/2024/v45i53931

Bhuyan, M. S., Venkatramanan, S., Selvam, S., Szabo, S., Hossain, M. M., Rashed-Un-Nabi, M., Paramasivam, C. R., Jonathan, M. P., & Islam, M. S. (2021). Plastics in marine ecosystem: A review of their sources and pollution conduits. *Regional Studies in Marine Science, 41*, 101539. https://doi.org/10.1016/j.rsma.2020.101539

Bouzid, N., Anquetil, C., Dris, R., Gasperi, J., Tassin, B., & Derenne, S. (2022). Quantification of microplastics by pyrolysis coupled with gas chromatography and mass spectrometry in sediments: Challenges and implications. *Microplastics, 1*(2), 229–239.

Brander, S. M., Renick, V. C., Foley, M. M., Steele, C., Woo, M., Lusher, A., Carr, S., Helm, P., Box, C., Cherniak, S., Andrews, R. C., & Rochman, C. M. (2020). Sampling and quality assurance and quality control: a guide for scientists investigating the occurrence of microplastics across matrices. *Applied Spectroscopy, 74*(9), 1099–1125.

Brandon, J., Goldstein, M., & Ohman, M. D. (2016). Long-term aging and degradation of microplastic particles: Comparing in situ oceanic and experimental weathering patterns. *Marine Pollution Bulletin, 110*(1), 299–308. https://doi.org/10.1016/j.marpolbul.2016.06.048

Chakraborty, I., Banik, S., Biswas, R., Yamamoto, T., Noothalapati, H., & Mazumder, N. (2022). Raman spectroscopy for microplastic detection in water sources: A systematic review. *International Journal of Environmental Science and Technology, 20*(9), 10435–10448. https://doi.org/10.1007/s13762-022-04505-0

Chen, X., Chen, X., Liu, Q., Zhao, Q., Xiong, X., & Wu, C. (2021). Used disposable face masks are significant sources of microplastics to environment. *Environmental Pollution, 285*, Article 117485.

Corcoran, P. L. (2022). Degradation of microplastics in the environment. In *Handbook of microplastics in the environment* (pp. 531–532). https://doi.org/10.1007/978-3-030-39041-9_10.pdf

Courtene-Jones, W., Quinn, B., Gary, S. F., Mogg, A. O., & Narayanaswamy, B. E. (2017). Microplastic pollution identified in deep-sea water and ingested by benthic invertebrates in the Rockall Trough, North Atlantic Ocean. *Environmental Pollution, 231*, 271–280.

Coyle, R., Hardiman, G., & Driscoll, K. O. (2020). Microplastics in the marine environment: A review of their sources, distribution processes, uptake and exchange in ecosystems. *Case Studies in Chemical and Environmental Engineering, 2*, Article 100010. https://doi.org/10.1016/j.cscee.2020.100010

Cózar, A., Echevarría, F., González-Gordillo, J. I., Irigoien, X., Úbeda, B., Hernández-León, S., Palma, Á. T., Navarro, S., García-De-Lomas, J., Ruiz, A., Fernández-De-Puelles, M. L., & Duarte, C. M. (2014). Plastic debris in the open ocean. *Proceedings of the National Academy of Sciences, 111*(28), 10239–10244. https://doi.org/10.1073/pnas.1314705111

Crawford, C. B., & Quinn, B. (2016). *Microplastic pollutants*. Elsevier Limited.

Ding, L., Yu, X., Guo, X., Zhang, Y., Ouyang, Z., Liu, P., Zhang, C., Wang, T., Jia, H., & Zhu, L. (2021). The photodegradation processes and mechanisms of polyvinyl chloride and polyethylene terephthalate microplastic in aquatic environments: Important role of clay minerals. *Water Research, 208*, Article 117879. https://doi.org/10.1016/j.watres.2021.117879

Du, H., Xie, Y., & Wang, J. (2021). Microplastic degradation methods and corresponding degradation mechanism: Research status and future perspectives. *Journal of Hazardous Materials, 418*, Article 126377. https://doi.org/10.1016/j.jhazmat.2021.126377

Dutta, S., Sethulekshmi, S., & Shriwastav, A. (2022). Abundance, morphology, and spatio-temporal variation of microplastics at the beaches of Mumbai, India. *Regional Studies in Marine Science, 56*, Article 102722. https://doi.org/10.1016/j.rsma.2022.102722

Elsayed, A. A., Erfan, M., Sabry, Y. M., Dris, R., Gaspéri, J., Barbier, J. S., Marty, F., Bouanis, F., Luo, S., Nguyen, B. T., Liu, A. Q., & Bourouina, T. (2021). A microfluidic chip enables fast analysis of water microplastics by optical spectroscopy. *Scientific reports, 11*(1), 10533.

Enfrin, M., Lee, J., Gibert, Y., Basheer, F., Kong, L., & Dumée, L. F. (2020). Release of hazardous nanoplastic contaminants due to microplastic fragmentation under shear stress forces. *Journal of Hazardous Materials, 384*, Article 121393.

Ephsy, D., & Raja, S. (2023). Characterization of microplastics and its Pollution load index in freshwater Kumaraswamy Lake of Coimbatore, India. https://doi.org/10.21203/rs.3.rs-2610640/v1

Ephsy, D., Raja, S.(2020). Sources and impact of microplastic pollution in indian aquatic ecosystem: a review. *Current World Environment*. Special Issue (Sustainable Mining). https://doi.org/10.12944/CWE.15.Special-Issue1.01

Gaylarde, C. C., Neto, J. A. B., & da Fonseca, E. M. (2021). Paint fragments as polluting microplastics: A brief review. *Marine Pollution Bulletin, 162*, Article 111847.

Gokul, T., Kumar, K. R., Veeramanikandan, V., Arun, A., Balaji, P., & Faggio, C. (2023). Impact of particulate pollution on aquatic invertebrates. *Environmental Toxicology and Pharmacology, 100*, Article 104146.

Gopinath, K., Seshachalam, S., Neelavannan, K., Anburaj, V., Rachel, M., Ravi, S., Bharath, M., & Achyuthan, H. (2020). Quantification of microplastic in Red Hills Lake of Chennai city, Tamil Nadu, India. *Environmental Science and Pollution Research International, 27*(26), 33297–33306. https://doi.org/10.1007/s11356-020-09622-2

Goßmann, I., Halbach, M., & Scholz-Böttcher, B. M. (2021). Car and truck tire wear particles in complex environmental samples–a quantitative comparison with "traditional" microplastic polymer mass loads. *Science of the Total Environment, 773*, Article 145667.

Goswami, P., Vinithkumar, N. V., & Dharani, G. (2020). First evidence of microplastics bioaccumulation by marine organisms in the Port Blair Bay, Andaman Islands. *Marine pollution bulletin, 155*, 111163.

Gupta, P., Saha, M., Rathore, C., Suneel, V., Ray, D., Naik, A. Unnikrishnan, K., Dhivya, M., & Daga, K. (2021). Spatial and seasonal variation of microplastics and possible sources in the estuarine system from central west coast of India. *Environmental Pollution, 288*, 117665.

Gurjar, U. R., Xavier, K. M., Shukla, S. P., Jaiswar, A. K., Deshmukhe, G., & Nayak, B. B. (2022). Microplastic pollution in coastal ecosystem off Mumbai coast, India. *Chemosphere, 288*, Article 132484.

Harshvardhan, K., & Jha, B. (2013). Biodegradation of low-density polyethylene by marine bacteria from pelagic waters, Arabian Sea, India. *Marine Pollution Bulletin, 77*(1–2), 100–106.

Hebner, T. S., & Maurer-Jones, M. A. (2020). Characterizing microplastic size and morphology of photodegraded polymers placed in simulated moving water conditions. *Environmental Science Processes & Impacts, 22*(2), 398–407. https://doi.org/10.1039/c9em00475k

Hu, K., Zhou, P., Yang, Y., Hall, T., Nie, G., Yao, Y., Duan, X., & Wang, S. (2021). Degradation of microplastics by a thermal fenton reaction. *ACS ES&T Engineering, 2*(1), 110–120. https://doi.org/10.1021/acsestengg.1c00323

Srinivasalu, S., Natesan, U., Ayyamperumal, R., Kalam, N., Anbalagan, S., Sujatha. K., & Alagarasan, C. (2021). Microplastics as an emerging threat to the freshwater ecosystems of Veeranam lake in south India: A multidimensional approach. *Chemosphere, 264*, 128502. https://doi.org/10.1016/j.chemosphere.2020.128502

Kalogeraki, N., Karkanorachaki, K., Kalogerakis, G. C., Triantafyllidi, E. I., Gotsis, A. D., Partsinevelos, P., & Fava, F. (2017). Microplastics generation: onset of fragmentation of polyethylene films in marine environment mesocosms. *Frontiers in Marine Science, 4*. https://doi.org/10.3389/fmars.2017.00084

Karthikeyan, P., & Subagunasekar, M. (2023). Microplastics pollution studies in India: A recent review of sources, abundances and research perspectives. *Regional Studies in Marine Science, 61*, Article 102863. https://doi.org/10.1016/j.rsma.2023.102863

Kataria, N., Yadav, S., Garg, V. K., Rene, E. R., Jiang, J.-J., Rose, P. K., Kumar, M., & Khoo, K. S. (2024). Occurrence, transport, and toxicity of microplastics in tropical food chains: perspectives view and way forward. *Environmental Geochemistry and Health, 46*(3). https://doi.org/10.1007/s10653-024-01862-2

Laursen, S. N., Fruergaard, M., & Andersen, T. J. (2022). Rapid flocculation and settling of positively buoyant microplastic and fine-grained sediment in natural seawater. *Marine Pollution Bulletin, 178*, Article 113619.

Lebreton, L. C., Van Der Zwet, J., Damsteeg, J. W., Slat, B., Andrady, A., & Reisser, J. (2017). River plastic emissions to the world's oceans. *Nature Communications, 8*(1), 15611.

Li, C., Wang, X., Liu, K., Zhu, L., Wei, N., Zong, C., & Li, D. (2021). Pelagic microplastics in surface water of the Eastern Indian Ocean during monsoon transition period: Abundance, distribution, and characteristics. *Science of the Total Environment, 755*, Article 142629. https://doi.org/10.1016/j.scitotenv.2020.142629

Li, Y., Zhang, H., & Tang, C. (2020). A review of possible pathways of marine microplastics transport in the ocean. *Anthropocene Coasts, 3*(1), 6–13.

Lin, Z., Jin, T., Zou, T., Xu, L., Xi, B., Xu, D., He, J., Xiong, L., Tang, C., Peng, J., Zhou, Y., & Fei, J. (2022). Current progress on plastic/microplastic degradation: Fact influences and mechanism. *Environmental Pollution, 304*, Article 119159. https://doi.org/10.1016/j.envpol.2022.119159

Liu, D., Guo, Z., Xu, Y., Chan, F. K. S., Xu, Y., Johnson, M., & Zhu, Y. (2022). Widespread occurrence of microplastics in marine bays with diverse drivers and environmental risk. *Environment International, 168*, Article 107483. https://doi.org/10.1016/j.envint.2022.107483

Luo, Z., Zhou, X., Su, Y., Wang, H., Yu, R., Zhou, S., Xu, E. G., & Xing, B. (2021). Environmental occurrence, fate, impact, and potential solution of tire microplastics: Similarities and differences with tire wear particles. *Science of the Total Environment, 795*, 148902.

Manickavasagam, S., Kumar, S., Kumar, K., Bhuvaneswari, G. R., Paul, T., & Shukla, S. (2020). Quantitative assessment of influx and efflux of marine debris in a water channel of South Juhu creek, Mumbai, India. *Regional Studies in Marine Science, 34*, 101095. https://doi.org/10.1016/j.rsma.2020.101095

Meyers, N., Catarino, A. I., Declercq, A. M., Brenan, A., Devriese, L., Vandegehuchte, M., De Witte, B., Janssen, C., & Everaert, G. (2022). Microplastic detection and identification by Nile red staining: Towards a semi-automated, cost-and time-effective technique. *Science of the Total Environment, 823*, 153441.

Nandal, M., Khyalia, P., Ghalawat, A., Jugiani, H., Kaur, M., & Laura, J. S. (2022). Review on the use of Microalgae Biomass for Bioplastics Synthesis: A Sustainable and Green approach to control Plastic Pollution. *Pollution, 8*(3), 844–859.

OSPAR. (2007). OSPAR Pilot Project on Monitoring Marine Beach Litter: Monitoring of marine litter in the OSPAR region.

Othman, A. R., Hasan, H. A., Muhamad, M. H., Ismail, N., & Abdullah, S. R. S. (2021). Microbial degradation of microplastics by enzymatic processes: A review. *Environmental Chemistry Letters, 19*(4), 3057–3073. https://doi.org/10.1007/s10311-021-01197-9

Patchaiyappan, A., Dowarah, K., Ahmed, S. Z., Prabakaran, M., Jayakumar, S., Thirunavukkarasu, C., & Devipriya, S. P. (2021a). Prevalence and characteristics of microplastics present in the street dust collected from Chennai metropolitan city, India. *Chemosphere, 269*, Article 128757.

Patchaiyappan, A., Zakiahmed, S., Dowarah, K., Khadanga, S. S., Singh, T., Jayakumar, S., Thirunavukkarasu, C., & Devipriya, S. (2021b). Prevalence of microplastics in the sediments of Odisha beaches, southeastern coast of India. *Marine Pollution Bulletin, 167*, Article 112265. https://doi.org/10.1016/j.marpolbul.2021.112265

Pavithran, V. A. (2021). Study on microplastic pollution in the coastal seawaters of selected regions along the northern coast of Kerala, southwest coast of India. *Journal of Sea Research, 173*, Article 102060. https://doi.org/10.1016/j.seares.2021.102060

Peng, X., Chen, M., Chen, S., Dasgupta, S., Xu, H., Ta, K., Du, M., Li, J., Guo, Z., & Bai, S. (2018). Microplastics contaminate the deepest part of the world's ocean. *Geochemical Perspectives Letters, 9*(1), 1–5.

Pielichowski, K., Njuguna, J., & Majka, T. M. (2023). Mechanisms of thermal degradation of polymers. In *Elsevier eBooks* (pp. 9–11). https://doi.org/10.1016/b978-0-12-823023-7.00001-0

Pirsaheb, M., Hossini, H., & Makhdoumi, P. (2020). Review of microplastic occurrence and toxicological effects in marine environment: Experimental evidence of inflammation. *Process Safety and Environmental Protection, 142*, 1–14.

Pothiraj, C., Gokul, T. A., Kumar, K. R., Ramasubramanian, A., Palanichamy, A., Venkatachalam, K., Pastorino, P., Barcelo, D., Balaji, P., & Faggio, C. (2023). Vulnerability of microplastics on marine environment: A review. *Ecological Indicators, 155*, 111058.

Prata, J. C., Da Costa, J. P., Duarte, A. C., & Rocha-Santos, T. (2019). Methods for sampling and detection of microplastics in water and sediment: A critical review. *TrAC Trends in Analytical Chemistry, 110*, 150–159.

Qi, H., Li, H., Meng, X., Peng, L., Zheng, H., Wang, L., Wang, W., Chen, K., Zhang, J., Zhang, H., & Cai, M. (2022). Fate of microplastics in deep-sea sediments and its influencing factors: Evidence from the Eastern Indian Ocean. *Science of the Total Environment, 828*, Article 154266. https://doi.org/10.1016/j.scitotenv.2022.154266

Qiu, Q., Peng, J., Yu, X., Chen, F., Wang, J., & Dong, F. (2015). Occurrence of microplastics in the coastal marine environment: First observation on sediment of China. *Marine Pollution Bulletin, 98*(1–2), 274–280. https://doi.org/10.1016/j.marpolbul.2015.07.028

Radhakrishnan, K., Sivapriya, V., Rajkumar, A., Akramkhan, N., Prakasheswar, P., Krishnakumar, S., & Hussain, S. (2021). Characterization and distribution of microplastics in estuarine surface sediments, Kayamkulam estuary, southwest coast of India. *Marine Pollution Bulletin, 168*, Article 112389. https://doi.org/10.1016/j.marpolbul.2021.112389

Ranjani, M., Veerasingam, S., Venkatachalapathy, R., Mugilarasan, M., Bagaev, A., Mukhanov, V., & Vethamony, P. (2021). Assessment of potential ecological risk of microplastics in the coastal sediments of India: A meta-analysis. *Marine Pollution Bulletin, 163*, Article 111969. https://doi.org/10.1016/j.marpolbul.2021.111969

Razeghi, N., Hamidian, A. H., Wu, C., Zhang, Y., & Yang, M. (2021). Microplastic sampling techniques in freshwaters and sediments: A review. *Environmental Chemistry Letters, 19*(6), 4225–4252.

Rødland, E. S., Samanipour, S., Rauert, C., Okoffo, E. D., Reid, M. J., Heier, L. S., Lind, O. C., Thomas, K. V., & Meland, S. (2022). A novel method for the quantification of tire and polymer-modified bitumen particles in environmental samples by pyrolysis gas chromatography mass spectroscopy. *Journal of Hazardous Materials, 423*, 127092.

Rogers, K. L., Carreres-Calabuig, J. A., Gorokhova, E., & Posth, N. R. (2020). Micro-by-micro interactions: How microorganisms influence the fate of marine microplastics. *Limnology and Oceanography Letters, 5*(1), 18–36. https://doi.org/10.1002/lol2.10136

Romano, E., Bergamin, L., Di Bella, L., & Provenzani, C. (2024). Microplastic pollution in marine caves. In *EGU General Assembly 2024*, Vienna, Austria, 14–19 Apr 2024, EGU24-16399. https://doi.org/10.5194/egusphere-egu24-16399

Rose, P. K., Yadav, S., Kataria, N., & Khoo, K. S. (2023). Microplastics and nanoplastics in the terrestrial food chain: Uptake, translocation, trophic transfer, ecotoxicology, and human health risk. *TrAC Trends in Analytical Chemistry, 167*, Article 117249. https://doi.org/10.1016/j.trac.2023.117249

Ryan, P. G., Moore, C. J., Van Franeker, J. A., & Moloney, C. L. (2009). Monitoring the abundance of plastic debris in the marine environment. *Philosophical Transactions of the Royal Society B: Biological Sciences, 364*(1526), 1999–2012.

Saha, M., Naik, A., Desai, A., Nanajkar, M., Rathore, C., Kumar, M., & Gupta, P. (2021). Microplastics in seafood as an emerging threat to marine environment: A case study in Goa, west coast of India. *Chemosphere, 270*, Article 129359.

Sathish, M. N., Jeyasanta, I., & Patterson, J. (2020). Microplastics in Salt of Tuticorin, southeast coast of India. *Archives of Environmental Contamination and Toxicology, 79*(1), 111–121. https://doi.org/10.1007/s00244-020-00731-0

Sathish, N., Jeyasanta, K. I., & Patterson, J. (2019). Abundance, characteristics and surface degradation features of microplastics in beach sediments of five coastal areas in Tamil Nadu, India. *Marine Pollution Bulletin, 142*, 112–118. https://doi.org/10.1016/j.marpolbul.2019.03.037

Selvam, S., Manisha, A., Venkatramanan, S., Chung, S., Paramasivam, C., & Singaraja, C. (2020). Microplastic presence in commercial marine sea salts: A baseline study along Tuticorin Coastal salt pan stations, Gulf of Mannar, South India. *Marine Pollution Bulletin, 150*, Article 110675. https://doi.org/10.1016/j.marpolbul.2019.110675

Sharma, S., Bhardwaj, A., Thakur, M., et al. (2024). Understanding microplastic pollution of marine ecosystem: A review. *Environmental Science and Pollution Research, 31*, 41402–41445. https://doi.org/10.1007/s11356-023-28314-1

Sharma, S., Bhardwaj, A., Thakur, M., & Saini, A. (2023). Understanding microplastic pollution of marine ecosystem: A review. *Environmental Science and Pollution Research.* https://doi.org/10.1007/s11356-023-28314-1

Sivadas, S. K., Mishra, P., Kaviarasan, T., Sambandam, M., Dhineka, K., Murthy, M. R., Nayak, S., Sivyer, D., & Hoehn, D. (2022). Litter and plastic monitoring in the Indian marine environment: A review of current research, policies, waste management, and a roadmap for multidisciplinary action. *Marine Pollution Bulletin, 176*, Article 113424. https://doi.org/10.1016/j.marpolbul.2022.113424

Song, Y. K., Hong, S. H., Jang, M., Han, G. M., Jung, S. W., & Shim, W. J. (2017). Combined effects of UV exposure duration and mechanical abrasion on microplastic fragmentation by polymer type. *Environmental Science and Technology, 18*, 4368–4376.

Sorolla-Rosario, D., Llorca-Porcel, J., Pérez-Martínez, M., Lozano-Castelló, D., & Bueno-López, A. (2023). Microplastics' analysis in water: Easy handling of samples by a new Thermal Extraction Desorption-Gas Chromatography-Mass Spectrometry (TED-GC/MS) methodology. *Talanta, 253*, Article 123829.

Sundar, S., Chokkalingam, L., Roy, P. D., & Usha, T. (2020). Estimation of microplastics in sediments at the southernmost coast of India (Kanyakumari). *Environmental Science and Pollution Research International, 28*(15), 18495–18500. https://doi.org/10.1007/s11356-020-10333-x

Tang, L., Feng, J., Li, C., Liang, J., Zhang, S., & Yang, Z. (2023). Global occurrence, drivers, and environmental risks of microplastics in marine environments. *Journal of Environmental Management, 329*, Article 116961. https://doi.org/10.1016/j.jenvman.2022.116961

Turner, A. (2021). Paint particles in the marine environment: An overlooked component of microplastics. *Water rEsearch x, 12*, Article 100110.

Tursi, A., Baratta, M., Easton, T., Chatzisymeon, E., Chidichimo, F., De Biase, M., & De Filpo, G. (2022). Microplastics in aquatic systems, a comprehensive review: Origination, accumulation, impact, and removal technologies. *RSC Advances, 12*(44), 28318–28340. https://doi.org/10.1039/d2ra04713f

Vahabi, H., Sonnier, R., & Ferry, L. (2014). Effects of ageing on the fire behaviour of flame-retarded polymers: A review. *Polymer International, 64*(3), 313–328. https://doi.org/10.1002/pi.4841

Vahabi, H., Sonnier, R., & Ferry, L. (2015). Effects of ageing on the fire behaviour of flame-retarded polymers: A review. *Polymer International, 64*(3), 313–328.

Veerasingam, S., Mugilarasan, M., Venkatachalapathy, R., & Vethamony, P. (2016). Influence of 2015 flood on the distribution and occurrence of microplastic pellets along the Chennai coast, India. *Marine Pollution Bulletin, 109*(1), 196–204. https://doi.org/10.1016/j.marpolbul.2016.05.082

Veerasingam, S., Ranjani, M., Venkatachalapathy, R., Bagaev, A., Mukhanov, V., Litvinyuk, D., Mugilarasan, M., Gurumoorthi, K., Guganathan, L., Aboobacker, V. M., & Vethamony, P. (2021). Contributions of Fourier transform infrared spectroscopy in microplastic pollution research: A review. *Critical Reviews in Environmental Science and Technology, 51*(22), 2681–2743.

Venkatramanan, S., Chung, S., Selvam, S., Sivakumar, K., Soundhariya, G., Elzain, H. E., & Bhuyan, M. S. (2022). Characteristics of microplastics in the beach sediments of Marina tourist beach, Chennai, India. *Marine Pollution Bulletin, 176*, Article 113409. https://doi.org/10.1016/j.marpolbul.2022.113409

Vidyasakar, A., Krishnakumar, S., Kasilingam, K., Neelavannan, K., Bharathi, V. A., Godson, P. S., Prabha, K., & Magesh, N. (2020). Characterization and distribution of microplastics and plastic debris along Silver Beach, Southern India. *Marine Pollution Bulletin, 158*, Article 111421. https://doi.org/10.1016/j.marpolbul.2020.111421

Vidyasakar, A., Krishnakumar, S., Kumar, K. S., Neelavannan, K., Anbalagan, S., Kasilingam, K., Srinivasalu, S., Saravanan, P., Kamaraj, S., & Magesh, N. S. (2021). Microplastic contamination in edible sea salt from the largest salt-producing states of India. *Marine Pollution Bulletin, 171*, 112728.

Yadav, S., Kataria, N., Khyalia, P., Rose, P. K., Mukherjee, S., Sabherwal, H., Chai, W. S., Rajendran, S., Jiang, J., & Khoo, K. S. (2023). Recent analytical techniques, and potential eco-toxicological impacts of textile fibrous microplastics (FMPs) and associated contaminates: A review. *Chemosphere, 326*, Article 138495. https://doi.org/10.1016/j.chemosphere.2023.138495

Yang, H., Chen, G., & Wang, J. (2021). Microplastics in the marine environment: Sources, fates, impacts and microbial degradation. *Toxics, 9*(2), 41. https://doi.org/10.3390/toxics9020041

Yang, S. J., Feng, W. W., Wang, Q., Cai, Z. Q., Liu, Q. Y., Hou, Y. B., & Zhang, Q. Q. (2020, October). Rapid identification of microplastic using portable Raman system and extra trees algorithm. In *Real-time photonic measurements, data management, and processing V* (vol. 11555, pp. 70–77). SPIE.

Yu, R., & Singh, S. (2023). Microplastic pollution: Threats and impacts on global marine ecosystems. *Sustainability, 15*(17), 13252. https://doi.org/10.3390/su151713252

Zhang, J., Gao, D., Li, Q., Zhao, Y., Li, L., Lin, H., Bi, Q., & Zhao, Y. (2020). Biodegradation of polyethylene microplastic particles by the fungus Aspergillus flavus from the guts of wax moth Galleria mellonella. *Science of the Total Environment, 704*, 135931.
Zhang, K., Hamidian, A. H., Tubić, A., Zhang, Y., Fang, J. K., Wu, C., & Lam, P. K. (2021). Understanding plastic degradation and microplastic formation in the environment: A review. *Environmental Pollution, 274*, Article 116554. https://doi.org/10.1016/j.envpol.2021.116554
Zhou, X. X., Liu, R., Hao, L. T., & Liu, J. F. (2021). Identification of polystyrene nanoplastics using surface-enhanced Raman spectroscopy. *Talanta, 221*, Article 121552.

Health Impacts of Microplastics on the Indian Terrestrial Ecosystem

Pooja Yadav, Manju Rani, and Sudesh Chaudhary

Abstract Microplastic was first discovered in the marine environment, but 4–23 times more plastic debris contaminates the terrestrial ecosystem yearly. Rapid industrialisation, urbanisation, growing demand for plastic consumption, ineffective agricultural practices, and illegal dumping lead to MP soil pollution. Therefore, MP pollution in terrestrial environment has started to raise serious concerns. Sampling, extraction, identification, characterisation and quantification are currently used to assess MPs. This chapter emphasises the ecotoxicological impacts of microplastics on the terrestrial environment and their significant effects on soil ecosystems, terrestrial plants, biota and humans. The transfer of related pollutants and degradation consequences of microplastic on the environment was also highlighted. Microplastic toxicity in the Indian terrestrial ecosystem in various regions was also described in this chapter. MPs have been investigated in Indian marine environment, freshwater ecosystem, marine plants, animals and terrestrial ecosystem. Microplastics are transported by various means and accumulate in soil profile, altering the functions of soil ecosystem by changing its biophysical properties. They also accumulate in the food chain and potentially risk living life including plants, organisms, and human beings by interfering with their reproduction rate, growth rate, seed germination, digestive system, and genetic structure.

Keywords Microplastics · Soil health impacts · Soil microbiota · Indian terrestrial ecosystem

---

P. Yadav · S. Chaudhary (✉)
Center of Excellence for Energy and Environmental Studies, Deenbandhu Chhotu Ram University of Science and Technology, Murthal, Haryana, India
e-mail: sudesh.energy@dcrustm.org

M. Rani
Department of Chemical Engineering, Deenbandhu Chhotu Ram University of Science and Technology, Murthal, Haryana, India
e-mail: manjurani.che@dcrustm.org

P. Khyalia and S. Yadav (eds.), *Microplastic Pollution in India and Its Environmental Impacts*, Earth and Environmental Sciences Library,
https://doi.org/10.1007/978-3-031-92530-6_7

Abbreviations

HDPE	High density polyethene
LDPE	Low density polyethene
MPs	Microplastics
MNPs	Micro-nano-plastic
PET	Polyethylene terephthalate
PES	Polyethersulfone
PHBV	polyhydroxy butyrate co hydroxy valerate
PLA	Polylactic Acid
PP	Polypropylene
PU	Polyurethane
PVC	Polyvinyl chloride
PS	Polystyrene

1 Introduction

In recent years, there has been ubiquitous amount of plastic accumulated in the environment due to increased plastic production, improper disposal techniques, and an extremely slow rate of degradation. Large plastic breaks into smaller particles due to various environmental conditions, including mechanical abrasions and continuous UV exposure. The resulting small particles of plastic known as microplastics, have a length of less than five millimetres (Veerasingam et al., 2020). They are made up of synthetic polymers that are typically identified as solid particles that are insoluble in water and have thermoplastic properties. At present, microplastic pollution exists in every kind of ecosystem globally, through different sources and migration in various modes. Terrestrial ecosystems are more exposed to microplastics than marine, but surprisingly, there's no adequate study has been conducted on terrestrial ecosystems. There are a variety of microplastic sources in soil, such as textile industries, farming practices, wastewater treatment plants, plastic industries, landfill stations, household waste, etc. (Xu et al., 2020a, 2020b). One of the primary sources of microplastics in soil is modern agricultural practices, which include the use of plastic mulching films. Plastic mulching films release plastic particles over time. Secondly, agricultural practices such as tilling, crop rotation, and ploughing break those fragments into microplastics (Huang et al., 2020). Majority of MPs remain in water and sludge even after efficient treatment techniques are used in plants. This contaminated sewage sludge and water released from waste-water treatment plants are applied as a source of compost and irrigation purposes respectively in agro fields (Collivignarelli et al., 2021). Additionally, road runoffs and illegal dumping are also among the major sources of soil microplastic (Gokdemir et al., 2023). Road tyre dust and atmospheric deposition move from one place to another and transport microplastics (Roychand et al., 2020). Urban and agricultural soil is mainly more prone to microplastic contamination because of more exposure to human activities, currently, there are approximately 4–5% of MP study data available for Asia, and only 2–3% of those studies address the terrestrial environment (Jenkins et al., 2022). Soil acts as a plane of contact between the atmosphere, hydrosphere, and lithosphere; once microplastics are introduced into soil, they can endure for a long period. Soil ecosystems may be a substantial long-term sink for microplastics because most of the plastic waste is generated and released on the land (Kim et al., 2020). The quantity of MPs in terrestrial ecosystems is expected to rise over time due to the reduced ability of MPs to migrate through soil. The long-term presence of microplastics in soil lowers the bulk density of the soil. Furthermore, it leads the soil structure to become more porous, which increases soil-water evaporation and causes the soil to dry out. A change in the soil's structure due to drying certainly will affect soil aggregation, which could harm root development and soil aeration. The accumulation of residual microplastics in soil may disintegrate the structure of soil, change its physicochemical characteristics, and impact the soil's ability to diffuse moisture and carry nutrients. They also lessen nutrient availability, change microbial activity, impact crop growth and development, and decrease seed germination. When ingested, MPs significantly impact the environment and soil organisms' development and reproduction at various trophic levels. They act as stressors in the terrestrial environment, which leads to global change (de Souza Machado et al., 2018). The main objective of this book chapter is to provide an overview of the current state of research on the impacts of microplastic (MPs) pollution on soil environments. It also aims to emphasize and explore MPs' transport behavior, the pollution level they cause, and the combined effects of MPs and other pollutants on soil ecology.

2 Ecotoxicological Impacts of Microplastics on Different Compartments of Soil Ecosystem

Microplastic particles move from one place to another and significantly negatively impact flora and fauna. Several factors are responsible for the migration of these plastic particles in the soil environment, such as soil crakes, bio pores, soil aggregates, and soil biota (Martínez Urreaga et al., 2020). Many agricultural practices such as ploughing, use of various fertilizers, irrigation, and agricultural runoff cause the transport of microplastics horizontally and vertically, as represented in Fig. 1 (Briassoulis, 2023). These particles may reach the aquifers, contaminate the freshwater resources, and disturb biodiversity. Soil organisms such as earthworms, collembola, etc., distribute the small plastic particles through body attachment and ingestion/egestion mechanisms (Rillig et al., 2017). These microplastics leach down into the soil and alter the normal functioning of the soil ecosystem by changing the pH, soil fertility, biomass, germination, chlorophyll content, reproduction rate, crop productivity, and nutrient cycle. Fig. 2 describes the possible impacts of MPs on soil environment (Li et al., 2023).

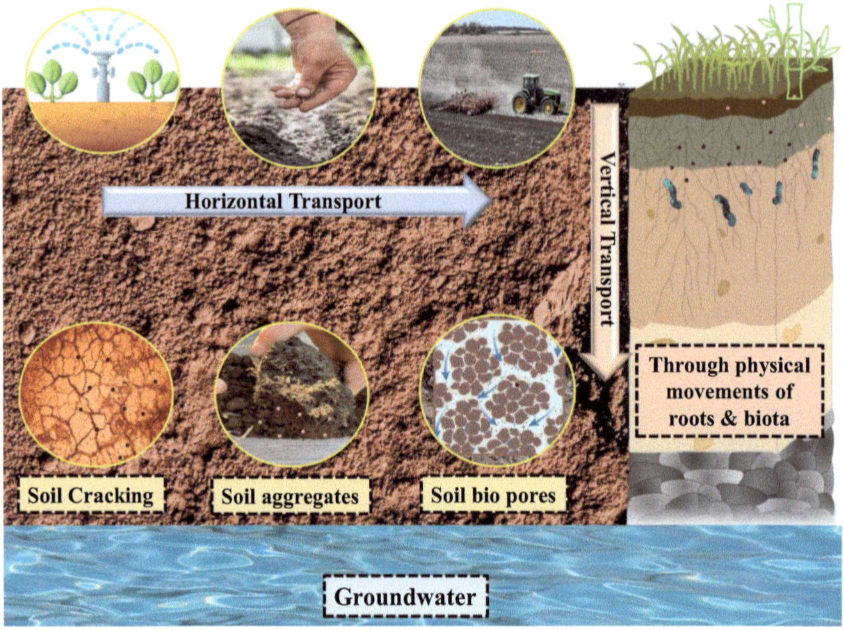

Fig. 1 Movement of microplastics in soil ecosystem

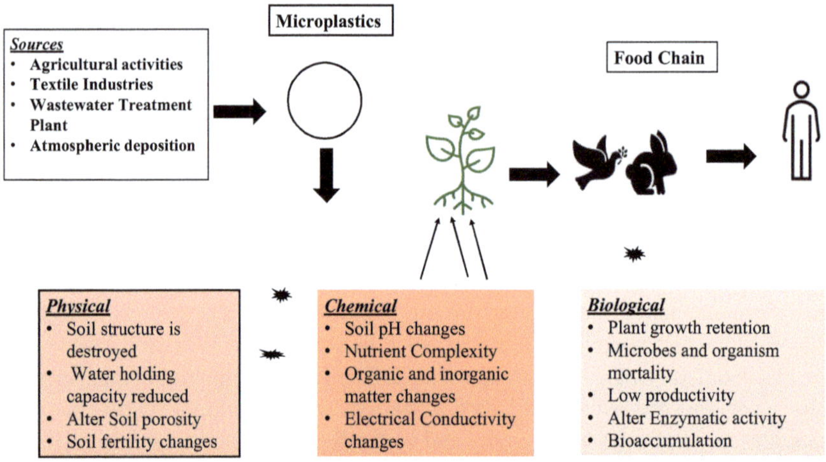

Fig. 2 Effects of microplastic in soil environment from sources passing through physical, chemical and biological processes in the soil to food chain

2.1 Physio-chemical properties of soil

MPs have a profound impact on the physical & chemical characteristics of the soil, including the capacity of soil to retain water, pH, pore size, pore availability, hydraulic properties, and soil conductivity (De Souza MacHado et al., 2018). Based on the available data, different kinds of microplastics have distinguished impacts on the soil characteristics. MPs can alter the pH of soil based on the shapes (fibres, films, foams, and fragments) and types of polymers; as shown in laboratory incubation research, the pH of the soil can first decrease and then increase the rise in the case of MPs, which are foam and fragment shaped. Adding MPs to the soil decreases the pore size distribution, which in turn decreases the hydraulic conductivity of saturated soils. Microplastics have potential to alter the distribution of soil's water-stable aggregates and hinder water infiltration by reducing soil stability. Microplastics have a less obvious impact on soil structure and the water cycle, because of their similar size and shape to soil particles. As an example, polyester fibres (0.4%, w/w) increased water-holding capacity and evapotranspiration and significantly decreased bulk density and water-stable aggregates in the soil (Zhao et al., 2021). Different forms of microplastics, like PE fragments or polyamide (PA) beads (2.0%, w/w), had less pronounced impacts (De Souza MacHado et al., 2018).

It is hypothesized that a key aspect in determining how MPs affect the soil attributes is soil texture. The bulk density of soil can decrease the presence of various plastic particles like PES, PS, PET, and PP (Guo et al., 2022). They can reduce soil aeration and water infiltration, making it less suitable for soil-dwelling organisms. This alteration of soil conditions can disrupt the habitats of soil organisms. The porosity and aeration of the soil were dramatically reduced by relatively smaller MPs when compared to larger MPs. The quantity of micropores in soil and the amount of MPs are inversely correlated. Microplastics can also serve as a means for the transmission of hazardous substances, such as toxic compounds, heavy metals such as Cd in plastic or other toxins absorbed from soil matrices to soil biota and posing a risk (Brennecke et al., 2016; Yadav et al., 2024) Once MPs are incorporated into the soil matrix, their characteristics change. MPs hetero aggregates with inorganic elements of soil like Al, Fe and Mg. Fe oxides are positively charged, and easily engage electrostatically with the negatively charged MPs. MPs' density and zeta potential are increased by their hetero aggregation with OM and soil minerals. In a sandy loam soil, the proportion of dissolved organic C, inorganic nitrogen, and total phosphorous were considerably enhanced by a 28% (w/w) microplastic amendment rate; however, these modifications were not seen at lower amendment rate of 7% (w/w) over 30 days. At 500 kg ha^{-1}, the existence of plastic film residue considerably reduced the amount of soil organic matter (SOM) and alkali-hydrolysable nitrogen (N), as well as the amount of accessible P and N by 55% and 60%, at 2000 kg ha^{-1} respectively and microplastics also interfere with the fluorescein diacetate hydrolase (FDAse) in soil (Iqbal et al., 2020; H. Liu et al., 2017; Yang et al., 2021). These results indicate that plastic residue reduces soil fertility and the growth of crops.

Future research should consider soil with varying textures and long-term impacts to understand these field phenomena.

2.2 Soil Microorganisms

Soil microorganisms help in many activities such as bioturbation, decomposition of the organic matter, nutrient cycle, and maintaining the physiochemical activities of soil. Based on MPs' surface roughness and hydrophobicity, microorganisms may adhere to one another, colonise, and form biofilms. Microplastics can influence or disrupt enzyme activity and soil microbial populations (Sajjad et al., 2022). Microplastic's addition rate, polymer composition, size, shape (linear or non-linear), and soil texture all have different impacts on soil microbial and enzyme activity (Mbachu et al., 2021). PLA and polyester microfibers (0.1%, w/w) significantly reduce microbial activities in soil. After 50 days of incubation, the bacterial community's richness and variety were decreased by LDPE and PVC at concentrations of 1–4% (Wang et al., 2020a, 2020b, Yan et al., 2021). On the other hand, phyla and families showed a community-dependent relationship with respect to relative bacterial abundance, with an increase in Burkholderiaceae and Xanthobacteriaceae (Aralappanavar et al., 2024). The effects of PU on bacterial diversity were not statistically significant within the 0.01–1% concentration range. On the other hand, PU enhances the abundance of Firmicutes, Bacteroidetes, and Verrucomicrobia (Liu et al., 2023). Polystyrene (PS) nanoplastics at 100 and 1000 ng g^{-1} dramatically decreased soil microbial biomass & increased basal respiration, respectively (Awet et al., 2018). E. coli *(Escherichia coli)* transport in quartz sand was essentially unaffected by PS particles under low ionic strength, but it was accelerated by plastic particles under high ionic strength conditions (Wang et al., 2021a, 2021b). The growth of antibiotic resistance genes is a growing concern because it could cause potential risks to human health. PS microplastics (0.1%) have been found to optimize the retention period of antibiotics and ARGs in soil ecosystems (Awe et al., 2018). According to Wang et al. 2016, plastic-film residues (67.5 kg ha^{-1}) have the potential to significantly reduce soil microbial community diversity, soil microbial C and N levels, and the activity of dehydrogenase and fluoroscein diacetate hydrolase by 10% and 20%, respectively. The fungus helps in the degradation of microplastic as some fungus-like *Zalerion maritimum* use microplastics as a nutrient (Vaksmaa et al., 2023). Microplastics can dramatically reduce *C. elegans* survival rates, maximum body length, and reproduction rate after being ingested and accumulated by nematodes. Moreover, virgin microplastics (generated from human daily life and have no uniform shape) are found more toxic to *C. elegans,* causing intestinal injury, and transgenerational toxicity (Lei et al., 2018). MPs also increase the expression of oxidative stress gene *gst 4* and decrease the calcium level in the intestines in *C. elegans* (Hu et al., 2020). The presence of PE microplastics at 0.5% and 1% (w/w) changed the microbial community in the collembolan stomach and inhibited the reproduction and growth of *F. candida* in soil, respectively (Ju et al., 2019, Azeem et al., 2021). Overall, anthropogenic microplastic pollution may change the variety of the soil microbial community, and the activity of soil microbiota and enzymes; this disrupts microbial ecosystems and impacts the cycles of soil nutrients (Li et al., 2020a, 2020b).

2.3 *Impacts on Terrestrial Animals*

Soil animals such as earthworms, ants, and springtails act as transformers in terrestrial ecosystems because they convert the organic matter into a form that plants can easily utilize. These activities can be altered by microplastic contamination. They ingest microplastics which accumulate in the digestive tracts of these organisms, potentially leading to physical blockages, reduced nutrient absorption, and long-term health problems (Campanale et al., 2022). Song et al. (2019) noticed the drastic impacts of microplastics on the proper functioning of snails (*Achatina fulica)*. Polyethylene terephthalate (PET) microfibers (0.71g kg^{-1}) significantly decreased intake of food and excretion and also induced oxidative stress in *Achatina fulica,* posing a potential ecological risk to terrestrial ecosystems. Earthworms are ecosystem engineers, and, when exposed to microplastics may exhibit changes in burrowing behaviour and reduced foraging activity. The movements of earthworms enhance the transfer and integration of MNPs into the soil matrix. MPs can inhibit earthworm growth and survival even at 0.10% (w/w). Many earthworm species, including *Lumbricus terrestris,* and *Eisenia fetida*, have been shown in numerous studies to be capable of consuming microplastics. It is revealed that, after exposure to PE or PS particles at the maximum rate (20%), Superoxide dismutase and glutathione S-transferase activity in *E. fetida* were suppressed, although catalase, peroxidase, and the degree of lipid peroxidation were considerably ($p < 0.05$) elevated indicating tissue damage from microplastic particles and oxidative stress. PE microplastics are reduced in size after digestion in *L. terrestris* cast indicating that the earthworms' digestive systems either broke down PE microplastics more effectively or that they only consumed smaller plastic particles (Wang et al., 2019; Yu et al., 2019; Meng et al., 2023). Furthermore, the growth rate of *L. terrestris* was significantly reduced in the presence of higher concentrations of PE (0.4, 0.5, and 1.2%, w/w), but not in the presence of lower concentrations of 0.2% (w/w) in sandy soil. Moreover, the reproduction of *L. terrestris* remained unaffected even at the highest tested concentration of 1.2% (w/w) (Huerta Lwanga et al., 2016). Compared to lower concentrations (0–7%), larger concentrations (30–60%) of MPs of PE were found to cause greater mortality and slower development rates in L. terrestris. Evidence showed that MPs affected a wide range of soil animal functions, either increasing or decreasing. Different microplastic particle sizes have distinct effects on the soil worm *Enchytraeus crypticus's* rate of reproduction. It was found that smaller particles (i.e., 20 m) had a bigger impact on survival and reproduction than larger particles (i.e., 160 m), which might be explained by the fact that soil worms ingested more smaller-sized MPs (Lahive et al., 2019). The fertility and life duration of rotifer species were more significantly affected by smaller PS particles (50 nm) than by bigger PS particles (0.50–5μm). Since smaller particles are more readily aggregated, consumed, and digested by organisms than larger particles, this accounts for the size-based inequalities in MP toxicities.

2.4 Impact on Terrestrial Plants

The soil-plant relation is the most significant part of the terrestrial ecosystem because it sustains the biosphere. Microplastics and nano plastics have impact on the growth of many plants, such as faba beans, sorghum, wheat, mung beans, lettuce, rice, vascular cress, and spring onions (Ullah et al., 2021). The results of the meta-analysis indicate that plants experienced notable negative impacts on their physical and molecular changes due to large plastic particles (>1 µm) (Huo et al., 2022). Plastics larger than 1 µm consistently produced unfavourable results, significantly lowering plant growth (i.e., above ground biomass, below ground biomass & above ground length), germination, and oxidative stress. These findings suggest that the main reason for the decreased plant performance could be a physical obstruction. It was discovered that for broad beans (Vicia faba L.), carrots (Daucus carota L.), and cress (Lepidium sativum L.), 5 µm PS MPs caused more growth inhibition than 100 nm PS NPs. Microplastics can transfer to distinct plant parts, such as stems, leaves and fruits. Once they have entered the root zone (Campanale et al., 2022). Microplastics mainly polystyrene accumulate and are absorbed by the roots of raw vegetables and reach up to shoots of plants (Azeem et al., 2021). Microplastics can impact a plant's vegetative and reproductive growth, delay germination, low productivity (Fig. 3) and cause genotoxicity and ecotoxicity (De Souza Machado et al., 2019; Huo et al., 2022). Plants exposed to various microplastics performed undergo distinct alterations. In onion bulbs *(Allium fistulosum)*, for example, the water content was doubled by PA beads (2.0%, w/w), whereas it was decreased by polyester fibres (0.2%, w/w), PET, and PP fragments (2.0%, w/w); root colonisation by arbuscular mycorrhizal fungi was roughly eight times higher by polyester (0.2%, w/w), but was decreased by roughly 50% by PET (2.0%, w/w). However, a recent study conducted by revealed that the growth of mung beans, lettuce, and rice in LUFA soil was unaffected by macro-sized extended PS (8.3 ± 0.5 mm) (Xu et al., 2020a, 2020b). The BY-2 cells in tobacco uptake PS nanoparticles, which range in size from 20 to 40 nm, have been linked to endocytosis. It is widely accepted that MPs in the terrestrial ecosystem are vulnerable to plant systems following root uptake and cause phytotoxicities. PS MPs accumulated at the root cap cells of wheat and Arabidopsis but did not enter the roots (Jia et al., 2023). MPs can impact the antioxidant system, photosynthesis, and morphology of a variety of plants, including wheat, corn, lettuce, and cress (Huo et al., 2022). The confocal microscopic studies verified that PS accumulated at the root surface and entered the cortex, stele, and epidermis, particularly in the xylem of rice root seedlings. Plants can either absorb or internalize MPs, depending on their size, charge, and the characteristics of their tissues, such as their sticky or hydrophobic surface layer (Yin et al., 2024) (Table 1).

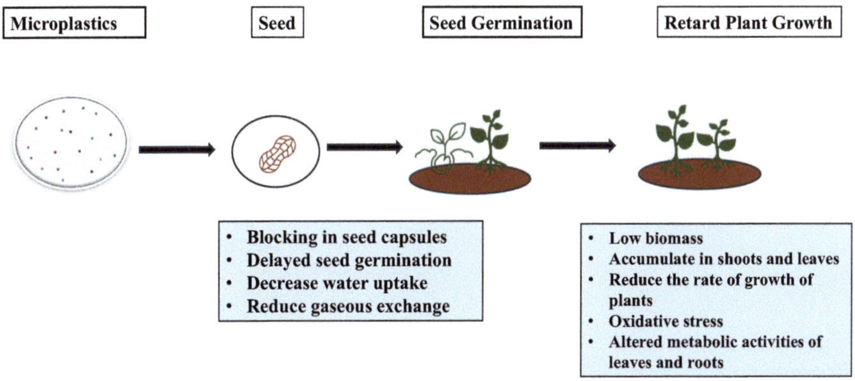

Fig. 3 Impacts of microplastic on seed, and plant growth

2.5 Impacts of microplastics on Terrestrial food chain

As MPs affect the marine biodiversity and food chain similarly contaminate the terrestrial food chain shown in Fig. 4. Although very little information available on the transfer of MPs in the terrestrial food chain, it is confirmed that they can from one tropic level to another from soil and through a prey-predator relationship. In the terrestrial food chain, producers constitute the first trophic level. MPs affect plants' roots, leaves and biomass. MPs transfer from the producer level to the consumer level. (Chae et al., 2020) investigated the movement of PS NPs (28 nm) throughout the terrestrial food chain from soil to plant (mung bean, Vigna radiata) to consumer (Achatina fulica, African giant snail). After that, the overall health of decomposers is significantly impacted by the existence of MPs, which may lead to the locking up of nutrients in the environment, which in turn reduces ecosystem productivity and hurts the food chain, web, and overall health of the environment (Rose et al., 2023).

A study by Huerta Lwanga et al. (2017), illustrated the tropic transfer as well as the process of biomagnification and bioaccumulation of MPs in chicken gizzards from soil via earthworms. Various studies suggest that being exposed to MPs can lead to reproductive problems in soil animals, including reduced fertility and altered sex ratios. These effects can have long-term consequences for population dynamics and soil animals are integral components of terrestrial food webs. Changes in the abundance and health of soil animals due to microplastics can disrupt these food webs, affecting predator-prey dynamics and the overall functioning of ecosystems (Van Raamsdonk et al., 2020). In a recent study *Hypoaspis aculeifer* was used as a predator to observe the transportation of commercial PVC and researchers found that transportation was enhanced due to the prey-predator relationship (Zhu et al., 2018). The intake of contaminated food and transfer of MPs at the tropical level also affect human health. The enormous surface area of M/NPs can potentially produce oxidative stress, cytotoxicity, and translocation of MPs to other tissues (Winiarska et al., 2024). They cause respiratory disease, fertility issues, pregnancy problems

Table 1 Impact of microplastics on plants in soil

MPs type	Concentration	Plant species	Impacts	Ref
HDPE, PS-Cd 5 mg/kg	0–10%	*Z. mays*	Reduced plant growth and decreased dry biomass.	(Boots et al., 2019)
PVC-a and b	0.5–2%	*Lactuva sativa L.*	PVC an enhanced the carotenoid synthesis on other hand PVC b inhibited it. There's no significant effect of both MPs on roots.	(Li et al., 2020a, 2020b)
Biodegrade-able PLA and HDPE	1g kg^{-1}	*Lolium perenne L.*	Seed germination and root biomass are reduced. A precise quantity of Chlorophyll does not vary significantly.	(Wang et al., 2020a, 2020b)
PE, PLA	0.1, 1 and 10%	*Z. mays L.*	PLA reduced the chlorophyll content and dry biomass. PE is not significantly toxic compared to PLA.	(Wang et al., 2020a, 2020b)
PVC	0.1and 1 %	Add with the bacterial community and N/P	It altered the availability of P content to the soil. However, it doesn't affect the bacterial community, but some species get declined or are enhanced.	(Yan et al., 2021)
PHBV	1%	*T. aestivum*	It enhanced the dissolved carbon content and soil biomass. It leads to death of plant after few days.	(Zhou et al., 2021)
PS	100 and 400mg/L	*Allium Cepa*	Reduce root development, mitotic index, induce large % of chromosomal and nuclear abnormalities	(Kaur et al., 2022)

and genotoxicity in human beings when inhaled and ingested (Amato-Lourenço et al., 2021; Çobanoğlu et al., 2021; Zurub et al., 2023) This shows that microplastic interferes with the food chain and affects biodiversity.

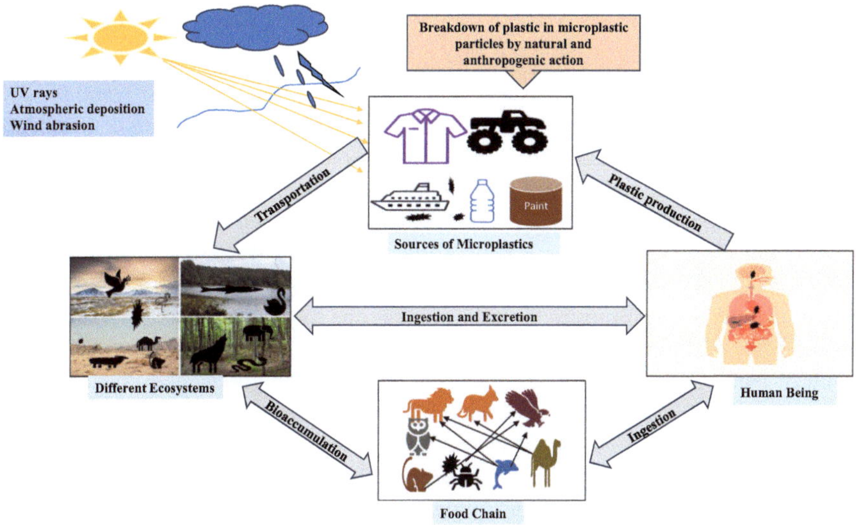

Fig. 4 Transportation of microplastics from the production source to the human being and in the environment

3 Microplastic and Associated Contaminants

Besides the ingestion toxicity of MPs, indirect toxicity is also a matter of concern in the soil ecosystem. MPs release PCB, propylene, BPA, and bis (2-ethylhexyl) phthalate because they are used in plastic manufacturing. Further research has also shown that several monomers, oligomers and other chemical additives such as di-n-octyl phthalate, di(2-ethylhexyl) phthalate, polybrominated diphenyl ethers, and tetrabromobisphenol A, show negative effects on living beings when exposed to them through air, food and water. Some chemicals like BPA are more dangerous to living beings than MPs themselves. These chemicals are released into the food chain after being ingested since they are not chemically bound by the plastic matrix (Wang et al., 2021a, 2021b). Conversely, degradable MPs have the greatest ability to adsorb triamefon and difenoconazole. The adsorption behaviour is almost unaffected by natural factors such as dissolved organic matter, pH and salinity. PBS and PLA microplastics were shown to have a higher concentration ability for the pesticide fipronil than traditional microplastics; the affinity is stronger than that of traditional PP/PE-MPs. Tetracycline (TC) and ciprofloxacin (CIP) have greater multilayer adsorption capacities than PVC-MP monolayer adsorption. So, the degradable MPs are more dangerous to the ecosystem (Qin et al., 2021). Apart from this, microplastics also act as vectors to hydrophobic organic pollutants and provide a greater surface area to inorganic pollutants such as heavy metals, they generate new complex compounds by associating with them and negatively affect the terrestrial environment and food chain (Yadav et al., 2023).

4 Health Impacts of Microplastics in Indian Ecosystem

India is located between latitudes 8°4′ and 37°6′ N and longitudes 68°7′ and 97°25′ E and has a 7517 km coastline (Kumar et al., 2006) . It is a developing country and is becoming more urbanized and industrialized rapidly, which has led to a high rate of plastic production and consumption (Chowdhury et al., 2023). Among the states that consume the most plastics (47%) is Western India, which includes Chhattisgarh, Daman and Diu, Dadra Nagar Haveli, Gujarat, Madhya Pradesh and Maharashtra (Vaid et al., 2021). Since India is the most populous country in the world, microplastic contamination is ubiquitous in the environment due to plastic waste generation (9.3 Mt per year) (Cottom et al., 2024). The vast majority of plastic trash produced is a result of packaging, synthetic fabrics, used tyres, plastic goods, and poor disposal and treatment of waste facilities which affects the terrestrial ecosystem severely (Kibria et al., 2023). However, Marine ecosystems received noticeably more attention from studies investigating microplastics in Indian ecosystems than freshwater, atmosphere, terrestrial habitats, and human consumables (Vaid et al., 2021). In addition to being a vital part of the terrestrial ecosystem, soil is also necessary for crop production and the nation's economic stability. Since India is an agricultural nation, soil pollution with microplastics must be addressed because it might harm the terrestrial ecosystem. According to Singh et al. (2023), in the investigation of microplastics in central India, microplastics not only contaminate the adjacent lake but also have an adverse effect on the health of the soil. Moreover, the micro- and mesofauna of the soil, including mites, enchytraeids, collembolans, and others may consume the microplastic particles. Since these species are crucial to the dynamics of organic matter and the preservation of the soil's physical structure, negative effects on the food chain of debris are expected. Consequently, pollution of the soil of Van Vihar National Park could have detrimental effects on ecological as well as animal development and wellness. Similar studies on the impacts of microplastics on terrestrial ecosystems in India are summarised in Table 2.

5 Conclusion

India is a developing country so microplastic pollution is ubiquitous in the environment. Microplastics (less than 5 mm) are present in terrestrial ecosystems as they are released from various sources, from household waste to large industries. These particles disperse and leach down in terrestrial ecosystems by several factors such as air, water runoff, soil cracks, soil aggregates, biopores, and bioturbation of plant roots and soil organisms. Microplastic contamination alters the physicochemical parameters such as water content availability, fertility, pH, bulk density, and gaseous exchange cycle in soil. It has been noticed that microplastic particles whose size and shape are similar to soil particles have lesser effects on soil properties. Along with this, they also interfere with the growth of plants, productivity of crops, delay

Table 2 Various studies related to microplastic toxicity in soil in different regions of India

Area	Microplastics present	Impacts	Ref
Kasargod municipal Town (Kerala)	PP, HDPE, LDPE	Microplastic accumulation in soil affects the soil health and nearby rivers. There are chances of depletion of groundwater in many areas.	(Reethu et al., 2023)
Bhopal	Polyethylene, PP, PVC, Polyethylene terephthalate, and Polyvinyl acetate	Loss of biodiversity by contaminating groundwater wetlands and national parks.	(Singh et al., 2023)
Mangalavanam, Kerala	Polypropylene, LDPE, HDPE, Polypropylene Isotactic, Polyethylene, 1-octene copolymer	Negatively impact on the biodiversity of mangroves and the health of birds.	(Kannankai et al., 2022)
Cooch Behar, Eastern India	Presence of microplastics and nano plastics with heavy metals.	High NO^{-3} –N content in soil, heavy metals and microplastics affected the soil health.	(Sarkar et al., 2022)

seed germination, biomass production, and accumulate into the root, and shoot of plants which leads them into food chain. Microplastics increase the mortality rate, reduce the reproduction rate, cause various diseases and oxidative stress, and alter soil organisms' behavior. For example, they change the burrowing pattern in earthworms. Some microplastics have detrimental effects on terrestrial environments such as Polyacrylic and polyester particles significantly reduce microbial activities, plastic films reduce the organic matter in soil, PET particles reduced the food uptake in soil worms and cause oxidative stress to them which is an ecological risk to biodiversity and environment. Degradable microplastics has more severe impacts on the environment as they release toxins and other chemical additives which bind with soil particles and go into food-chain. There is enormous microplastic contamination in the Indian terrestrial environment, significantly higher than the average level worldwide

6 Recommendations

As microplastic is an emergent pollutant, more comprehensive and detailed research is needed to monitor plastic pollution and minimise its impacts. If this contamination is not addressed in a way that is appropriate and grounded in science, there will be serious risks to both India's terrestrial resources and the global environment. So, there is a need to reduce plastic pollution in India, develop more effective strategies to control it and raise public awareness about the potential effects, which is crucial.

Mandatory enforcement of strict regulations and laws is to be adopted on plastic production, use and disposal. Government, stakeholders and researchers must work in tandem to address this issue and researchers must be promoted to develop alternative material which is less harmful to the environment.

References

Amato-Lourenço, L. F., Carvalho-Oliveira, R., Júnior, G. R., dos Santos Galvão, L., Ando, R. A., & Mauad, T. (2021). Presence of airborne microplastics in human lung tissue. *Journal of Hazardous Materials, 416*. https://doi.org/10.1016/j.jhazmat.2021.126124

Aralappanavar, V. K., Mukhopadhyay, R., Yu, Y., Liu, J., Bhatnagar, A., Praveena, S. M., Li, Y., Paller, M., Adyel, T. M., Rinklebe, J., Bolan, N. S., & Sarkar, B. (2024). Effects of microplastics on soil microorganisms and microbial functions in nutrients and carbon cycling—A review. *In Science of the Total Environment, 924*. https://doi.org/10.1016/j.scitotenv.2024.171435

Awet, T. T., Kohl, Y., Meier, F., Straskraba, S., Grün, A. L., Ruf, T., Jost, C., Drexel, R., Tunc, E., & Emmerling, C. (2018). Effects of polystyrene nanoparticles on the microbiota and functional diversity of enzymes in soil. *Environmental Sciences Europe, 30*(1). https://doi.org/10.1186/s12302-018-0140-6

Azeem, I., Adeel, M., Ahmad, M. A., Shakoor, N., Jiangcuo, G. D., Azeem, K., Ishfaq, M., Shakoor, A., Ayaz, M., Xu, M., & Rui, Y. (2021). Uptake and accumulation of nano/microplastics in plants: A critical review. *In Nanomaterials, 11*(11). https://doi.org/10.3390/nano11112935

Boots, B., Russell, C. W., & Green, D. S. (2019). Effects of microplastics in soil ecosystems: Above and below ground. *Environmental Science and Technology, 53*(19), 11496–11506. https://doi.org/10.1021/acs.est.9b03304

Brennecke, D., Duarte, B., Paiva, F., Caçador, I., & Canning-Clode, J. (2016). Microplastics as vector for heavy metal contamination from the marine environment. *Estuarine, Coastal and Shelf Science, 178*, 189–195. https://doi.org/10.1016/j.ecss.2015.12.003

Briassoulis, D. (2023). Agricultural plastics as a potential threat to food security, health, and environment through soil pollution by microplastics: Problem definition. *Science of The Total Environment, 892*, Article 164533. https://doi.org/10.1016/j.scitotenv.2023.164533

Campanale, C., Galafassi, S., Savino, I., Massarelli, C., Ancona, V., Volta, P., & Uricchio, V. F. (2022). Microplastics pollution in the terrestrial environments: Poorly known diffuse sources and implications for plants. *In Science of the Total Environment, 805*. https://doi.org/10.1016/j.scitotenv.2021.150431

Chae, Y., & An, Y. J. (2020). Nanoplastic ingestion induces behavioral disorders in terrestrial snails: Trophic transfer effects: Via vascular plants. *Environmental Science: Nano, 7*(3), 975–983. https://doi.org/10.1039/c9en01335k

Chowdhury, S., Tiwari, M., Mishra, P., Parihar, R. S., Verma, A., Mehrotra, R., Punj, N., & Sharma, A. (2023). Recent trends of plastic waste management for sustainable environment in Indian context. *Materials Today: Proceedings*. https://doi.org/10.1016/j.matpr.2023.06.063

Çobanoğlu, H., Belivermiş, M., Sıkdokur, E., Kılıç, Ö., & Çayır, A. (2021). Genotoxic and cytotoxic effects of polyethylene microplastics on human peripheral blood lymphocytes. *Chemosphere, 272*. https://doi.org/10.1016/j.chemosphere.2021.129805

Collivignarelli, M. C., Carnevale Miino, M., Caccamo, F. M., & Milanese, C. (2021). Microplastics in sewage sludge: A known but underrated pathway in wastewater treatment plants. *In Sustainability (Switzerland), 13*(22). https://doi.org/10.3390/su132212591

Cottom, J. W., Cook, E., & Velis, C. A. (2024). A local-to-global emissions inventory of macroplastic pollution. *Nature, 633*(8028), 101–108. https://doi.org/10.1038/s41586-024-07758-6

de Souza Machado, A. A., Kloas, W., Zarfl, C., Hempel, S., & Rillig, M. C. (2018). Microplastics as an emerging threat to terrestrial ecosystems. *In Global Change Biology, 24*(4), 1405–1416. https://doi.org/10.1111/gcb.14020

De Souza MacHado, A. A., Lau, C. W., Till, J., Kloas, W., Lehmann, A., Becker, R., & Rillig, M. C. (2018). Impacts of Microplastics on the soil biophysical environment. *Environmental Science and Technology, 52*(17), 9656–9665. https://doi.org/10.1021/acs.est.8b02212

De Souza Machado, A. A., Lau, C. W., Kloas, W., Bergmann, J., Bachelier, J. B., Faltin, E., Becker, R., Görlich, A. S., & Rillig, M. C. (2019). Microplastics can change soil properties and affect plant performance. *Environmental Science and Technology, 53*(10), 6044–6052. https://doi.org/10.1021/acs.est.9b01339

Gokdemir, O., Randenikumara, S., Floss, M., Rochfort, A., & Astier Peña, M. P. (2024). Illegal waste dumping practices: Where does all the garbage go? *Environmental Forensics 25*(5):1–4. https://doi.org/10.1080/15275922.2023.2297421

Guo, Z. Q., Li, P., Yang, X. M., Wang, Z. H., Lu, B. B., Chen, W. J., Wu, Y., Li, G. W., Zhao, Z. W., Liu, G. Bin., Ritsema, C., Geissen, V., & Xue, S. (2022). Soil texture is an important factor determining how microplastics affect soil hydraulic characteristics. *Environment International, 165*, Article 107293. https://doi.org/10.1016/J.ENVINT.2022.107293

Hu, J., Li, X., Lei, L., Cao, C., Wang, D., & He, D. (2020). The Toxicity of (Nano)Microplastics on C. elegans and Its Mechanisms. *In Handbook of Environmental Chemistry* (Vol. 95, pp. 259–278). Springer Science and Business Media Deutschland GmbH. https://doi.org/10.1007/698_2020_452

Huang, Y., Liu, Q., Jia, W., Yan, C., & Wang, J. (2020). Agricultural plastic mulching as a source of microplastics in the terrestrial environment. *Environmental Pollution, 260*. https://doi.org/10.1016/j.envpol.2020.114096

Huerta Lwanga, E., Gertsen, H., Gooren, H., Peters, P., Salánki, T., Van Der Ploeg, M., Besseling, E., Koelmans, A. A., & Geissen, V. (2016). Microplastics in the terrestrial ecosystem: Implications for lumbricus terrestris (oligochaeta, lumbricidae). *Environmental Science and Technology, 50*(5), 2685–2691. https://doi.org/10.1021/acs.est.5b05478

Huerta Lwanga, E., Mendoza Vega, J., Ku Quej, V., Chi, J. de los A., Sanchez del Cid, L., Chi, C., Escalona Segura, G., Gertsen, H., Salánki, T., van der Ploeg, M., Koelmans, A. A., & Geissen, V. (2017). Field evidence for transfer of plastic debris along a terrestrial food chain. *Scientific Reports, 7*(1). https://doi.org/10.1038/s41598-017-14588-2

Huo, Y., Dijkstra, F. A., Possell, M., & Singh, B. (2022). Ecotoxicological effects of plastics on plants, soil fauna and microorganisms: A meta-analysis. *In Environmental Pollution, 310*. https://doi.org/10.1016/j.envpol.2022.119892

Iqbal, S., Xu, J., Allen, S. D., Khan, S., Nadir, S., Arif, M. S., & Yasmeen, T. (2020). Unraveling consequences of soil micro- and nano-plastic pollution on soil-plant system: Implications for nitrogen (N) cycling and soil microbial activity. *In Chemosphere, 260*. https://doi.org/10.1016/j.chemosphere.2020.127578

Jenkins, T., Persaud, B. D., Cowger, W., Szigeti, K., Roche, D. G., Clary, E., Slowinski, S., Lei, B., Abeynayaka, A., Nyadjro, E. S., Maes, T., Thornton Hampton, L., Bergmann, M., Aherne, J., Mason, S. A., Honek, J. F., Rezanezhad, F., Lusher, A. L., Booth, A. M., Van Cappellen, P. (2022). Current State of Microplastic Pollution Research Data: Trends in availability and sources of open data. *Frontiers in Environmental Science, 10*. https://doi.org/10.3389/fenvs.2022.912107

Jia, L., Liu, L., Zhang, Y., Fu, W., Liu, X., Wang, Q., Tanveer, M., & Huang, L. (2023). Microplastic stress in plants: effects on plant growth and their remediations. *In Frontiers in Plant Science* (Vol. 14). Frontiers Media SA. https://doi.org/10.3389/fpls.2023.1226484

Wang, Jie, Coffin, S., Sun, C., Schlenk, D., & Gan, J. (2019). Negligible effects of microplastics on animal fitness and HOC bioaccumulation in earthworm Eisenia fetida in soil. *Environmental Pollution, 249*, 776–784. https://doi.org/10.1016/j.envpol.2019.03.102

Wang, Jun, Lv, S., Zhang, M., Chen, G., Zhu, T., Zhang, S., Teng, Y., Christie, P., & Luo, Y. (2016). Effects of plastic film residues on occurrence of phthalates and microbial activity in soils. *Chemosphere, 151*, 171–177. https://doi.org/10.1016/j.chemosphere.2016.02.076

Ju, H., Zhu, D., & Qiao, M. (2019). Effects of polyethylene microplastics on the gut microbial community, reproduction and avoidance behaviors of the soil springtail, Folsomia candida. *Environmental Pollution, 247*, 890–897. https://doi.org/10.1016/j.envpol.2019.01.097

Kannankai, M. P., Alex, R. K., Muralidharan, V. V., Nazeerkhan, N. P., Radhakrishnan, A., & Devipriya, S. P. (2022). Urban mangrove ecosystems are under severe threat from microplastic pollution: a case study from Mangalavanam, Kerala, India. *Environmental Science and Pollution Research, 29*(53), 80568–80580. https://doi.org/10.1007/s11356-022-21530-1

Kaur, M., Xu, M., & Wang, L. (2022). Cyto–genotoxic effect causing potential of polystyrene microplastics in terrestrial plants. *Nanomaterials, 12*(12). https://doi.org/10.3390/nano12122024

Kibria, M. G., Masuk, N. I., Safayet, R., Nguyen, H. Q., & Mourshed, M. (2023). Plastic Waste: Challenges and Opportunities to Mitigate Pollution and Effective Management. *In International Journal of Environmental Research, 17*(1). https://doi.org/10.1007/s41742-023-00507-z

Kim, Y. N., Yoon, J. H., & Kim, K. H. (2020). Microplastic contamination in soil environment—A review. *In Soil Science Annual, 71*(4), 300–308. https://doi.org/10.37501/soilsa/131646

Kumar, V. S., Pathak, K. C., Pednekar, P., Raju, N. S. N., & Gowthaman, R. (2006). *Coastal processes along the Indian coastline* (Vol. 91, Issue 4). Accessed https://www.jstor.org/stable/24093957

Lahive, E., Walton, A., Horton, A. A., Spurgeon, D. J., & Svendsen, C. (2019). Microplastic particles reduce reproduction in the terrestrial worm Enchytraeus crypticus in a soil exposure. *Environmental Pollution, 255*. https://doi.org/10.1016/j.envpol.2019.113174

Lei, L., Wu, S., Lu, S., Liu, M., Song, Y., Fu, Z., Shi, H., Raley-Susman, K. M., & He, D. (2018). Microplastic particles cause intestinal damage and other adverse effects in zebrafish Danio rerio and nematode Caenorhabditis elegans. *Science of the Total Environment, 619–620*, 1–8. https://doi.org/10.1016/j.scitotenv.2017.11.103

Li, L., Luo, Y., Li, R., Zhou, Q., Peijnenburg, W. J. G. M., Yin, N., Yang, J., Tu, C., & Zhang, Y. (2020). Effective uptake of submicrometre plastics by crop plants via a crack-entry mode. *Nature Sustainability, 3*(11), 929–937. https://doi.org/10.1038/s41893-020-0567-9

Li, J., Song, Y., & Cai, Y. (2020). Focus topics on microplastics in soil: Analytical methods, occurrence, transport, and ecological risks. *In Environmental Pollution, 257*. https://doi.org/10.1016/j.envpol.2019.113570

Li, X., Wang, R., Dai, W., Luan, Y., & Li, J. (2023). Impacts of Micro(nano) plastics on terrestrial plants: germination, growth, and litter. *In Plants, 12*(20). https://doi.org/10.3390/plants12203554

Liu, H., Yang, X., Liu, G., Liang, C., Xue, S., Chen, H., Ritsema, C. J., & Geissen, V. (2017). Response of soil dissolved organic matter to microplastic addition in Chinese loess soil. *Chemosphere, 185*, 907–917. https://doi.org/10.1016/j.chemosphere.2017.07.064

Liu, C., Chen, L., Xu, A., & Song, Z. (2023). Evaluation of polyurethane sponge biocarrier effects on Litopenaeus vannamei cultivation in zero water exchange systems based on water quality, shrimp performance and bacterial community analysis. *Frontiers in Environmental Science, 11*. https://doi.org/10.3389/fenvs.2023.1326194

Martínez Urreaga, J., Beltrán, F. R., Acosta, J., Aguinaco, T., Fonseca, C., Ochoa, A., Oliet Palá, J. A., González-Sánchez, C., & de la Orden, M. U. (2020). Tube shelters from agricultural plastic waste: An example of circular economy. *Journal of Cleaner Production, 268*. https://doi.org/10.1016/j.jclepro.2020.122401

Mbachu, O., Jenkins, G., Kaparaju, P., & Pratt, C. (2021). The rise of artificial soil carbon inputs: Reviewing microplastic pollution effects in the soil environment. *In Science of the Total Environment, 780*. https://doi.org/10.1016/j.scitotenv.2021.146569

Meng, K., Lwanga, E. H., van der Zee, M., Munhoz, D. R., & Geissen, V. (2023). Fragmentation and depolymerization of microplastics in the earthworm gut: A potential for microplastic bioremediation? *Journal of Hazardous Materials, 447*. https://doi.org/10.1016/j.jhazmat.2023.130765

Qin, M., Chen, C., Song, B., Shen, M., Cao, W., Yang, H., Zeng, G., & Gong, J. (2021). A review of biodegradable plastics to biodegradable microplastics: Another ecological threat to soil environments? *In Journal of Cleaner Production, 312*. https://doi.org/10.1016/j.jclepro.2021.127816

Reethu, M., Biswajit, R., Aravind, G. H., Rafaz, A. K., Sandeep, K., Sijinkumar, A. V., & Warrier, A. K. (2023). A first report on the spatial and temporal variability of microplastics in coastal soils of an urban town in south-western India: Pre–and post-COVID scenario. *Marine Pollution Bulletin, 190*. https://doi.org/10.1016/j.marpolbul.2023.114888

Rillig, M. C., Ziersch, L., & Hempel, S. (2017). Microplastic transport in soil by earthworms. *Scientific Reports, 7*(1). https://doi.org/10.1038/s41598-017-01594-7

Rose, P. K., Yadav, S., Kataria, N., & Khoo, K. S. (2023). Microplastics and nanoplastics in the terrestrial food chain: Uptake, translocation, trophic transfer, ecotoxicology, and human health risk. *In TrAC—Trends in Analytical Chemistry, 167*. https://doi.org/10.1016/j.trac.2023.117249

Roychand, R., & Pramanik, B. K. (2020). Identification of micro-plastics in Australian road dust. *Journal of Environmental Chemical Engineering, 8*(1). https://doi.org/10.1016/j.jece.2019.103647

Sajjad, M., Huang, Q., Khan, S., Khan, M. A., Liu, Y., Wang, J., Lian, F., Wang, Q., & Guo, G. (2022). Microplastics in the soil environment: A critical review. *In Environmental Technology and Innovation, 27*. https://doi.org/10.1016/j.eti.2022.102408

Sarkar, A., Deb, S., Ghosh, S., Mandal, S., Quazi, S. A., Kushwaha, A., Hoque, A., & Choudhury, A. (2022). Impact of anthropogenic pollution on soil properties in and around a town in Eastern India. *Geoderma Regional, 28*. https://doi.org/10.1016/j.geodrs.2021.e00462

Singh, S., Chakma, S., Alawa, B., Kalyanasundaram, M., & Diwan, V. (2023). Identification, characterization, and implications of microplastics in soil—A case study of Bhopal, central India. *Journal of Hazardous Materials Advances, 9*. https://doi.org/10.1016/j.hazadv.2022.100225

Song, Y., Cao, C., Qiu, R., Hu, J., Liu, M., Lu, S., Shi, H., Raley-Susman, K. M., & He, D. (2019). Uptake and adverse effects of polyethylene terephthalate microplastics fibers on terrestrial snails (Achatina fulica) after soil exposure. *Environmental Pollution, 250*, 447–455. https://doi.org/10.1016/j.envpol.2019.04.066

Ullah, R., Tsui, M. T. K., Chen, H., Chow, A., Williams, C., & Ligaba-Osena, A. (2021). Microplastics interaction with terrestrial plants and their impacts on agriculture. *Journal of Environmental Quality, 50*(5), 1024–1041. https://doi.org/10.1002/jeq2.20264

Vaid, M., Mehra, K., & Gupta, A. (2021). Microplastics as contaminants in Indian environment: a review. *In Environmental Science and Pollution Research* (Vol. 28, Issue 48, pp. 68025–68052). Springer Science and Business Media Deutschland GmbH. https://doi.org/10.1007/s11356-021-16827-6

Van Raamsdonk, L. W. D., Van Der Zande, M., Koelmans, A. A., Hoogenboom, P. L. A., Peters, R. J. B., Groot, M. J., Peijnenburg, M. A. C., & Weesepoel, Y. J. A. (2020). Current insights into monitoring, bioaccumulation, and potential health effects of microplastics present in the food chain. *In Foods, 9*(1). https://doi.org/10.3390/foods9010072

Vaksmaa, A., Guerrero-Cruz, S., Ghosh, P., Zeghal, E., Hernando-Morales, V., & Niemann, H. (2023). Role of fungi in bioremediation of emerging pollutants. *In Frontiers in Marine Science* (Vol. 10). Frontiers Media S.A. https://doi.org/10.3389/fmars.2023.1070905

Veerasingam, S., Ranjani, M., Venkatachalapathy, R., Bagaev, A., Mukhanov, V., Litvinyuk, D., Verzhevskaia, L., Guganathan, L., & Vethamony, P. (2020). Microplastics in different environmental compartments in India: Analytical methods, distribution, associated contaminants and research needs. *In TrAC–Trends in Analytical Chemistry, 133*. https://doi.org/10.1016/j.trac.2020.116071

Wang, F., Zhang, X., Zhang, S., Zhang, S., Adams, C. A., & Sun, Y. (2020). Effects of Co-contamination of microplastics and Cd on plant growth and Cd accumulation. *Toxics, 8*(2). https://doi.org/10.3390/TOXICS8020036

Wang, J, Huang, M., Wang, Q., Sun, Y., Zhao, Y., & Huang, Y. (2020). LDPE microplastics significantly alter the temporal turnover of soil microbial communities. *Science of the Total Environment, 726*. https://doi.org/10.1016/j.scitotenv.2020.138682

Wang, C., Zhao, J., & Xing, B. (2021). Environmental source, fate, and toxicity of microplastics. *In Journal of Hazardous Materials, 407*. https://doi.org/10.1016/j.jhazmat.2020.124357

Wang, L., Wu, W. M., Bolan, N. S., Tsang, D. C. W., Li, Y., Qin, M., & Hou, D. (2021). Environmental fate, toxicity and risk management strategies of nanoplastics in the environment: Current status and future perspectives. *Journal of Hazardous Materials, 401*. https://doi.org/10.1016/j.jhazmat.2020.123415

Winiarska, E., Jutel, M., & Zemelka-Wiacek, M. (2024). The potential impact of nano- and microplastics on human health: Understanding human health risks. *In Environmental Research, 251*. https://doi.org/10.1016/j.envres.2024.118535

Xu, B., Liu, F., Cryder, Z., Huang, D., Lu, Z., He, Y., Wang, H., Lu, Z., Brookes, P. C., Tang, C., Gan, J., & Xu, J. (2020). Microplastics in the soil environment: Occurrence, risks, interactions and fate–A review. *Critical Reviews in Environmental Science and Technology, 50*(21), 2175–2222. https://doi.org/10.1080/10643389.2019.1694822

Xu, C., Zhang, B., Gu, C., Shen, C., Yin, S., Aamir, M., & Li, F. (2020). Are we underestimating the sources of microplastic pollution in terrestrial environment? *In Journal of Hazardous Materials, 400*. https://doi.org/10.1016/j.jhazmat.2020.123228

Yadav, S., Kataria, N., Khyalia, P., Rose, P. K., Mukherjee, S., Sabherwal, H., & Khoo, K. S. (2023). Recent analytical techniques, and potential eco-toxicological impacts of textile fibrous microplastics (FMPs) and associated contaminates: A review. *Chemosphere, 326*, Article 138495.

Yadav, P., Dahiya, S., Yadav, S., Dahiya, D., Rani, M., & Chaudhary, S. (2024). Examining the ecological footprint of microplastics: a holistic exploration from genesis to demise. *Oriental Journal Of Chemistry, 40*(3), 788–793. https://doi.org/10.13005/ojc/400321

Yan, Y., Chen, Z., Zhu, F., Zhu, C., Wang, C., & Gu, C. (2021). Effect of polyvinyl chloride microplastics on bacterial community and nutrient status in two agricultural soils. *Bulletin of Environmental Contamination and Toxicology, 107*(4), 602–609. https://doi.org/10.1007/s00128-020-02900-2

Yang, L., Zhang, Y., Kang, S., Wang, Z., & Wu, C. (2021). Microplastics in soil: A review on methods, occurrence, sources, and potential risk. *In Science of the Total Environment, 780*. https://doi.org/10.1016/j.scitotenv.2021.146546

Yin, L., Zhang, S., Liu, B., Zheng, Q., Wang, Z., & Qu, R. (2024). Investigation of the photolysis process of benzo(a)anthracene (BaA) on polyvinyl chloride (PVC) and polystyrene (PS) microplastics: Plastics aging effect, transformation products and toxicity assessment. *Science of the Total Environment, 929*. https://doi.org/10.1016/j.scitotenv.2024.172394

Yu, M., Van Der Ploeg, M., Lwanga, E. H., Yang, X., Zhang, S., Ma, X., Ritsema, C. J., & Geissen, V. (2019). Leaching of microplastics by preferential flow in earthworm (Lumbricus terrestris) burrows. *Environmental Chemistry, 16*(1), 31–40. https://doi.org/10.1071/EN18161

Zhao, T., Lozano, Y. M., & Rillig, M. C. (2021). Microplastics increase soil pH and decrease microbial activities as a function of microplastic shape, polymer type, and exposure time. *Frontiers in Environmental Science, 9*. https://doi.org/10.3389/fenvs.2021.675803

Zhou, J., Gui, H., Banfield, C. C., Wen, Y., Zang, H., Dippold, M. A., Charlton, A., & Jones, D. L. (2021). The microplastisphere: Biodegradable microplastics addition alters soil microbial community structure and function. *Soil Biology and Biochemistry, 156*. https://doi.org/10.1016/j.soilbio.2021.108211

Zhu, D., Bi, Q. F., Xiang, Q., Chen, Q. L., Christie, P., Ke, X., Wu, L. H., & Zhu, Y. G. (2018). Trophic predator-prey relationships promote transport of microplastics compared with the single Hypoaspis aculeifer and Folsomia candida. *Environmental Pollution, 235*, 150–154. https://doi.org/10.1016/j.envpol.2017.12.058

Zurub, R. E., Cariaco, Y., Wade, M. G., & Bainbridge, S. A. (2023). Microplastics exposure: implications for human fertility, pregnancy and child health. *In Frontiers in Endocrinology* (Vol. 14). Frontiers Media SA. https://doi.org/10.3389/fendo.2023.1330396

Health Impact of Microplastic on Indian Aquatic System

Shikha Kumari, Alka Rao, Manjeet Kaur, and Geeta Dhania

Abstract The amount of plastic waste being generated has skyrocketed over the past few decades, and 79% of it is being dumped without effective waste management in landfills or nearby aquatic bodies. As plastic waste degrades due to abiotic factors, microplastics are created. MPs are synthetic or semi-synthetic polymeric particles. The aquatic ecosystem is particularly vulnerable to microplastic pollution as most plastic waste is dumped there. Persistent organic pollutants and microplastics interact to increase the harm posed by the pollutants. Microplastics are directly ingested by organisms from polluted water or indirectly through the contaminated food web. The impacts of microplastics are wide-ranging, impacting marine life, fisheries, economics, tourism, plants, marine aesthetics, and human health. This study involves the adverse impacts of microplastic pollution on the Indian aquatic ecosystem.

Keywords Plastic waste · Microplastic · Freshwater ecosystem · Marine ecosystem · Toxic effects

1 Introduction

Plastics are synthetic materials and they are highly flexible, durable, easy to handle and can be moulded into different shapes. Thus, they are widely used for various purposes. Polyethylene, polystyrene, polythylene terephthalate, polypropylene and

S. Kumari · A. Rao · G. Dhania (✉)
Department of Environmental Science, Maharshi Dayanand University, Rohtak, Haryana, India
e-mail: Geeta.env.sc@mdurohtak.ac.in

S. Kumari
e-mail: shikha.rs.envsc@mdurohtak.ac.in

M. Kaur
University Institute of Engineering and Technology, Maharshi Dayanand University, Rohtak, Haryana, India
e-mail: manjeetkaurbiotech.uiet@mdurohtak.ac.in

P. Khyalia and S. Yadav (eds.), *Microplastic Pollution in India and Its Environmental Impacts*, Earth and Environmental Sciences Library,
https://doi.org/10.1007/978-3-031-92530-6_8

polyvinylchloride form about 90% of global plastic production (Andrady & Neal, 2009). Abiotic factors like UV ray exposure and temperature lead to photodegradation of plastic waste (Browne, 2015). Degradation and fragmentation of plastic waste result in the formation of small-sized particles called microplastic, having size less than 5 mm (Geyer et al., 2017; Sighicelli et al., 2018). Besides the small size MPs possessed a large specific area with not easily degradable, chemically durable and water-insoluble properties (Gasperi et al., 2018).

There are two kinds of MP (Fig. 1) first is Primary MP which includes microsized plastic materials and microbeads. They are used in sandblasting media, manufacturing of synthetic clothes, chemical formulations and cosmetic products (Andrady, 2015). Secondary MP consists of the fragmented plastic waste degraded due to environmental factors (Costa et al., 2016). The MP pollutants enter the aquatic system either through a primary source or a secondary source. Anthropogenic activities like wastewater from industries, polyester fibre from textile industries and dumping of garbage into aquatic environments are the biggest contributing factor to the availability of MPs in aquatic environments (Chatterjee & Sharma, 2019; Morritt et al., 2014). The other sources of MP in the aquatic environment are fishing activity and aqua tourism (Nace, 2017).

Plastic waste firstly comes into contact with freshwater ecosystem and from there, they are introduced into the marine ecosystem (Browne et al., 2011; Murphy et al., 2016). The Color, shape, and size of MPs affect their ingestion by the aquatic organism. Both physical and chemical effects of MPs pollution have been observed in aquatic organisms (Meng et al., 2020). Aquatic animals such as mammals, turtles, and birds, namely green turtles, seagulls, and sea lions, are highly vulnerable to MPs toxicity (Thushari and Senevi rathma, 2020). Seafood like oyster, crab, sea cucumber, Mussels were identified to consume MPs (Li et al., 2018a; Sharma & Chatterjee, 2017). Bioaccummulation of MPs disturbs the digestive and nervous

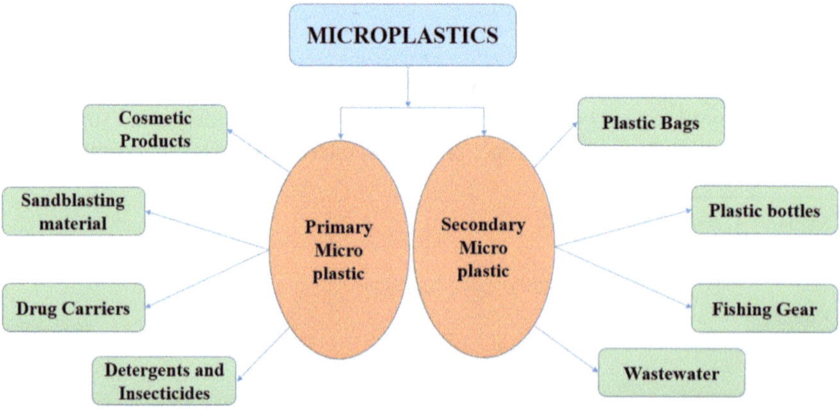

Fig. 1 Types and sources of microplastics

system, damages organ, impede the growth of organisms and causes infertility (Guo et al., 2020; Kumar et al., 2018; Yang et al., 2022).

The present study encloses the adverse impacts of microplastics on the freshwater and marine aquatic ecosystems in India, with a special focus on the impact of microplastic pollution on water quality, aquatic plants, and other aquatic creatures, including phytoplankton, zooplankton, crustaceans, water birds and other large marine animals.

2 MPs Pollution in Indian Aquatic Ecosystem

MPs are abundant in aquatic ecosystem. Routes for transportation of these MPs in aquatic ecosystem has been shown through Fig. 2. The Indian aquatic system is mainly divided into the freshwater aquatic system and marine aquatic system. MPs, as a pollutant, are widely distributed throughout the freshwater and marine environment. The growing amount of MPs will result in a noticeable increase in MP pollution in the Indian aquatic system (Urbanek et al., 2018). This issue is more noticeable in developing countries like India due to the mismanagement of waste. Presently to the best of the author knowledge, no studies are available and most of them are based on the presence of Mps on beaches, sediments and water samples. However, very few researches have been done on their impact on the health of aquatic biota. Freshwater pollution due to plastic waste is an emerging hazard. Freshwater sources like rivers are greatly linked to the marine ecosystems. The occurrence of MPs in freshwater and marine water in different parts of India are discussed further.

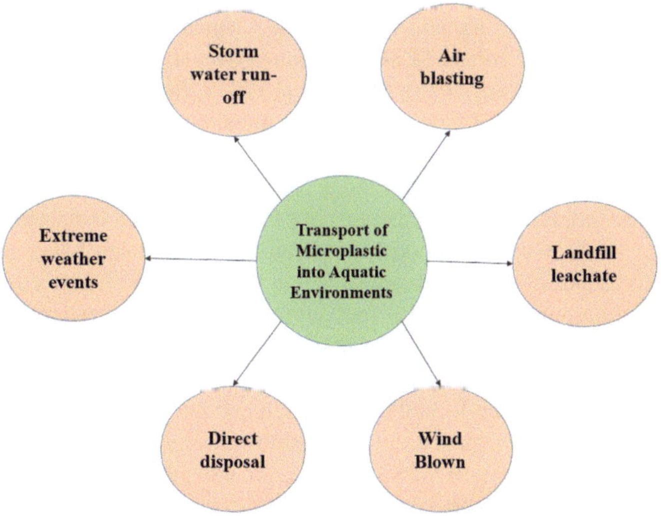

Fig. 2 Different routes of transport of microplastics into aquatic environment

A no. of samples of Indian marine water were collected using different techniques and were found to be contaminated with MPs by Ganesan et al. (2019) and Sathish et al. (2020). Marine debris pollution was noticed along marina beach, Chennai (Kumar et al., 2016). Three separate beaches, Aksa, Versova, and Girgaon Chowpatty, along the Mumbai coastline, were chosen for the study. Seawater and marine sediment samples were taken from each, and mp was found in all. In both the sediments and the saltwater column, beads were the most common kinds of MPs, whereas fibres contributed the least. The majority of the particles that were observed belonged to the 0.45 to 500 m size range (Dutta et al., 2022). Valsan et al., 2023, reported finding microplastics in four areas in southwest India's pore water, groundwater, and seawater. The most common type of MPs were fibres. Kerubo et al., 2020 conducted a study on the presence of microplastics (MPs) in Kenyan creeks (Tudor, Port-Reitz, and Mida). Small-size (20–250 m) MPs were found in large numbers, while Tudor and Port Reitz had higher amounts than Mida Creek. The study gives information on the levels of microplastic in the streams and suggests concentrating on small-sized microplastics for monitoring because of their high concentrations, which can be harmful to creatures.

Rivers, streams, lakes, ponds and ditches are the major freshwater sources in India. Their size, habitat and location may vary. Lakes and rivers form an important part of the freshwater aquatic system in India. Based on various investigations carried out by the researchers it was found that both the rivers and lakes are contaminated with MPs. Maximun research on MPs pollution was conducted in southern India (Vaid et al., 2021); and the major focus of researchers was on MPs ingestion by fishes. Evidence of MPs was found in the water of three rivers, namely, the Adyar River, Chennai; the Kosasthalaiyar River, Chennai; and the Multhirappuzhayar River, Munnar, Kerala. Among the three rivers Adyar River was highly polluted. The MP were mainly in the forms of fibre and fragments (Lechthaler et al., 2021). MP abundance was found in the different freshwater aquatic systems like Karman river, Kerala, Kodaikamal Lake, Sabarmati river, Vembanad Lake, and ganga river (Laju et al., 2022; Owens & Kamil, 2020; Ram & Kumar, 2020; Sarkar et al., 2019; Sruthy & ramasamy, 2017).

PP and PE were widely found in the Indian aquatic system (Veerasingam et al., 2016). Fishing nets and ecotourism discards resulted in the introduction of MPs fragments in the Periyar River, South India. MPs in the form of Films, foams, fibres and fragments were found in the sediments of kaveri river, south India, (Maheswaran et al., 2022). A study conducted on MPs pollution in Veeranam Lake, Tamilnadu, showed the presence of different kinds of MP in the lake water. These compositions were mainly PE (polyethylene), PS (polystyrene), PP (polypropylene), and PVC (polyvinyl chloride) and nylon. Napper et al., 2021, recognized the presence of MP in the Ganga river. Further, they found that the Brahmaputra, Ganga and Meghna rivers collectively introduce 1–3 billion MP each day into the Bay of Bengal.

Very few studies were carried out on groundwater contamination due to MPs. Leaching is the major source of MPS's introduction into groundwater. Ganesan et al., (2019) and Selvam et al., studied the effect of MPs in groundwater and surface water resources in Tamilnadu and depicted the presence of MPS in groundwater based on their research. PP, PVC, and Polyamide were found widely in the groundwater samples. The no. of MPs fragments found in surface water and groundwater were 12 and 96, respectively.

3 Impact of MPs on Indian Aquatic Ecosystem

3.1 Impact of MP on Water Quality

MPs possess good hydrophobicity and thus act as a carrier of pollutants (Yuan et al., 2020). MPs consist of Monomers like vinyl monomers, Functional monomers, acrylic and methacrylic acid monomers, MPs also contain additives like phthalates, aliphatic binary esters, phosphate epoxy compounds, dyes and pigments. MPs absorb heavy metals like zinc, lead, copper, chromium, and cadmium (Godoy et al., 2019). Along with heavy metals they also absorb polychlorinated biphenyl, organochlorine pesticides, polybrominated diphenyl ether, petroleum hydrocarbon and bisphenol A (Wang et al., 2020a, 2020b). The introduction of these pollutants into water alters the water's physicochemical properties, thus deteriorating its quality (Yadav et al., 2023).

3.2 Impact of MP on Aquatic Plants

Microplastics are abundant in the aquatic environment and they easily interact with aquatic plants like seagrass (Goss et al., 2018), Duckweed (Dovidat et al., 2020) and Mangrove (Li et al., 2018a, 2018b). Aquatic plants could trap microplastics from the surrounding water and accumulate them to plant surface by phoyostabilization (Goss et al., 2018; Nolte et al., 2017; Yuan et al., 2019). The microplastics contaminated plants are ingested by herbivorous species and thus they are transferred from one trophic level to another (Dovidat et al., 2020; Goss et al., 2018; Gutow et al, 2016). Such trophic transfer of contaminants is a serious concern but very less research has been done in this field. Microplastics affect the plants adversely; for instance, the cell viability and root growth were found to be reduced in duckweed Lemna minor, as per Kalcikova et al. Yuan et al. (2019) also reported adverse effects of microplastic accumulation on the growth and reproduction of aquatic plants and ferns.

3.3 Impact of MP on Aquatic Biota

Studies related to the impact of microplastic on aquatic biota have been compiled in Table 1.

Table 1 Impact of microplastic on aquatic biota

Aquatic Biota	Impact	Ref
Phytoplankton	Inhibited growth	Bellingeri et al., 2019
Zooplankton	Inhibited growth, Increased mortality	Yu and Chan, 2020
Polychaete	Neurotoxicity, Oxidative damage	Urban-malinga et al.; Silva et al., (2020a, 2020b)
Crustaceans	Reduced reproductive ability, Increased mortality	Au et al., 2015
Fish	Oxidative stress; Liver inflammation, Brain Damage, Tissue damage	Naidoo and glasom, (2019); Mattson et al.
Sea birds	Delayed growth, Abnormal blood calcium level	Lavers et al., (2019); Liu et al., (2019)
Seals	Increased Mortality	Alimba and Faggio, (2019)
Sea turtle	Kidney Damage, Severe anaemia	Pagano et al., (2019)

3.3.1 Phytoplankton

Aquatic phytoplankton is the foundation of aquatic food webs and is crucial for ecosystem services and ecosystem functioning on a global scale. Although phytoplankton make up less than 1% of the planet's photosynthetic biomass. They produce nearly 50 percent of the world's net primary production and serve as the main source of energy for aquatic ecosystems (Winder & Sommer, 2012). A number of studies have documented the negative impact of MPs on Phytoplankton. These impact includes reduced growth of Phytoplankton tissues, genetic toxicity, lysosome alteration, peroximal expansions, irritation of antioxidant system (Anbumani & Kakkar, 2018; Khoshnamvand et al., 2021; Li et al., 2018a, 2018b; Paul-pont et al., 2016).

The toxicity of MPs depends greatly on the size and type of the microplastics. With the increase in the size of MPs, negative growth-related impacts exist (Khoshnamvand et al., 2021; Mao et al., 2018). Loss of juvenility in algal species *Rhodmonas baltica* and *Oxyrrhis marina* was noticed on performing an expanded uptake assay using 10 μm and fluorescent polystyrene particle by Lyakurwa, (2017). *Chlorella vulgaris, Dunaliella tertiolecta* and *Thalassiosira pseudonana* showed a decrease in photosynthetic activity on exposure to Polystyrene microplastic for about 72 H (Sjollema et al., 2016a, 2016b).

In another study, the growth rate, chlorophyll A content and photosynthetic activity was reduced in Cenedesmus obliqus due to interaction with microsized PS particles (Besseling et al., 2013). The algal species *Chlorella pyrenoidosa, Chlorell* Sp., *Scenedesmus obliqus* and *Dunaliella tertiolecta* showed destruction of algal cells, blocked photosynthesis due to blocked light and air, decreased chlorophyll content respectively (Besseling et al., 2013; Bhattacharya et al., 2010; Yi et al., 2019).

MPs affect the chlorophyll level in algae by impacting the flow of light and air, which ultimately decreases the photosynthetic activity of algae (Sjollema et al., 2016a, 2016b). In *Tigriupus japonicas*, fatality rate increased with enhanced exposure to MPs (Lee et al., 2013). MPs could destroy the filteration process in Daphnia magna, because they ingest MP and store them in the form of fat droplet (Jemec et al., 2016).

3.3.2 Zooplankton

Zooplanktons are a group of free-floating invertebrates (Cozar et al., 2014). They are considered as primary consumers in the aquatic system and thus play an important role in the aquatic food chain. Zooplanktons are widely present in the aquatic system and have different feeding mechanisms, which make them interact with different kinds of microplastic (Cole et al., 2013).

Different species of zooplankton including shrimps, worms, ciliates, cladocerans, and copepods etc. found to ingets microplastics. A number of secondary consumers feed on the zooplankton, which increases their susceptibility to the transfer of these MPs from zooplankton to different levels of the food chain. On ingestion of MPs, the feeding ability of Centropages typicus get lost (Cole et al., 2013). MPs namely Polymethylmethaacrylate (PMMA) and Polyhydroxybutyrate (PHB) exposure leads to decreased growth of *Gammarus fossarum* as per Straub et al. (2017). A study by Besseling et al. (2013) concluded that MPs ingestion in marine lugworm Arenicpla marian, resulted in loss in weight and feeding activity.

3.3.3 Polychaete

Polychaete are widely found in the aquatic ecosystem and play an important role in the aquatic food chain (Rodrigues et al., 2011). They serve as a food source for aquatic animals including seabirds, fish, crustaceans etc. They also act as a vector for many contaminants through bioaccumulation. Decreased feeding response was observed in A. marina at high dose of MPs by Besseling et al. (2013). In *Perineresis albuhitensis*, a four-week exposure to PSMP's reduced the posterior segment generation ration and increased mortality (Leung & Chan, 2018). Another species of Polychaete namely *Marenzelleri*a sp. Showed an increased consumption ratio on ingestion of PS MPs.

An investigation of *Acaryochloris marina*, exposed to PVC for a duration of 48 Hours and four weeks, showed that worm exposed to PVC for 48 h displayed extended gut residence time and those exposed for four weeks exhibited lower eating activity (Wright et al., 2013). Muller Karanassos et al. (2021), investigated acute and chronic toxicity of MPs on *H. Diversicolor* and the results showed loss in body weight, reduced feeding response and burrowing activity. At higher exposure to PVC, the activity of catalase (CAT) and glutathione S-transferase were inhibited and regeneration capacity was reduced. Smaller size MPs greatly affect the oxidative status, behaviour defence efficiency and regeneration ability (Silva et al., 2020a; Silva et al., 2020b). Smaller size MPs greatly affect the oxidative status, behaviour defence efficiency and regeneration ability (Silva et al., 2020a, 2020b).

3.3.4 Crustaceans

They include shrimps, crayfish, lobsters, crabs, krills etc. MPs cause reduced fecundity, alteration in behavior, impaired enzyme activity and reduced food uptake (Cole et al., 2015). Crustacean species namely *Hyalella Azteca* showed an increase in the mortality rate and a decrease in reproductive ability and growth with a decrease in the size of MPs (Au et al., 2015). *Eriocheir sinensis, Carcinus maenus* showed a decrease in Na^+ content and increase in Ca^+ Concentration and haemocyanin content, and a significant reduction in the oxygen consumption in haemolyph. It was also observed that male crabs consume more MP as compared to female crabs (Mak et al., 2019). Increased exposure to MPs, destroys crustacean filtration system and obstruct the intestines and even lead to mortality. The consumption of MP is higher in free swimming crustaceans than the sessile ones (Jemec et al., 2016; Au et al., 2015). Delayed molting, decreased fecundity and retarded development are other consequences of MPs pollution.

3.3.5 Fish

A noumber of studies are availalable on the impact of MPs on Zebrafish. For instance, accummulation of MPs in the gill, liver and gut of zebrafish was observed after 7 days of exposure to MPs by Naidoo and Glasom, (2019). Further fat buildup, inflammation in liver, oxidative stress and change in metabolism were the most probably occurring toxic effects. MPs greatly influence the growth rate and increase mortality in Zebrafish. The adverse effects on zebrafish include MP accumulation, intestinal inflammation, tissue damage, development and reproductive impact immune dysfunction (Lei et al., 2018a, 2018b; Pitt et al., 2018; Qiao et al., 2019a, 2019b). In Zebrafish larvae, the catalytic capacity of acetylcholinesterase was suppressed when the larvae came into contact of 1 mg/l of PS MPs. The suppressed activity of Ach can lead to paralysis and death of the zebrafish (Chen et al., 2017). The normal metabolism in zebrafish was disturbed and oxidative stress was induced on exposure to MP for a duration of 2 days (Qiao et al., 2019a, 2019b).

The impact of PS MPs includes oxidative stress and disturbed metabolism in red tilapia (Ding et al., 2020). PS MPs accumulation in the digestive tract of goldfish larvae *carassius auratus* caused oxidative stress, organ damage and inhibition of movement and growth (Yang et al., 2020). Brain damage and behavioural changes may occur in crucian carp fish, if the nanoplastic passes through blood to the brain barrier (Mattsson et al., 2017).

MPs affect the antioxidant response, which lead to increased cellular oxidative stress which in turn causes lipid peroxidation (LPO). LPO has the ability to damage liver, muscles, and gills. These harmfull effects of LPO are documented in different species of fish and other aquatic organisms (Hilfiker et al., 1999; Wen et al., 2018). On exposure to the mixture of MPs and Hg for 96 h, LPO damage in the brain of *Dicentrachus labra* was noticed (Barboza et al.,2018b). A six week exposure to MPs causes loss in weight, inflammation, microgranulomas in the liver and disruption in the buccal cavity in marine fish, *Carassius carassius* (Jabeen et al., 2018; Wang et al., 2020a, 2020b).

PE MPs effects on Danio rerio were intercepted development of eggs and neurotoxic effects (Mak et al., 2019). In another study on Danio rerio PS MP ingestion led to its accumulation in liver intestine and gills (Lu et al., 2016). An increase in exposure dose increased mortality rate by 14% in Dicentrachus labrax (Mazurais et al., 2015). Schiniri et al., 2020 found polypropylene and polystyrene MPs in the digestive tract of medium pelagic fish. Cracking of Microvilli in the intestine, tissue alteration, change in gills and digestive glands (Faggio et al., 2018; Lei et al., 2018a, 2018b; Macmillan, 2018).

Jabeen et al. (2018) analysed that fibers of MPs were ingested by *Carassius auratus* and caused severe damage to intestinal lining of goldfish. Accummulation of MPs in the gastrointestinal tract of fish could induce normal behavior in fish. Long-term exposure to MP influences their nutrition and affects their health and growth (Yin et al., 2018).

MPs can easily interact with other organic pollutants and thus they affect the absorption, distribution, metabolism and excreation of these pollutants in different aquatic creatures. Due to MPs exposure, *Seabastes schelegelli* lost its weight gross energy and growth rate at 65.4%, 9.5% and 65.9%, respectively (Yin et al., 2018).

Fragments and pellets form of MPs were chewed and expelled by fish which caused adverse impacts on their jaws. Most of the small MPs are easily ingested by the aquatic animals instead of the large ones (Jabeen et al., 2018).

Fishes namely Blue musel *mytilus edulis L. Granulocytoma, Calanus helogolandicus, Mytilus Spp., mytilus galloprovinciallis Macerel squid*, tuber, yellow squid, shark, red squid and sankara were found to have MPs (Avio et al., 2015; Cole et al., 2013, 2015; Lechner et al., 2014; Moos et al., 2012; Sussarellu et al., 2016a, 2016b; Vandermeersch et al., 2015).

3.3.6 Large Marine Animals

Large marine animals like shark, whale, sea turtle, seals and polar bears either engulf MP directly or they ingest them by eating other maine creatures. MP were found in the intestine and stomach of habor seal and *Phoca vitulina.* A high concentration of phthaletes in Mediterranean fin whale (*Balaenoptera Physalus*), indicates the severity of MP pollution in the oceans (Fossi et al., 2012). Approx. 261 MPs particles were found in 7 dolphins, 2 seals, and 1 whale at the coast of Britain.

A study was conducted on sea turtles in Brazil and the results showed the accumulation of MPs in the digestive tract of approx. 60% of seaturtles (Nerland et al., 2014). Investigations have reported the presence of MP in the GI tract of harbor seal *Phaco Vitulina* in the Netherlands (Rebolledo et al., 2013). MPs were also found in south America fur seals Halichoerus *Grypus atlantica* (Hudak & Sette, 2019). Harbor seal *Phoca vitulina* had mps in their stomach and intestine (Rebolledo et al., 2013). A feeding study conducted by Nelms et al., (2018), proved the trophic transfer of MPs from Atlantic mackerel to gray seals *Halichoerus Grypus.*

Besides seals MPs were detected in Dolphins, for instance Short beaked common dolphin *Delphinus delphis* (Hernandez- Ganzalez et al., 2018), East Asian finless *Propiscs veophocaena asiaurientalis sunaueri* (Xiong et al., 2018), *Delphinus delphis* (sala et al., 2019), harbor porpoise *Phocoena phocoena* (Van Franneker et al., 2018) *Souura chinesis* (Zhu et al., 2019).

The Mediteranean sea turtle have cellular necrosis in different tissue, severe anaemia and kidney damage (Pagano et al., 2019). The marine C*opepocalanus helgolandicus* was fed with polystyrene which resulted in decreased fecundity (Cole et al., 2015). The acetylcholine esterase activity was reduced in *Pomatoschistus microps* due to MPs ingestion (Alimba et al., 2019).

3.3.7 Waterbirds

Both freshwater bird and sea bird are susceptible to the negative impacts of MPs. Besides this, the migratory behavior of waterbirds increase the chances of their transfer to different aquatic sources (Reynolds & Ryan, 2018; Basto et al., 2019). There are many evidences of MPs ingestion by waterbirds (Reynolds & Ryan, 2018). These ingested MPs remain in the gut and GI tract of the waterbirds (Basto et al., 2019; Brookson et al., 2019; Cartraud et al., 2019).

MPs fragments were found in the stomach of six species of seabirds *namely Phalecrocorax bougainvillli, larus dominicanus, Palecanoides garnotti, Pelenanoides urinatrix, Peelcamrs thagus, Spheniscus humboldti.* The highest MPs ingestion was detected in *Larus dominicams* (Theil et al., 2018). MPs block the digestive tract, in turn feeding rate is reduced (Wright et al., 2013).

The ingestion and presence of MPs were observed in GI tract of seabird like ducks, geese, loons, African duck, Phalacrocorax auritus (Brookson et al., 2019; Holland et al., 2016; Reynolds & Ryan, 2018). Bird species like great skua, *Eurasian dipers* and gulls ingest the prey containing MPs and the level of ingestion in seabirds varies from region to region (Hammer et al., 2016; Furtado et al., 2016; Basto et al., 2019; Catraud et al., 2019). The effects of MPs ingestion in seabirds are Delayed growth and sexual maturity, abnormal blood calcium level (Lavers et al., 2019).

3.4 MPs Pollution Impact on Indian Aquatic Biota

In Zebrafish, exposure to MPs resulted in reduction of branched chain amino acid. These amino acids play an important role in prevention of fatty acid accumulation and also in fatty acid metabolism. So, MPs disturb the metabolism in zebrafish. PS MPs affects the liver in zebrafish adversely. They induced accumulation and disturbed the lipid metabolism. The synthesis and transportation of phospholipid is also affected (Mahamud et al., 2022). Fish species namely *Sardiella longiceps* was collected from Indian coast had MP. MPs were evidenced in different species of shellfish and bottom feeding fishes, namely *Arius maculatus, E, Maculatus, Etroplus Suratensis* and *Villorita* sp. Found in vembanad lake (Nikki et al., 2021).

MPs were traced in the gut of red line torpedo barb collected from periyar river. In India, Fishes caught from the southeast coast and southwest coast were found to ingest MP. These were marine fishes, namely Mackerel, sardine, ribbon fish, anchovy, tunas and dolphin fish (Sulochanan et al., 2011; Sajikumar et al., 2013; Kripa et al., 2018). Around 365 shellfish were collected from adayar estuary and MPs were found in the majority of Shellfish (Jeyasanta et al., 2023).

MPs in Fish cause histopathological modifications in the intestine, leading to detachment of mucosa epithelial lining from the lamina propia and causing its widening, reduction and puffing of villi, increase in no. of goblet cells and certain alteration in serosa of fish. Apart from this, MPs ingestion in fish causes metabolic alteration like up-regulation and down-regulation of fatty acids and amino acids, respectively (Peda et al., 2016). The uptake of nano and microplastic altered the ratio of cholesterol and triglycerides in the blood serum level of fish. MPs also cause reduced growth and feeding imbalance in fishes (Jabeen et al., 2018). Endocrine disorders, oxidative stress, altered immune and gene response, reduced growth and reproduction (Jin et al., 2018; Ma et al., 2020; Qiao et al., 2019a, 2019b).

MPs exposure in glassfish lead to an increase in juvenile mortality and a reduction in growth performance (Naidoo & Glasson, 2019). MPs were detected in shellfish and finfish samples collected from the river Sal Estuary, Goa. The fish samples collected from ulhar river and MPs were present in them (Kumkar et al., 2021). At Kerala coast, the MPs ingestion rate in fish species was found to be 21% and around 653 species of commercial fish were found to have MPs (James et al., 2020). Fourty fish belonging to two species, namely *Epinephalus merra* and *Rastrilleger Kanagurta* were collected from Tuticoriin and twelve out of them were found to have MPs in their intestines (Kumar et al., 2018). MPs were found in the gastrointestinal tract of Bombay duck, white sardine, kadal shrimp and Malabar sole fish. The adverse impacts of MPs ingestion differ in different species. Feeding, development, reproduction and survival are mainly affected in aquatic invertebrates (Sussarellu et al., 2016a, 2016b; Trestrail et al., 2020). MP toxicity in fish mainly affects the tissue an organ health, intestinal permeability, metabolism, behavioural and neurological function and intestinal microbiome diversity (Jacob et al., 2020).

MPs trapped in the gills of fish can result in respiratory disease like gill infection and hypoxia (Mahamud et al., 2022). PE based MPs affect the shape and size of larvae and also reduces the fertilization and survival rate of larvae in Zebrafish. MPs may cause damage to digestive organs by altering villi height and width, disturbing functional patterns of villi, and damages the small intestine (Mahamud et al., 2022; Mbugani et al., 2022).

MPs were found in the Indian edible oyster caught from the tuticorin coast, Tamilnadu (Patterson et al., 2019). The sperm speed and egg diameter dwindled in oysters due to polystyrene particles (Sangkham et al., 2022). Mussels were collected from the coast of Kerala and Tamilnadu. MPs contamination was observed in the soft tissue of two mussel species, namely P. virdis and P. perna (Patterson et al., 2021). MPs particles were detected in the benthic invertebrates such as Magelona cinta,

Tellina sp. and Strenaspis scutata collected from the Kochi coast of India (Naidu et al., 2018).

4 Conclusion

Large-scale plastic production and usage have permeated every aspect of our daily life. Because there is so much plastic debris around us, the ecology has been severely harmed. Freshwater and marine waters are contaminated, and aquatic life is harmed by the microplastics created by the breakdown of plastic debris. The abundance and different compositions of microplastic were reported in Indian water bodies as pellet, fiber, fragment, film, and foam forms that accumulate on organisms make consequences. When these microplastics are found in aquatic species, the aquatic biota suffers. From the smallest phytoplankton to the greatest whale, MPs have an adverse impact. In India the research studies are limited to certain locations and many areas are still not noticed. More in-depth studies of these emerging pollutants are required for effective management action and risk assessment to reduce the negative social, ecological, and economic impacts in light of the potential threat and subsequent impacts of plastic pollution on human populations.

5 Recommendations

To address the growing concern of microplastic pollution in India's aquatic systems, future research should focus on long-term studies assessing the bioaccumulation and biomagnification of microplastics in the food chain and their interactions with other pollutants, such as heavy metals and pesticides. Standardised microplastic detection and toxicity assessment methods in Indian water bodies must be developed to ensure accurate monitoring. Additionally, studies on the impact of microplastic ingestion on human health through seafood consumption are crucial. For policy planners and decision-makers, implementing stricter regulations on plastic waste management, enforcing extended producer responsibility (EPR), and promoting eco-friendly alternatives can help reduce microplastic contamination. Establishing nationwide monitoring programs and integrating microplastic pollution control measures into existing water quality management policies are necessary. Stakeholders, including industries, researchers, and environmental organisations, should collaborate to raise public awareness and promote sustainable practices to mitigate the health risks associated with microplastic pollution in India's aquatic environment.

References

Alimba, C. G., & Faggio, C. (2019). Microplastics in the marine environment: Current trends in environmental pollution and mechanisms of toxicological profile. *Environmental Toxicology and Pharmacology, 68*, 61–74.

Anbumani, S., & Kakkar, P. (2018). Ecotoxicological effects of microplastics on biota: A review. *Environmental Science and Pollution Research, 25*, 14373–14396.

Andrady, A. L. (2015). Persistence of plastic litter in the oceans. In M. Bergmann, L. Gutow, & M. Klages (Eds.), *Marine anthropogenic litter* (pp. 57–72). Springer.

Avio, C. G., Gorbi, S., & Regoli, F. (2015). Experimental development of a new protocol for extraction and characterization of microplastics in fish tissues: First observations in commercial species from Adriatic Sea. *Marine Environmental Research, 111*, 18–26.

Au, S. Y., Bruce, T. F., Bridges, W. C., & Klaine, S. J. (2015). Responses of Hyalella azteca to acute and chronic microplastic exposures. *Environmental Toxicology and Chemistry, 34*(11), 2564–2572.

Andrady, A. L., & Neal, M. A. (2009). Applications and societal benefits of plastics. *Philosophical Transactions of the Royal Society b: Biological Sciences, 364*(1526), 1977–1984.

Anju, P., Sekharan, M. N., & Harikrishnan, M. (1999). *The occurrence of microplastics in gut contents of endemic barb Sahyadria chalakkudiensis inhabiting river systems of Western Ghats* (p. 2019). World Aquaculture Society, Chennai Trade Center.

Basto, M. N., Nicastro, K. R., Tavares, A. I., McQuaid, C. D., Casero, M., Azevedo, F., & Zardi, G. I. (2019). Plastic ingestion in aquatic birds in Portugal. *Marine Pollution Bulletin, 138*, 19–24.

Brookson, C. B., De Solla, S. R., Fernie, K. J., Cepeda, M., & Rochman, C. M. (2019). Microplastics in the diet of nestling double-crested cormorants (Phalacrocorax auritus), an obligate piscivore in a freshwater ecosystem. *Canadian Journal of Fisheries and Aquatic Sciences, 76*(11), 2156–2163.

Barboza, L. G. A., Vieira, L. R., Branco, V., Figueiredo, N., Carvalho, F., Carvalho, C., & Guilhermino, L. (2018). Microplastics cause neurotoxicity, oxidative damage and energy-related changes and interact with the bioaccumulation of mercury in the European seabass, Dicentrarchus labrax (Linnaeus, 1758). *Aquatic Toxicology, 195*, 49–57.

Bellingeri, A., Bergami, E., Grassi, G., Faleri, C., Redondo-Hasselerharm, P., Koelmans, A. A., & Corsi, I. (2019). Combined effects of nanoplastics and copper on the freshwater alga Raphidocelis subcapitata. *Aquatic toxicology, 210*, 179-187.

Browne, M. A. (2015). Sources and pathways of microplastics to habitats. *Marine anthropogenic litter*, 229–244.

Bhat, G. S., Lavanya, S., Dineshbabu, A. P., Kaladharan, P., & Sulochanan, B. (2014). Preliminary assessment of ecosystem process and marine litter in the beaches of Mangalore.

Besseling, E., Wegner, A., Foekema, E. M., Van Den Heuvel-Greve, M. J., & Koelmans, A. A. (2013). Effects of microplastic on fitness and PCB bioaccumulation by the lugworm Arenicola marina (L.). *Environmental science & technology, 47*(1), 593–600.

Bhattacharya, P., Lin, S., Turner, J. P., & Ke, P. C. (2010). Physical adsorption of charged plastic nanoparticles affects algal photosynthesis. *The journal of physical chemistry C, 114*(39), 16556–16561. Cole, M., Lindeque, P., Fileman, E., Halsband, C., Goodhead, R., Moger, J., & Galloway, T. S. (2013). Microplastic ingestion by zooplankton. *Environmental science & technology, 47*(12), 6646–6655.

Browne, M. A., Crump, P., Niven, S. J., Teuten, E., Tonkin, A., Galloway, T., & Thompson, R. (2011). Accumulation of microplastic on shorelines woldwide: Sources and sinks. *Environmental Science & Technology, 45*(21), 9175–9179.

Chinglenthoiba, C., Pukhrambam, B., Chanu, K. T., Devi, K. S., Meitei, N. J., Devika, Y., & Valiyaveettil, S. (2022). A review on microplastic pollution research in India. *Regional Studies in Marine Science*, 102777.

Chatterjee, S., & Sharma, S. (2019). Microplastics in our oceans and marine health. *Field Actions Science Reports. The Journal of Field Actions*, (Special Issue 19), 54–61.

Cartraud, A. E., Le Corre, M., Turquet, J., & Tourmetz, J. (2019). Plastic ingestion in seabirds of the western Indian Ocean. *Marine Pollution Bulletin, 140*, 308–314.

Chen, Q., Gundlach, M., Yang, S., Jiang, J., Velki, M., Yin, D., & Hollert, H. (2017). Quantitative investigation of the mechanisms of microplastics and nanoplastics toward zebrafish larvae locomotor activity. *Science of the Total Environment, 584*, 1022–1031.

Cole, M., Lindeque, P., Fileman, E., Halsband, C., & Galloway, T. S. (2015). The impact of polystyrene microplastics on feeding, function and fecundity in the marine copepod Calanus helgolandicus. *Environmental Science & Technology, 49*(2), 1130–1137.

Cozar, A., Echevarria, F., González-Gordillo, J. I., Irigoien, X., Ubeda, B., Hernandez-Leon, S., Palma, A. T., Navarro, S., Garcia-de-Lomas, J., Ruiz, A., Fernandez-depuelles, M. L., & Duarte, C. M. (2014). Plastic debris in the open ocean. *The Proceedings of the National Academy of Science., 111*, 10239–10244.

Cole, M., Lindeque, P., Fileman, E., Halsband, C., Goodhead, R., Moger, J., & Galloway, T. S. (2013). Microplastic ingestion by zooplankton. *Environmental Science & Technology, 47*(12), 6646–6655.

Cedervall, T., Hansson, L. A., Lard, M., Frohm, B., & Linse, S. (2012). Food chain transport of nanoparticles affects behaviour and fat metabolism in fish. *PLoS ONE, 7*(2), Article e32254.

Dutta, S., Sethulekshmi, S., & Shriwastav, A. (2022). Abundance, morphology, and spatio-temporal variation of microplastics at the beaches of Mumbai, India. *Regional Studies in Marine Science, 56*, Article 102722.

Dovidat, L. C., Brinkmann, B. W., Vijver, M. G., & Bosker, T. (2020). Plastic particles adsorb to the roots of freshwater vascular plant Spirodela polyrhiza but do not impair growth. *Limnology and Oceanography Letters, 5*, 37–45.

Ding, J., Huang, Y., Liu, S., Zhang, S., Zou, H., Wang, Z., Zhu, W., & Geng, J. (2020). Toxicological effects of nano- and micro-polystyrene plastics on red tilapia: Are larger plastic particles more harmless? *Journal of Hazardous Materials, 396*, Article 122693.

D'Souza, J. M., Windsor, F. M., Santillo, D., & Ormerod, S. J. (2020). Food web transfer of plastics to an apex riverine predator. *Global Change Biology, 26*, 3846–3857.

De Sá, L. C., Oliveira, M., Ribeiro, F., Rocha, T. L., & Futter, M. N. (2018). Studies of the effects of microplastics on aquatic organisms: What do we know and where should we focus our efforts in the future? *Science of the Total Environment, 645*, 1029–1039.

da Costa, J. P., Santos, P. S., Duarte, A. C., & Rocha-Santos, T. (2016). (Nano) plastics in the environment–sources, fates and effects. *Science of the Total Environment, 566*, 15–26.

Faggio, C., Tsarpali, V., & Dailianis, S. (2018). Mussel digestive gland as a model tissue for assessing xenobiotics: An overview. *Science of the Total Environment, 636*, 220–229.

Furtado, R., Menezes, D., Santos, C. J., & Catry, P. (2016). White-faced storm-petrels Pelagodroma marina predated by gulls as biological monitors of plastic pollution in the pelagic subtropical Northeast Atlantic. *Marine Pollution Bulletin, 112*, 117–122.

Fossi, M. C., Panti, C., Guerranti, C., Coppola, D., Giannetti, M., Marsili, L., & Minutoli, R. (2012). Are baleen whales exposed to the threat of microplastics? A case study of the Mediterranean fin whale (Balaenoptera physalus). *Marine Pollution Bulletin, 64*(11), 2374–2379.

Guo, J. J., Huang, X. P., Xiang, L., Wang, Y. Z., Li, Y. W., Li, H., & Wong, M. H. (2020). Source, migration and toxicology of microplastics in soil. *Environment International, 137*, Article 105263.

Godoy, V., Blázquez, G., Calero, M., Quesada, L., & Martín-Lara, M. A. (2019). The potential of microplastics as carriers of metals. *Environmental Pollution, 255*, Article 113363.

Ganesan, M., Nallathambi, G., & Srinivasalu, S. (2019). Fate and transport of microplastics from water sources. *Current Science, 117*(11), 1879–1885.

Gelatt, T., Ream, R., Baker, J. D., Faulhaber, K., & Lerner, D. T. (2019). Evaluating exposure of northern fur seals, Callorhinus ursinus, to microplastic pollution through fecal analysis. *Marine Pollution Bulletin, 138*, 213–221.

Gasperi, J., Wright, S. L., Dris, R., Collard, F., Mandin, C., Guerrouache, M., & Tassin, B. (2018). Microplastics in air: Are we breathing it in? *Current Opinion in Environmental Science & Health, 1*, 1–5.

Goss, H., Jaskiel, J., & Rotjan, R. (2018). Thalassia testudinum as a potential vector for incorporating microplastics into benthic marine food webs. *Marine Pollution Bulletin, 135*, 1085–1089.

Geyer, R., Jambeck, J. R., & Law, K. L. (2017). Production, use, and fate of all plastics ever made. *Science Advances, 3*(7), Article e1700782.

Gutow, L., Eckerlebe, A., Gim´enez, L., Saborowski, R. (2016). Experimental evaluation of seaweeds as a vector for microplastics into marine food webs. *Environmental Science & Technology 50*, 915–923.

Hudak, C. A., & Sette, L. (2019). Opportunistic detection of anthropogenic micro debris in harbor seal (Phoca vitulina vitulina) and gray seal (Halichoerus grypus atlantica) fecal samples from haul-outs in southeastern Massachusetts, USA. *Marine Pollution Bulletin, 145*, 390–395.

Hernandez-Gonzalez, A., Saavedra, C., Gago, J., Covelo, P., Santos, M. B., & Pierce, G. J. (2018). Microplastics in the stomach contents of common dolphin (Delphinus delphis) stranded on the Galician coasts (NW Spain, 2005–2010). *Marine Pollution Bulletin, 137*, 526–532.

Hammer, S., Nager, R. G., Johnson, P. C. D., Furness, R. W., & Provencher, J. F. (2016). Plastic debris in great skua (Stercorarius skua) pellets corresponds to seabird prey species. *Marine Pollution Bulletin, 103*, 206–210.

Holland, E. R., Mallory, M. L., & Shutler, D. (2016). Plastics and other anthropogenic debris in freshwater birds from Canada. *Science of the Total Environment, 571*, 251–258.

Hilfiker, S., Pieribone, V. A., Czernik, A. J., Kao, H. T., Augustine, G. J., Greengard, P. (1999). Synapsins as regulators of neurotransmitter release. *Philosophical Transactions of the Royal Society of London. Series B: Biological Sciences 354*(1381), 269–279.

Jeyasanta, K. I., Laju, R. L., Patterson, J., Jayanthi, M., Bilgi, D. S., Sathish, N., & Edward, J. P. (2023). Microplastic pollution and its implicated risks in the estuarine environment of Tamil Nadu India. *Sciences of the Total Environment, 861*, Article 160572.

James, K., Vasant, K., Padua, S., Gopinath, V., Abilash, K. S., Jeyabaskaran, R., & John, S. (2020). An assessment of microplastics in the ecosystem and selected commercially important fishes off Kochi, south eastern Arabian Sea India. *Marine Pollution Bulletin, 154*, Article 111027.

Jacob, H., Besson, M., Swarzenski, P. W., Lecchini, D., & Metian, M. (2020). Effects of virgin micro- and nanoplastics on fish: Trends, meta-analysis, and perspectives. *Environmental Science and Technology, 54*, 4733–4745.

Jabeen, K., Li, B., Chen, Q., Su, L., Wu, C., Hollert, H., & Shi, H. (2018). Effects of virgin microplastics on goldfish (Carassius auratus). *Chemosphere, 213*, 323–332.

Jin, Y., Xia, J., Pan, Z., Yang, J., Wang, W., & Fu, Z. (2018). Polystyrene microplastics induce microbiota dysbiosis and inflammation in the gut of adult zebrafish. *Environmental Pollution, 235*, 322–329.

Jemec, A., Horvat, P., Kunej, U., Bele, M., & Kržan, A. (2016). Uptake and effects of microplastic textile fibers on freshwater crustacean Daphnia magna. *Environmental Pollution, 219*, 201–209.

Khoshnamvand, M., Hanachi, P., Ashtiani, S., & Walker, T. R. (2021). Toxic efects of polystyrene nanoplastics on microalgae Chlorella vulgaris: Changes in biomass, photosynthetic pigments and morphology. *Chemosphere, 280*, Article 130725. https://doi.org/10.1016/j.chemosphere.2021.130725

Kumkar, P., Gosavi, S. M., Verma, C. R., Pise, M., & Kalous, L. (2021). Big eyes can't see microplastics: Feeding selectivity and eco-morphological adaptations in oral cavity affect microplastic uptake in mud-dwelling amphibious mudskipper fish. *Science of the Total Environment, 786*, Article 147445.

Kerubo, J. O., Muthumbi, A. W., Onyari, J. M., Kimani, E. N., & Robertson-Andersson, D. (2020). Microplastic pollution in the surface waters of creeks along the Kenyan coast, Western Indian Ocean (WIO). *Western Indian Ocean Journal of Marine Science, 19*(2), 75–88.

Kalčíková, G., Skalar, T., Marolt, G., & Kokalj, A. J. (2020). An environmental concentration of aged microplastics with adsorbed silver significantly affects aquatic organisms. *Water Research, 175*, Article 115644.
Kumar, V. E., Ravikumar, G., & Jeyasanta, K. I. (2018). Occurrence of microplastics in fishes from two landing sites in Tuticorin, South east coast of India. *Marine Pollution Bulletin, 135*, 889–894.
Kaza, S., Yao, L., Bhada-Tata, P., & Van Woerden, F. (2018). What a Waste 2.0: A Global Snapshot of Solid Waste Management to 2050. The World Bank.
Kripa, V. (2018). Marine debris-a threat to sustainable fisheries.
Kripa, V., Prema, D., Varghese, M., Padua, S., Jeyabaskaran, R., Sumithra, T. G., & Shylaja, G. (2018). *Book of Abstracts & Success Stories National Conference on Marine Debris COMAD 2018.*
Kumar, A., Sivakumar, R., Reddy, Y. S. R., Raja, B., Nishanth, T., & Revanth, V. (2016). Preliminary study on marine debris pollution along Marina beach, Chennai, India. *Regional Studies in Marine Science, 5*, 35–40.
Laju, R. L., Jayanthi, M., Jeyasanta, K. I., Patterson, J., Asir, N. G. G., Sathish, M. N., & Edward, J. P. (2022). Spatial and vertical distribution of microplastics and their ecological risk in an Indian freshwater lake ecosystem. *Science of the Total Environment, 820*, Article 153337.
Lechthaler, S., Waldschläger, K., Sandhani, C. G., Sannasiraj, S., Sundar, V., Schwarzbauer, J., & Schüttrumpf, H. (2021). Baseline study on microplastics in Indian rivers under diferent anthropogenic infuences. *Water, 13*, 1648. https://doi.org/10.3390/w13121648
Lavers, J. L., Hutton, I., & Bond, A. L. (2019). Clinical pathology of plastic ingestion in marine birds and relationships with blood chemistry. *Environmental Science and Technology, 53*, 9224–9231.
Liu, S. L., Jian, M. F., Zhou, L. Y., Li, W. H., Wu, X. E., Rao, D., (2019d). Pollution characteristics of microplastics in migratory bird habitats located within Poyang lake wetlands. *Huan jing ke xue= Huanjing kexue 40*, 2639–2646.
Li, J., Green, C., Reynolds, A., Shi, H., & Rotchell, J. M. (2018a). Microplastics in mussels sampled from coastal waters and supermarkets in the United Kingdom. *Environmental Pollution, 241*, 35–44.
Lei, L., Wu, S., Lu, S., Liu, M., Song, Y., Fu, Z., & He, D. (2018a). Microplastic particles cause intestinal damage and other adverse effects in zebrafish Danio rerio and nematode Caenorhabditis elegans. *Science of the Total Environment, 619*, 1–8.
Leung, J., & Chan, K. Y. K. (2018). Microplastics reduced posterior segment regeneration rate of the polychaete Perinereis aibuhitensis. *Marine Pollution Bulletin, 129*(2), 782–786. https://doi.org/10.1016/j.marpolbul.2017.10.072
Lei, L., Wu, S., Lu, S., Liu, M., Song, Y., Fu, Z., Shi, H., Raley-Susman, K. M., & He, D. (2018b). Microplastic particles cause intestinal damage and other adverse effects in zebrafish Danio rerio and nematode Caenorhabditis elegans. *Science of the Total Environment, 619–620*, 1–8.
Li, J., Liu, H., & Chen, J. P. (2018b). Microplastics in freshwater systems: A review on occurrence, environmental efects, and methods for microplastics detection. *Water Research, 137*, 362–374. https://doi.org/10.1016/j.watres.2017.12.056
Lyakurwa, D. J. (2017). Uptake and efects of microplastic particles in selected marine microalgae species; Oxyrrhis marina and Rhodomonas baltica (Master's thesis, NTNU).
Lu, Y., Zhang, Y., Deng, Y., Jiang, W., Zhao, Y., Geng, J., & Ren, H. (2016). Uptake and accumulation of polystyrene microplastics in zebrafish (Danio rerio) and toxic effects in liver. *Environmental Science & Technology, 50*(7), 4054–4060.
Lechner, A., Keckeis, H., Lumesberger-Loisl, F., Zens, B., Krusch, R., Tritthart, M., & Schludermann, E. (2014). The Danube so colourful: A potpourri of plastic litter outnumbers fish larvae in Europe's second largest river. *Environmental Pollution, 188*, 177–181.
Lee, K. W., Shim, W. J., Kwon, O. Y., & Kang, J. H. (2013). Size-dependent effects of micro polystyrene particles in the marine copepod Tigriopus japonicus. *Environmental Science & Technology, 47*(19), 11278–11283.

Mahamud, A. S. U., Anu, M. S., Baroi, A., Datta, A., Khan, M. S. U., Rahman, M., & Rahman, T. (2022). Microplastics in fishmeal: A threatening issue for sustainable aquaculture and human health. *Aquaculture Reports, 25*, Article 101205.

Mbugani, J. J., Machiwa, J. F., Shilla, D. A., Kimaro, W., Joseph, D., & Khan, F. R. (2022). Histomorphological damage in the small intestine of wami tilapia (Oreochromis urolepis)(Norman, 1922) exposed to microplastics remain long after depuration. *Microplastics, 1*(2), 240–253.

Maheswaran, B., Karmegam, N., Al-Ansari, M., Subbaiya, R., Al-Humaid, L., Raj, J. S., & Govarthanan, M. (2022). Assessment, characterization, and quantification of microplastics from river sediments. *Chemosphere, 298*, Article 134268.

Muller-Karanassos, C., Arundel, W., Lindeque, P. K., Vance, T., Turner, A., & Cole, M. (2021). Environmental concentrations of antifouling paint particles are toxic to sediment-dwelling invertebrates. *Environmental Pollution, 268*, Article 115754.

Meng, Y., Kelly, F. J., & Wright, S. L. (2020). Advances and challenges of microplastic pollution in freshwater ecosystems: A UK perspective. *Environmental Pollution, 256*, Article 113445.

Ma, H., Pu, S., Liu, S., Bai, Y., Mandal, S., & Xing, B. (2020). Microplastics in aquatic environments: Toxicity to trigger ecological consequences. *Environmental Pollution, 261*, Article 114089.

Malafaia, G., de Souza, A. M., Pereira, A. C., Gonçalves, S., da Costa Araújo, A. P., Ribeiro, R. X., & Rocha, T. L. (2020). Developmental toxicity in zebrafish exposed to polyethylene microplastics under static and semi-static aquatic systems. *Science of the Total Environment, 700*, Article 134867.

Mak, C. W., Yeung, K. C. F., & Chan, K. M. (2019). Acute toxic effects of polyethylene microplastic on adult zebrafish. *Ecotoxicology and Environmental Safety, 182*, Article 109442.

Mao, Y., Ai, H., Chen, Y., Zhang, Z., Zeng, P., Kang, L., & Li, H. (2018). Phytoplankton response to polystyrene microplastics: Perspective from an entire growth period. *Chemosphere, 208*, 59–68.

MacMillan, A. (2018). There's plastic in your poop: study warns microplastics may be hurting us all.

Mattsson, K., Johnson, E. V., Malmendal, A., Linse, S., Hansson, L. A., & Cedervall, T. (2017). Brain damage and behavioural disorders in fish induced by plastic nanoparticles delivered through the food chain. *Scientific Reports, 7*(1), 11452.

Murphy, F., Ewins, C., Carbonnier, F., & Quinn, B. (2016). Wastewater treatment works (WwTW) as a source of microplastics in the aquatic environment. *Environmental Science & Technology, 50*(11), 5800–5808.

Mazurais, D., Ernande, B., Quazuguel, P., Severe, A., Huelvan, C., Madec, L., & Zambonino-Infante, J. (2015). Evaluation of the impact of polyethylene microbeads ingestion in European sea bass (Dicentrarchus labrax) larvae. *Marine Environmental Research, 112*, 78–85.

Morritt, D., Stefanoudis, P. V., Pearce, D., Crimmen, O. A., & Clark, P. F. (2014). Plastic in the Thames: A river runs through it. *Marine Pollution Bulletin, 78*(1–2), 196–200.

Nikki, R., Jaleel, K. A., Ragesh, S., Shini, S., Saha, M., & Kumar, P. D. (2021). Abundance and characteristics of microplastics in commercially important bottom dwelling finfishes and shellfish of the Vembanad Lake, India. *Marine Pollution Bulletin, 172*, Article 112803.

Nelms, S. E., Galloway, T. S., Godley, B. J., Jarvis, D. S., & Lindeque, P. K. (2018). Investigating microplastic trophic transfer in marine top predators. *Environmental Pollution, 238*, 999–1007.

Nolte, T. M., Hartmann, N. B., Kleijn, J. M., Garnæs, J., van de Meent, D., Jan Hendriks, A., & Baun, A. (2017). The toxicity of plastic nanoparticles to green algae as influenced by surface modification, medium hardness and cellular adsorption. *Aquatic Toxicology, 183*, 11–20.

Napper, I. E., Baroth, A., Barrett, A. C., Bhola, S., Chowdhury, G. W., Davies, B. F., & Koldewey, H. (2021). The abundance and characteristics of microplastics in surface water in the transboundary Ganges River. *Environmental Pollution, 274*, Article 116348.

Naidoo, T., & Glassom, D. (2019). Decreased growth and survival in small juvenile fish, after chronic exposure to environmentally relevant concentrations of microplastic. *Marine Pollution Bulletin, 145*, 254–259. https://doi.org/10.1016/j.marpolbul.2019.02.037

Naidu, S. A., Ranga Rao, V., & Ramu, K. J. E. G. (2018). Microplastics in the benthic invertebrates from the coastal waters of Kochi, Southeastern Arabian Sea. *Environmental Geochemistry and Health, 40*, 1377–1383.

Nace, T. (2017). Your Table Salt Is Likely Sprinkled with Microplastics. *Research Reveals.*

Nerland, I. L., Halsband, C., Allan, I., & Thomas, K. V. (2014). Microplastics in marine environments: Occurrence, distribution and effects.

Owens, K. A., & Kamil, P. I. (2020). Adapting coastal collection methods for river assessment to increase data on global plastic pollution: Examples from India and Indonesia. *Frontiers in Environmental Science, 7*, 208.

Pires, A., Cuccaro, A., Sole, M., Freitas, R., 2022. Micro (nano) plastics and plastic additives effects in marine annelids: a literature review. *Environmental Resource 214*, 113642. https://doi.org/10.1016/j.envres.2022.113642.

Patterson, J., Jeyasanta, K. I., Laju, R. L., & Edward, J. P. (2021). Microplastic contamination in Indian edible mussels (Perna perna and Perna viridis) and their environs. *Marine Pollution Bulletin, 171*, Article 112678.

Patterson, J., Jeyasanta, K. I., Sathish, N., Booth, A. M., & Edward, J. P. (2019). Profiling microplastics in the Indian edible oyster, Magallana bilineata collected from the Tuticorin coast, Gulf of Mannar, Southeastern India. *Science of the Total Environment, 691*, 727–735.

Pagano, M., Vazzana, I., Gentile, A., Caracappa, G., & Faggio, C. (2019). Hematological and biochemical parameters in Sea turtles (Caretta caretta) after stranding. *Regional Studies in Marine Science, 32*, Article 100832.

Pitt, J. A., Kozal, J. S., Jayasundara, N., Massarsky, A., Trevisan, R., Geitner, N., Wiesner, M., Levin, E. D., & Di Giulio, R. T. (2018). Uptake, tissue distribution, and toxicity of polystyrene nanoparticles in developing zebrafish (Danio rerio). *Aquatic Toxicology, 194*, 185–194.

Paul-Pont, I., Lacroix, C., Fernández, C. G., Hégaret, H., Lambert, C., Le Goïc, N., & Soudant, P. (2016). Exposure of marine mussels Mytilus spp. to polystyrene microplastics: toxicity and infuence on fuoranthene bioaccumulation. *Environmental pollution, 216*, 724–737. https://doi.org/10.1016/j.envpol.2016.06.039.

Peda, C., Caccamo, L., Fossi, M. C., Gai, F., Andaloro, F., Genovese, L., & Maricchiolo, G. (2016). Intestinal alterations in European sea bass Dicentrarchus labrax (Linnaeus, 1758) exposed to microplastics: Preliminary results. *Environmental Pollution, 212*, 251–256.

Qiao, R., Sheng, C., Lu, Y., Zhang, Y., Ren, H., & Lemos, B. (2019a). Microplastics induce intestinal inflammation, oxidative stress, and disorders of metabolome and microbiome in zebrafish. *Science of the Total Environment, 662*, 246–253.

Qiao, R., Lu, K., Deng, Y., Ren, H., & Zhang, Y. (2019b). Combined effects of polystyrene microplastics and natural organic matter on the accumulation and toxicity of copper in zebrafish. *Science of the Total Environment, 682*, 128–137.

Ram, B., & Kumar, M. (2020). Correlation appraisal of antibiotic resistance with fecal, metal and microplastic contamination in a tropical Indian river, lakes and sewage. *NPJ Clean Water, 3*(1), 1–12.

Reynolds, C., & Ryan, P. G. (2018). Micro-plastic ingestion by waterbirds from contaminated wetlands in South Africa. *Marine Pollution Bulletin, 126*, 330–333.

Rebolledo, E. L. B., Van Franeker, J. A., Jansen, O. E., & Brasseur, S. M. (2013). Plastic ingestion by harbour seals (Phoca vitulina) in The Netherlands. *Marine Pollution Bulletin, 67*(1–2), 200–202.

Rodrigues, A. M., Quintino, V., Sampaio, L., et al. (2011). Benthic biodiversity patterns in ria de Aveiro, Western Portugal: Environmental-biological relationships. *Estuarine, Coastal and Shelf Science, 95*(2–3), 338–348. https://doi.org/10.1016/j.ecss.2011.05.019

Sangkham, S., Faikhaw, O., Munkong, N., Sakunkoo, P., Arunlertaree, C., Chavali, M., & Tiwari, A. (2022). A review on microplastics and nanoplastics in the environment: Their occurrence, exposure routes, toxic studies, and potential effects on human health. *Marine Pollution Bulletin, 181*, Article 113832.

Selvam, S., Jesuraja, K., Venkatramanan, S., Roy, P. D., & Kumari, V. J. (2021). Hazardous microplastic characteristics and its role as a vector of heavy metal in groundwater and surface water of coastal south India. *Journal of Hazardous Materials, 402*, Article 123786.

Srinivasalu, S., Natesan, U., Ayyamperumal, R., Kalam, N., Anbalagan, S., Sujatha, K., & Alagarasan, C. (2021). Microplastics as an emerging threat to the freshwater ecosystems of Veeranam lake in south India: A multidimensional approach. *Chemosphere, 264*, Article 128502.

Silva, M. S. S., Oliveira, M., Lopéz, D., et al. (2020a). Do nanoplastics impact the ability of the polychaeta Hediste diversicolor to regenerate? *Ecological Indicators 110*, 105921. https://doi.org/10.1016/j.ecolind.2019.105921.

Silva, M. S. S., Oliveira, M., Valente, P., et al. (2020b). Behavior and biochemical responses of the polychaeta Hediste diversicolor to polystyrene nanoplastics. *Science Total Environment 707*, 134434. https://doi.org/10.1016/j.scitotenv.2019.134434.

Schirinzi, G. F., Pedà, C., Battaglia, P., Laface, F., Galli, M., Baini, M., & Romeo, T. (2020). A new digestion approach for the extraction of microplastics from gastrointestinal tracts (GITs) of the common dolphinfish (Coryphaena hippurus) from the western Mediterranean Sea. *Journal of Hazardous Materials, 397*, Article 122794.

Sathish, M. N., Jeyasanta, K. I., & Patterson, J. (2020). Monitoring of microplastics in the clam Donax cuneatus and its habitat in Tuticorin coast of Gulf of Mannar (GoM) India. *Environmental Pollution, 266*, Article 115219.

Sarkar, D. J., Sarkar, S. D., Das, B. K., et al. (2019). Spatial distribution of meso and microplastics in the sediments of river ganga at eastern India. *Science of the Total Environment, 694*, Article 133712.

Sala, B., Gim´enez, J., de Stephanis, R., Barcelo, D., Eljarrat, E. (2019). First determination of high levels of organophosphorus flame retardants and plasticizers in dolphins from Southern European waters. *Environmental Resources 172*, 289–295.

Sighicelli, M., Pietrelli, L., Lecce, F., Iannilli, V., Falconieri, M., Coscia, L., & Zampetti, G. (2018). Microplastic pollution in the surface waters of Italian Subalpine Lakes. *Environmental Pollution, 236*, 645–651.

Straub, S., Hirsch, P. E., & Burkhardt-Holm, P. (2017). Biodegradable and petroleum-based microplastics do not differ in their ingestion and excretion but in their biological effects in a freshwater invertebrate Gammarus fossarum. *International Journal of Environmental Research and Public Health, 14*(7), 774.

Sharma, S., & Chatterjee, S. (2017). Microplastic pollution, a threat to marine ecosystem and human health: A short review. *Environmental Science and Pollution Research, 24*, 21530–21547.

Sruthy, S., & Ramasamy, E. V. (2017). Microplastic pollution in Vembanad Lake, Kerala, India: The first report of microplastics in lake and estuarine sediments in India. *Environmental Pollution, 222*, 315–322.

Sussarellu, R., Suquet, M., Thomas, Y., Lambert, C., Fabioux, C., Pernet, M. E. J., Le Goïc, N., Quillien, V., Mingant, C., Epelboin, Y., Corporeau, C., Guyomarch, J., Robbens, J., Paul-Pont, I., Soudant, P., & Huvet, A. (2016a). Oyster reproduction is affected by exposure to polystyrene microplastics. *Proceedings of the National Academy of Sciences, 113*, 2430.

Setälä, O., Norkko, J., Lehtiniemi, M. (2016). Feeding type affects microplastic ingestion in a coastal invertebrate community. *Marine Pollution Bulletin 102*(1), 95–101. https://doi.org/10.1016/j.marpolbul.2015.11.053.

Sjollema, S. B., Redondo-Hasselerharm, P., Leslie, H. A., Kraak, M. H., & Vethaak, A. D. (2016). Do plastic particles afect microalgal photosynthesis and growth? *Aquatic Toxicology, 170*, 259–261. https://doi.org/10.1016/j.aquatox.2015.12.002.

Setälä, O., Norkko, J., & Lehtiniemi, M. (2016b). Feeding type affects microplastic ingestion in a coastal invertebrate community. *Marine Pollution Bulletin, 102*(1), 95–101.

Sjollema, S. B., Redondo-Hasselerharm, P., Leslie, H. A., Kraak, M. H., & Vethaak, A. D. (2016b). Do plastic particles affect microalgal photosynthesis and growth? *Aquatic Toxicology, 170*, 259–261.

Sussarellu, R., Suquet, M., Thomas, Y., Lambert, C., Fabioux, C., Pernet, M. E. J., & Huvet, A. (2016b). Oyster reproduction is affected by exposure to polystyrene microplastics. *Proceedings of the National Academy of Sciences, 113*(9), 2430–2435.

Setälä, O., Fleming-Lehtinen, V., & Lehtiniemi, M. (2014). Ingestion and transfer of microplastics in the planktonic food web. *Environmental Pollution, 185*, 77–83.

Sajikumar, K. K., Ragesh, N., Remya, R. & Mohamed, K. S. (2013). Occurrence of plastic debris in the stomach of yelowfin tuna (Thunnus albacares) from the Arabian sea: A cause of concern. *Marine Fisheries Inforormation Service Technical & Extension Services, 217*, 13.

Sulochanan, B., Bhat, G. S., & Lavanya, R. (2011). Marine litter in the coastal environment of Mangalore. *Marine Fisheries Information Service, 208*, 18–19.

Thushari, G. G. N., & Senevirathna, J. D. M. (2020). Plastic pollution in the marine environment. *Heliyon, 6*(8).

Trestrail, C., Nugegoda, D., & Shimeta, J. (2020). Invertebrate responses to microplastic ingestion: Reviewing the role of the antioxidant system. *Science of the Total Environment, 734*, Article 138559.

Thiel, M., Luna-Jorquera, G., Álvarez-Varas, R., Gallardo, C., Hinojosa, I. A., Luna, N., & Zavalaga, C. (2018). Impacts of marine plastic pollution from continental coasts to subtropical gyres—fish, seabirds, and other vertebrates in the SE Pacific. *Frontiers in Marine Science, 238.*

Urbanek, A. K., Rymowicz, W., & Mirończuk, A. M. (2018). Degradation of plastics and plastic-degrading bacteria in cold marine habitats. *Applied Microbiology and Biotechnology, 102*, 7669–7678.

Valsan, G., Warrier, A. K., Amrutha, K., Anusree, S., & Rangel-Buitrago, N. (2023). Exploring the presence and distribution of microplastics in subterranean estuaries from southwest India. *Marine Pollution Bulletin, 190*, Article 114820.

Vaid, M., Mehra, K., & Gupta, A. (2021). Microplastics as contaminants in Indian environment: a review. *Environmental Science and Pollution Research*, 1–28.

Veerasingam, S., Saha, M., Suneel, V., Vethamony, P., Rodrigues, C. A., Bhattacharyya, S., & Naik, B. G. (2016). Characteristics, seasonal distribution and surface degradation features of microplastic pellets along the Goa coast India. *Chemosphere, 159*, 496–505. https://doi.org/10.1016/j.chemosphere.2016.06.056

Vandermeersch, G., Lourenço, H. M., Alvarez-Muñoz, D., Cunha, S., Diogène, J., Cano-Sancho, G., & Robbens, J. (2015). Environmental contaminants of emerging concern in seafood–European database on contaminant levels. *Environmental Research, 143*, 29–45.

Von Moos, N., Burkhardt-Holm, P., & Köhler, A. (2012). Uptake and effects of microplastics on cells and tissue of the blue mussel Mytilus edulis L. after an experimental exposure. *Environmental science & technology, 46*(20), 11327–11335.

van Franeker, J. A., Blaize, C., Danielsen, J., Fairclough, K., Gollan, J., Guse, N., Hansen, P.-L., Heubeck, M., Jensen, J.-K., Le Guillou, G., Olsen, B., Olsen, K.-O., Pedersen, J., Stienen, E. W. M., & Turner, D. M. (2011). Monitoring plastic ingestion by the northern fulmar Fulmarus glacialis in the North Sea. *Environmental Pollution, 159*, 2609–2615.

Wen, B., Jin, S. R., Chen, Z. Z., Gao, J. Z., Liu, Y. N., Liu, J. H., & Feng, X. S. (2018). Single and combined effects of microplastics and cadmium on the cadmium accumulation, antioxidant defence and innate immunity of the discus fish (Symphysodon aequifasciatus). *Environmental Pollution, 243*, 462–471.

Wang, C., Zhao, J., & Xing, B. (2021). Environmental source, fate, and toxicity of microplastics. *Journal of Hazardous Materials, 407*, Article 124357.

Wang, F., Zhang, M., Sha, W., Wang, Y., Hao, H., Dou, Y., & Li, Y. (2020a). Sorption behavior and mechanisms of organic contaminants to nano and microplastics. *Molecules, 25*(8), 1827.

Wang, W., Ge, J., & Yu, X. (2020b). Bioavailability and toxicity of microplastics to fish species: A review. *Ecotoxicology and Environmental Safety, 189*, Article 109913.

Wright, S. L., Thompson, R. C., & Galloway, T. S. (2013). The physical impacts of microplastics on marine organisms: A review. *Environmental Pollution, 178*, 483–492. https://doi.org/10.1016/j.envpol.2013.02.031

Winder, M., & Sommer, U. (2012). Phytoplankton response to a changing climate. *Hydrobiologia, 698*, 5–16.

Wirtz, K. W. (2012). Who is eating whom? Morphology and feeding type determine the size relation between planktonic predators and their ideal prey. *Marine Ecology Progress Series, 445*, 1–12.

Xiong, X., Chen, X., Zhang, K., Mei, Z., Hao, Y., Zheng, J., Wu, C., Wang, K., Ruan, Y., Lam, P. K. S., & Wang, D. (2018). Microplastics in the intestinal tracts of East Asian finless porpoises (Neophocaena asiaeorientalis sunameri) from Yellow Sea and Bohai Sea of China. *Marine Pollution Bulletin, 136*, 55–60.

Yadav, S., Kataria, N., Khyalia, P., Rose, P. K., Mukherjee, S., Sabherwal, H., & Khoo, K. S. (2023). Recent analytical techniques, and potential eco-toxicological impacts of textile fibrous microplastics (FMPs) and associated contaminates: A review. *Chemosphere, 326*, Article 138495.

Yang, S., Zhou, M., Chen, X., Hu, L., Xu, Y., Fu, W., & Li, C. (2022). A comparative review of microplastics in lake systems from different countries and regions. *Chemosphere, 286*, Article 131806.

Yang, H., Xiong, H., Mi, K., Xue, W., Wei, W., & Zhang, Y. (2020). Toxicity comparison of nano-sized and micron-sized microplastics to Goldfish Carassius auratus Larvae. *Journal of Hazardous Materials, 388*, Article 122058.

Yuan, W., Zhou, Y., Chen, Y., Liu, X., & Wang, J. (2020). Toxicological effects of microplastics and heavy metals on the Daphnia magna. *Science of the Total Environment, 746*, Article 141254.

Yuan, W., Zhou, Y., Liu, X., & Wang, J. (2019). New perspective on the nanoplastics disrupting the reproduction of an endangered fern in artificial freshwater. *Environmental Science and Technology, 53*, 12715–12724.

Yi, X., Chi, T., Li, Z., Wang, J., Yu, M., Wu, M., & Zhou, H. (2019). Combined effect of polystyrene plastics and triphenyltin chloride on the green algae Chlorella pyrenoidosa. *Environmental Science and Pollution Research, 26*, 15011–15018.

Yin, L., Chen, B., Xia, B., Shi, X., & Qu, K. (2018). Polystyrene microplastics alter the behavior, energy reserve and nutritional composition of marine jacopever (Sebastes schlegelii). *Journal of Hazardous Materials, 360*, 97–105.

Yu, S. P., & Chan, B. K. K. (2020). Effects of polystyrene microplastics on larval development, settlement, and metamorphosis of the intertidal barnacle Amphibalanus amphitrite. *Ecotoxicology and Environmental Safety, 194*, 110362.

Zhu, J., Yu, X., Zhang, Q., Li, Y., Tan, S., Li, D., Yang, Z., & Wang, J. (2019). Cetaceans and microplastics: First report of microplastic ingestion by a coastal delphinid. *Sousa Chinensis. Sci. Total Environ., 659*, 649–654.

Human Health Impact of Microplastics

Manjulata Yadav, Soumya Chatterjee, and Pankaj Kumar

Abstract The overuse of plastics in daily life results micro plastics (MPs) as global issue due to their toxic effect on human. Exposure to micro plastics has been associated with a number of respiratory disorders. Furthermore; the widespread presence of plastic pollution has an adverse effect on wildlife and animal life in addition to human health. Micro plastics cause a vicious cycle of environmental and human health degradation by getting into food supplies and contaminating water with different chemical substances. The ubiquitous plastic wrapping found in food stores contributes the micro plastics into human body. Research indicates that the production of plastic creates serious health dangers for those working in the industry and contributes to environmental contamination. Because they are more likely to contract crippling illnesses including mesothelioma, brain cancer, lymphoma, and leukemia, workers must strictly adhere to laws in order to protect public health. Plastic pollution endangers biodiversity in addition to human health by upsetting ecosystems and bringing in new pests and illnesses. Wildlife, in particular, suffers greatly when animals ingest waste plastics and get entangled in them or mistake them for food, which frequently has deadly results. Given these complex issues, reducing the manufacturing of plastic and supporting environmental conservation initiatives become essential tactics to reduce the harmful effects of micro plastics on human and environmental health. To solve this global catastrophe holistically and safeguard the health and well-being of current and future generations, immediate action is needed.

Keywords Micro plastics · Plastic pollution · Respiratory disorders · Environmental degradation · Biodiversity

M. Yadav (✉)
Department of Physics, S.S.S.T.S.R. Government Degree College Nainidanda, Pauri Garhwal, India
e-mail: manjuyadav187@gmail.com

S. Chatterjee
Department of Physics, Brainware University, Barasat, India

P. Kumar
Department of Physics, S.V. College, Aligarh, India
e-mail: pankaj198911@gmail.com

P. Khyalia and S. Yadav (eds.), *Microplastic Pollution in India and Its Environmental Impacts*, Earth and Environmental Sciences Library,
https://doi.org/10.1007/978-3-031-92530-6_9

1 Introduction

In our modern era, the pervasive presence of plastic has become an undeniable aspect of daily life. From packaging to textiles, this versatile material has woven itself into the fabric of society, offering convenience and durability like never before. However, the consequences of our plastic dependence extend far beyond its utility (Koelmans et al., 2016). Among the most concerning repercussions is the proliferation of microplastics—tiny plastic particles less than 5 millimeters in size—that now pervade in our environment, in air, water and food.

As the prevalence of microplastics continues to rise, so too does our understanding of their potential impact on human health (Rochman et al., 2013). This chapter delves into the complex network of interactions between the human body and microplastics, exploring the myriad ways in which these minuscule particles may exert their influence. From ingestion to inhalation, the pathways through which microplastics can enter our bodies are numerous and varied, raising pressing questions about their physiological consequences.

Drawing upon the latest research findings and expert insights, we will embark on a journey to unravel the complex relationship between microplastics and human health. We will examine the mechanisms by which these synthetic particles interact with biological systems, probing the potential for adverse effects ranging from inflammation and oxidative stress to the disruption of endocrine function (Hartmann et al., 2017; Yang et al., 2015). Moreover, we will explore the emerging field of microplastics toxicology, seeking to elucidate the long-term implications of chronic exposure on human well-being.

Yet, amidst the growing concern surrounding microplastics, there remains a glimmer of hope. Through proactive measures and interdisciplinary collaboration, we have the opportunity to reduce the risks posed by these widespread pollutants and pave the way towards a healthier, more sustainable future. By shedding light on the effect of microplastics on human health, this chapter seeks to empower readers with knowledge and insight, catalyzing informed action in the face of this pressing environmental challenge (Wright et al., 2013).

1.1 Classification of Micro-plastics

Microplastics are now a pervasive environmental problem, contaminating ecosystems worldwide. To fully comprehend their impacts on ecosystems, organisms, and human being health, understanding their classification is important. This includes categorization by size, shape, polymer type, source, and environmental fate.

a. **Size-based Classification**: Microplastics are divided into primary and secondary types. Primary microplastics are deliberately manufactured small particles, while secondary microplastics result from the breakdown of larger plastic items. These categories further include nanoplastics (less than 100 nm), microplastics (100 nm to 5 mm), and mesoplastics (5 mm to 1 cm).

b. **Shape-based Classification**: Microplastics exhibit a variety of shapes influenced by manufacturing processes, environmental degradation, and fragmentation. Shapes may include fibers, fragments, pellets, films, foams, and spheres, which help understand their behavior and interactions within ecosystems.
c. **Polymer-based Classification**: Microplastics are composed of different polymers such as polyethylene, polypropylene, polystyrene, and polyethylene terephthalate. Classifying them by polymer type aids in identifying sources and evaluating their persistence and toxicity in various environments.
d. **Source-based Classification**: Microplastics originate from diverse sources, including industrial pellets, microbeads in cosmetics, and fibers from textiles. Secondary sources involve the breakdown of larger plastic items. Classifying microplastics by source helps prioritize mitigation strategies and regulatory measures.
e. **Environmental Fate Classification**: Microplastics undergo intricate interactions in the environment, leading to various outcomes such as ingestion by organisms, sedimentation, adsorption of pollutants, and transport across ecosystems via air and water currents. Understanding their environmental fate is essential for evaluating ecological effects and developing effective mitigation strategies.

In conclusion, classifying microplastics based on size, shape, polymer type, source, and environmental fate offers a comprehensive framework for studying their distribution, behavior, and impacts on ecosystems. Continued research, analyzing and monitoring efforts are essential to deepen our understanding and mitigate their detrimental effects on the environment and human health. The major sources of microplastics have been divided as primary and secondary sources in Table 1.

In comparing primary microplastics to secondary microplastics, several key distinctions emerge. Primary microplastics are deliberately manufactured particles, often utilized in products such as cosmetics and industrial abrasives, intentionally released into the environment. On the other hand, secondary microplastics are generated from the degradation and disintegration of larger plastic items, such as bottles and bags, caused by environmental factors like UV radiation and mechanical wear and tear. Regarding size, primary microplastics are typically intentionally

Table 1 Sources of primary and secondary microplastics

Sl. No.	Primary sources	Secondary sources
1	Synthetic textiles	Plastic packaging
2	Tires	Tire wear particles
3	City dust	Synthetic textiles
4	Road making	Fishing gears
5	Marine coating	Tea bags
6	Personal care products	Personal care products
7	Plastics pellets	Paints and coatings

small, ranging from nanometers to a few millimeters, encompassing microbeads, pellets, and fibers. Conversely, the size of Secondary microplastics varies based on the extent of degradation, ranging from millimeters to micrometers. Shapes of primary microplastics vary widely, including spheres, fibers, fragments, films, and irregular shapes, influenced by their intended use and manufacturing process, while Secondary microplastics often exhibit irregular and heterogeneous shapes due to degradation processes. In terms of composition, Primary Microplastics consist of specific polymers chosen for their properties, like polyethylene and polypropylene, whereas secondary microplastics retain the composition of the original plastic item. The environmental impact of primary microplastics lies in their ingestion and bioaccumulation, posing risks to marine life and ecosystems, while Secondary Microplastics contribute to widespread pollution. Regulation and mitigation efforts for primary microplastics focus on banning or restricting their use and promoting biodegradable alternatives, while for secondary microplastics, it involves reducing plastic consumption, improving waste management, and preventing plastic debris from entering the environment. In summary, being differ in various aspects, both primary and secondary microplastics significantly contribute to global plastic pollution crisis, emphasizing the necessity of addressing their sources and pathways to mitigate their adverse effects on ecosystems and human health.

1.2 Transportation Mechanism of Micro-plastics into an Environment

Microplastics enter the environment through multiple transportation pathways, including atmospheric deposition, surface runoff, wastewater discharge, and marine transport. These mechanisms subscribe to the widespread dispersal of microplastics in terrestrial, freshwater, and marine environments. Here's a detailed overview of these transportation mechanisms:

1. **Atmospheric Deposition**
   - Microplastics can travel long distances through the atmosphere before being deposited onto land or water surfaces via atmospheric deposition.
   - Sources of airborne microplastics include urban areas, industrial activities, agricultural activities, and atmospheric transport from marine environments.
   - Atmospheric deposition of microplastics can occur through dry deposition (direct settling) or wet deposition (precipitation scavenging).
2. **Surface Runoff**
   - Microplastics present on surfaces such as roads, pavements, and urban areas can be transported by surface runoff during rainfall events.
   - Rainwater washes microplastics from these surfaces into stormwater drains, rivers, streams, and eventually into oceans and other water bodies.
   - Surface runoff can carry microplastics along with other pollutants, leading to the contamination of aquatic environments and coastal areas.

3. **Wastewater Discharge**
   - Microplastics enter in the environment via the flow of wastewater from various sources, e.g. domestic sewage, industrial effluents, wastewater treatment plants etc.
   - Microplastic pollution in wastewater is contributed by synthetic fibers from laundry effluents, micro beads from personal care products, and fragmented plastics from industrial processes.
   - Not all microplastics are effectively captured by wastewater treatment plants, allowing their discharge into surrounding water bodies.
4. **Marine Transport**
   - Microplastics are transported within marine environments through ocean currents, waves, and tides.
   - Coastal areas act as sinks for microplastics transported from upstream sources via rivers and estuaries, accumulating large concentrations of microplastics in sediments and beach environments.
   - Microplastics can also be transported across ocean basins through long-distance transport mechanisms such as gyres and oceanic currents, leading to their widespread distribution in remote marine ecosystems.
5. **Biological Transport**
   - Microplastics can be ingested by organisms at various trophic levels, from plankton to larger marine mammals.
   - Once ingested, microplastics can be transported within organisms through biological processes such as digestion, excretion, and biomagnification.
   - Bioaccumulation and biomagnification of microplastics can occur within food webs, leading to potential impacts on ecosystem health and biodiversity.

Understanding the transportation mechanisms of microplastics into the environment is essential for formulating effective strategies to reduce pollution and minimize ecological impacts. This requires multidisciplinary research and joint efforts to tackle the intricate pathways through which microplastics enter and pass within ecosystems.

1.3 Properties of Micro-plastics

Microplastics, possess a wide range of characteristics that significantly affect their behavior, distribution, and environmental consequences. These properties are essential to comprehend in order to evaluate the impact on ecosystems and the health of mankind. Firstly, microplastics vary greatly in size, spanning from nanometers to a few millimeters, rendering them easily ingestible by various organisms, from aquatic microbes to larger marine species. Secondly, their shapes can vary, including fibers, fragments, films, spheres, and irregular shapes, influencing their movement, deposition, and interactions within ecosystems. Thirdly, microplastics are composed of diverse polymers, which affect their density, buoyancy, durability, and potential toxicity. Additionally, their density varies based on composition, with some sinking in water while others float at the surface. Their large surface area relative to their volume enables them to effectively adsorb and transport contaminants like persistent organic pollutants (POPs) and heavy metals in aquatic environments. Furthermore, microplastics are highly durable and resistant to degradation, particularly in marine environments, where they may fragment into smaller particles rather than degrade fully. They also come in various colors, depending on the polymer and additives used, which can influence their visibility and identification in environmental samples. Lastly, their surface chemistry, including factors like charge, hydrophobicity, and functional groups, impacts their interactions with organisms, pollutants, and environmental constituents. Understanding these properties is important for evaluating their fate, and ecological impacts, and designing constructive mitigation policies against plastic pollution. Continued multidisciplinary data collection and analyses is necessary for exploring diverse properties of microplastics and their impacts on the ecosystems.

1.4 Effects of Micro-plastic

The effects of microplastics on human and animal life are of growing concern due to their widespread distribution in ecosystems and their potential to harm. Here's a discussion of the impacts of microplastics on animal and human life:

1. **Impacts on animal life**
   a. **Ingestion**: Marine animals, including fish, seabirds, turtles, and mammals, frequently confuse microplastics with food because of their small size and similarity to natural prey. Ingestion of microplastics can lead to blockages in the digestive tract, malnutrition, internal injuries, and reduced feeding efficiency, ultimately affecting marine wildlife's survival and health.
   b. **Bioaccumulation**: Microplastics can bioaccumulate in the tissues of marine organisms, leading the transfer of plastic particles and associated contaminants the food chain. This bioaccumulation can result in higher concentrations of toxins in predators at the top of the food web, posing risks to apex predators and potentially impacting entire ecosystems.
   c. **Habitat Alteration**: Microplastics can accumulate in coastal sediments, benthic habitats, and coral reefs, altering habitat structure, disrupting ecosystem functioning, and affecting the abundance and diversity of marine organisms. Microplastic pollution may also introduce non-native species and invasive organisms to marine environments, further disrupting ecological balance.

Overall, the detrimental effects of microplastics on human life and animal life highlight the urgent need for concerted efforts to mitigate microplastic and reduce the release of microplastics into the environment. Strategies such as improved waste management, regulation of plastic production and use, development of biodegradable alternatives, and public awareness campaigns are essential for protecting both human health and marine biodiversity from the impacts of microplastics.

2. **Impacts on human health**
   a. **Ingestion**: Microplastics can be ingested in the human body by consuming contaminated food and water. Seafood, salt, bottled water, and air have been found to contain microplastics, leading to potential human exposure.
   b. **Health Risks**: Microplastics act as carrier for pollutants, including toxic chemicals such as polychlorinated biphenyls (PCBs), polycyclic aromatic hydrocarbons (PAHs), and heavy metals. When ingested, these pollutants can bioaccumulate in human tissues and organs, potentially causing adverse health effects such as hormonal disruption, reproductive problems, immune system disorders, and even cancer.
   c. **Microbial Adhesion**: Microplastics in aquatic environments can serve as surfaces for microbial colonization, potentially facilitating the transfer of pathogenic bacteria and harmful microorganisms to humans through contaminated seafood and water.

2 Pathways of human exposure to micro plastics

According to the literature analysis, microplastics are regularly discovered in both environmental and human samples. Microplastics can be ingested, inhaled, or come into contact with the skin (Yee et al., 2021). We describe the harmful consequences of microplastics in several experimental models, including cells, organoids, and animals.

Previously considered inert and non-toxic, microplastics are now recognized as potentially harmful to organisms, with effects varying based on exposure and susceptibility (Galloway, 2015). Their high surface area can cause oxidative stress, cytotoxicity, and translocation to other tissues, while their persistent nature hinders removal from the body, potentially leading to chronic inflammation and an increased cancer risk. Microplastics, as part of particulate matter, may potentially contribute to an increase in the incidence of immunological or neurological illnesses. Furthermore, microplastics may release chemicals from their matrixes or adsorbed from the environment, or act as vectors for dangerous microorganisms (Kirstein et al., 2016). The major natural transportation route of microplastics to human beings has been shown in Fig. 1.

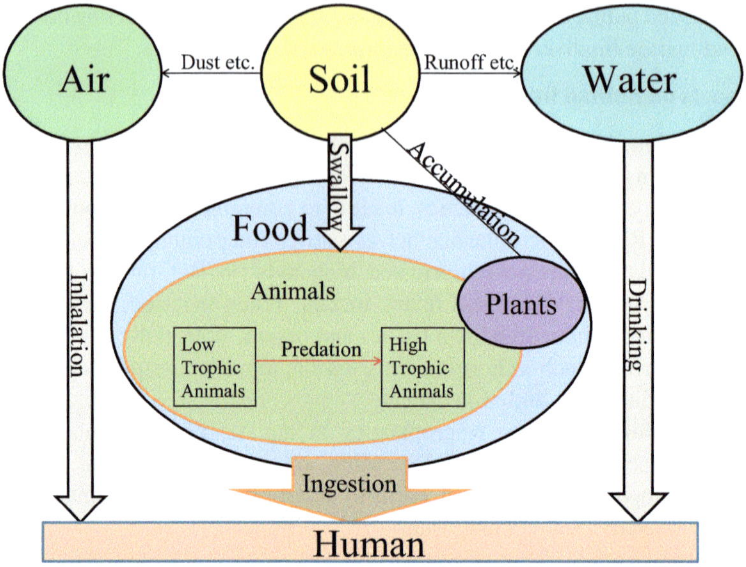

Fig. 1 The major natural transportation route of microplastics to human being

2.1 Inhalation

Microplastics are emitted into the air from multiple sources, e.g., synthetic fabrics, material abrasion, and the resuspension of microplastics from surfaces. Microplastics can enter the body through inhalation of airborne particles. Once inside the body, they can potentially accumulate in various organs and tissues. Early measurements of airborne microplastics reported outdoor concentrations ranging from 0.3 to 1.5 particles per m^3 and indoor concentrations from 0.4 to 56.5 particles per m^3 (33% of which were polymers), including particles small enough to be inhaled (Dris et al., 2017). Individual inhalation of airborne microplastics is estimated to range from 26 to 130 particles per day (Prata, 2018). Air sampling using a mannequin suggests that a male engaging in light activity may inhale around 272 microplastic particles daily (Vianello et al., 2019). Estimates of inhaled microplastics vary due to differences in sampling methods and environmental factors like cleaning schedules, activities, furniture materials, and seasons. Particle size and density affect how deeply they penetrate the lungs, with smaller and lighter particles reaching deeper. After deposition in the lungs, microplastics may be cleared by macrophages or enter the bloodstream or lymphatic system. However, small particles with large surface areas can trigger the release of chemotactic factors, hindering macrophage movement and increasing permeability, which may lead to chronic inflammation, a condition known as dust overload (Tran et al., 2000). Polystyrene (PS) nanospheres (64 nm) have been shown to cause neutrophil influx and lung inflammation in rats, as well as proinflammatory gene expression in epithelial cells, due to their high oxidant activity from the large surface area (Brown et al., 2001). In vitro studies also demonstrated that PVC particles (2 mm) produced via emulsion polymerization caused significant cytotoxicity in rat and human lung cells, along with hemolysis (Xu et al., 2002). Workers in industries like synthetic textiles, flock, and vinyl chloride have been exposed to airborne microplastics, which are associated with respiratory symptoms and lung disease. These conditions have been replicated in animal studies (Agarwal et al., 1977; Xu et al., 2004). Although a direct cause has not been confirmed, fibers up to 250 mm have been found in human lung biopsies, including cancer cases (Pauly et al., 1998). This suggests that airborne microplastics may pose a risk to the respiratory system, particularly in high-exposure environments or for vulnerable individuals.

2.2 *Ingestion*

The injection of microplastics directly into the human body is not a typical route of exposure, but it can occur under certain circumstances, such as medical procedures, accidental injuries involving plastic materials or via ingestion of contaminated food and water. Injection of microplastics can have detrimental effects on human health. Studies have shown that microplastics, including nano-sized particles, can be released from plastic products like feeding and water bottles, accumulating in the environment and potentially entering the human body through various exposure routes (Kazi et al., 2023). Microplastics injected into the body may migrate to other tissues or organs through the bloodstream or lymphatic system. This could potentially lead to systemic health effects or localized accumulation in certain organs. Once inside the body, microplastics can trigger oxidative processes, mitochondrial dysfunction, and induce cytotoxicity, leading to inflammation, tissue irritation, organ damage, the formation of granulomas or scar tissue around the injected particles, and systemic immunological conditions (Yasmin et al., 2023). The presence of microplastics in various human organs raises concerns about their harmful effects, impacting vital organs like the lungs, gastrointestinal system, and blood (Patwa et al., 2023). The sharp edges or irregular shapes of some microplastics may exacerbate tissue damage and inflammation. Furthermore, microplastics may interact with immune cells, induce pro-inflammatory mediators, and suppress immune regulation, potentially leading to autoimmune diseases. Strict control measures for plastic production and safe disposal strategies are crucial to prevent microplastics-mediated toxicity in humans (Iheanacho et al., 2023). However, the long-term health consequences of microplastic injection are not well understood. Chronic exposure to microplastics in the body could potentially contribute to the development or exacerbation of certain health conditions over time.

Humans are exposed to microplastics by inhalation, ingestion and dermal contact, potentially are exposed to chronic inflammatory lesions. Estimates, such the ones given for microplastic ingestion and inhalation, can be used to calculate the amount of microplastics that humans are exposed to (Kirstein et al., 2016; Prata, 2018). It is also possible to determine human exposure with greater accuracy by customizing standard diagnostic protocols. Stool samples can be used to assess intestinal exposure, while bronchial lavages can reveal details about the respiratory system. The histology of biopsies, autopsied tissues, and blood can all be utilized to determine the presence of microplastics. It is still necessary to develop techniques for recognizing plastics in these materials, which may involve staining histological sections or digesting tissues. The availability of tissue and the possibility of sample contamination (due to, for example, the extensive use of plastics in medicine) will always place restrictions on these techniques. The impact of microplastics on human health may be ascertained by research using cell cultures or animal models, such as mice or rats, while also taking various endpoints into account (Rubio et al., 2020). It may be possible to try observational studies that link exposure to unfavorable outcomes, but doing so may need large sample sizes and careful consideration of cofounder effects.

2.3 Skin Contact

While it is hypothesized that nanoplastics (smaller than 100 nm) could potentially penetrate the skin barrier, dermal contact with microplastics is considered a less significant route of exposure (Revel et al., 2018). This route is more frequently linked to exposure to plastic monomers and additives, such as phthalates and bisphenol A, which are known endocrine disruptors, via regular usage of household appliances. However, without further evidence, the hypothesis that nanoplastics could penetrate the skin barrier and be harmful should not be discounted. Plastics are recognized in medicine to cause fibrous encapsulation and mild inflammatory reactions in response to foreign bodies. For example, after 21 days, surgical sutures made of braided polyester and monofilament polypropylene caused less inflammation than silk and fibrous encapsulation (Salthouse & Matlaga, 1975). When a plastic of less than 10 mm in size particle was introduced subcutaneously into mice, an in vivo study found that after 98 days, Polyethylene (PE) disks caused encapsulation with minimal inflammation, whereas polyvinyl chloride (PVC) containing plasticizers or organotin led to encapsulation with moderate tissue degeneration and necrosis, likely due to leachate toxicity (Tienhoven et al., 2006). While both micro and nanoplastics have the potential to cause foreign body reactions and inflammation, variations in their surface qualities may potentially have different results. Exposure to micro and nanoplastics also causes oxidative stress in human epithelial cells (Schirinzi et al., 2017). Therefore, there is an urgent need for an hour to perform more research in the field due to the possible consequences of nanoplastics as well as the ubiquitous topical exposure to plastic particles (such as from dust, synthetic fibers, and microbeads in cosmetics).

3 Human Health Toxicity of Microplastic

Human food and air have both been found to contain microplastics. As a result, they could have an impact on human health through ingested or inhaled in. Human health may suffer if microplastic particles are consumed. The estimated total mass of ingested microplastic particles is equivalent to one credit card per week, fifty plastic bags annually, or 4.1 µg/week for people (Bai et al., 2024). To what extent are we consuming plastic? Some studies announced that, a typical human being may consume 5 g of microplastics, which is roughly equivalent to the weight of a credit card per week (Pletz, 2022). That reduces to as much as 52,000 particles each year from different food sources, according to another study (Cox et al., 2019). Prata carried out research in 2018 to investigate the effects of breathing in microplastics on human respiratory systems. According to the study, people who were exposed to microplastics developed airway inflammation, which made breathing difficult or caused shortness of breath (Amato-Lourenço et al., 2021). A variety of immediate symptoms, mostly of a respiratory character, might result from breathing in plastic vapors. People may have breathing difficulties, wheezing, and coughing. Skin and ocular irritation are also frequent neurological symptoms, along with headaches and vertigo. The degree of exposure and individual sensitivity were major determinants of these negative consequences (Sajadimajd et al., 2019). Studies have also suggested that microplastic exposure may result in cytotoxicity (Rochman et al., 2013; Yang et al., 2015), oxidative stress (Amato-Lourenço et al., 2021; Wright & Kelly, 2017), and translocation to other tissues (Kalb et al., 2015). Microplastics are long-lasting in the environment and do not degrade inside of living things. Consequently, the organisms are exposed to Microplastics for an extended period of time, which may cause chronic irritation, inflammation, cellular proliferation, necrosis, and immune cell impairment (Smith et al., 2018). Moreover, Microplastics may serve as carriers of other microbes (Kirstein et al., 2016). They can absorb materials from the environment or release compounds, both organic and inorganic, from their matrixes (Crawford & Quinn, 2017).

3.1 Oxidative stress DNA damage

Studies on inhalation exposure showed that the most effective model for Microplastics toxicity was oxidative stress, followed by inflammation and cytotoxicity (Nel et al., 2006). Microplastics may trigger oxidative stress by releasing oxidizing substances, such as metals adsorbed on their surface and reactive oxygen radicals produced by the host body during the inflammatory response (Kelly & Fussell, 2012). Reactive oxygen species (ROS) are a byproduct of the polymerization and processing of microplastics. But when these free radicals come into contact with UV radiation or reactive metals, their concentrations may rise dramatically. Additionally, as

Microplastics weathered, free radicals were produced, which oxidized the target tissues (Gewert et al., 2015).

3.2 Metabolic disorder

Research on the metabolic effects of microplastics and nanoplastics in humans is still in its early stages, but it could potentially contribute to metabolic disorders. Chemicals used in plastic manufacturing, such as phthalates and Bisphenol A (BPA), have endocrine-disrupting properties, potentially causing hormone dysfunction, obesity, insulin resistance, and dyslipidemia (Naomi et al., 2021). Exposure to these particles has been linked to chronic inflammation and oxidative stress, contributing to metabolic disorders like type 2 diabetes and metabolic syndrome. Microplastics and nanoplastics can interact with the gut microbiota, disrupting its composition and function, leading to metabolic disorders like obesity and insulin resistance (Haldar et al., 2023). Ingestion of these particles may physically obstruct nutrient absorption in the gastrointestinal tract, leading to nutrient deficiencies and metabolic imbalances. Adipose tissue can also be affected by these particles, leading to metabolic disorders such as obesity and insulin resistance. Environmental pollutants, including microplastics and nanoplastics, can lead to epigenetic changes that affect gene expression and metabolic function and lead to the development of metabolic disorders (Poma et al., 2023).

3.3 Neurotoxicity

A research was done in 2011, to assess the possible neurological consequences of styrene exposure on 21,962 Gulf Coast inhabitants in the United States. This region is thought to account for more than half of the US's total styrene output. The study assessed styrene exposure by measuring ambient styrene levels and evaluating blood samples from 874 participants. Data on neurological symptoms, including headaches, fatigue, dizziness, numbness, vision disturbances, and nausea, were collected. About 33% of the participants reported, experiencing at least one of aforesaid symptoms. A significant correlation was found between higher levels of environmental styrene exposure and the occurrence of multiple neurological symptoms, particularly in those in the highest exposure quartile. Although the biomarker findings were less conclusive, elevated blood styrene levels were associated with nausea. One important finding was that non-white participants reported more severe neurological symptoms, suggesting that environmental styrene exposure may contribute to neurotoxicity (Werder et al., 2018). Additionally, a retrospective study using Danish national registries examined the risk of encephalopathy and nonspecific dementia in individuals with confirmed styrene exposure. Among 72,465 workers in the reinforced plastics industry, there were 228 cases of encephalopathy and 565 cases of nonspecific dementia between 1977 and 2011 (Broadwell et al., 1995).

3.4 Gastrointestinal Tract

The concentration of fecal microplastics correlates positively with the intensity of Inflammatory Bowel Disease (IBD). Additionally, people with IBD had greater concentration of microplastics in feces (41.8 pcs/g) than a healthy person (28.0 pcs/g) (Nurk et al., 2022). Studies on IBD persons indicated a higher buildup of microplastics in rectal mucosal ulcers. However, nanoplastics were only detected in small levels on mucosal surfaces, indicating the possibility of designing precise medication delivery systems that target the inflamed intestinal mucosa (Schmidt et al., 2013). The possible influence of nano- and microplastics on carcinogenesis is particularly interesting. Scientific literature and meta-analyses were evaluated in response to reports of environmental variables that influence the start of colorectal cancer. The results indicate an increased risk of colorectal cancer among those working in a variety of industries, including plastics and rubber (Oddone et al., 2014). There has also been a significant rise in pancreatic cancer mortality among workers involved in the production of reinforced plastic (Kolstad et al., 1995). Another interesting topic is the impact of nano and microplastics on the microbiome. There is a scarcity of life in the existing literature. It has been hypothesized that changes in mammals gut microbiota following microplastic exposure might induce gut barrier breach and be connected with renal illness, liver damage, and neurological disturbances (Yong et al., 2019).

3.5 Respiratory System Disorders

Research into the potential toxicity of styrene began in the late 20th century. In 1999, the Styrene Information and Research Centre (SIRC) commissioned the Harvard Center to investigate the effects of styrene on human. The exposure of styrene varied, still remaining below 10 ppm, but the workers of industries of fiber-reinforced polymer composites levels reported more than 20 ppm. Studies conducted on animals, especially in rodents revealed possibilities of carcinogenic effects, neurological issues, and respiratory toxicity, but these findings did not establish a clear link to human health concerns (Cohen et al., 2002). A retrospective study of 3,704 boat building workers exposed to styrene for more than a year between 1959 and 1993 identified 516 cases of invasive cancer. Compared to the general population, there was a slight increase in cancers of the trachea, bronchus, and lungs. However, the absence of data on lifestyle factors such as smoking and other jobs weakened the findings, particularly for lung cancer, where the increase was modest relative to the reference population. The authors noted that the participants were relatively young (with a median age of 44 in 1991), whereas the average age of cancer diagnosis in the U.S. is 65 years (Coggon et al., 2015). Similar results were observed in a British cohort exposed to styrene (Mazzei et al., 2019), though other studies found no link between styrene exposure and lung cancer (Winiarska et al., 2024). Occupational Hypersensitivity Pneumonitis (OHP) can occur due to antigen inhalation in the workplace. In 2020, a case report detailed OHP in a 66 year-old who worked in the PET manufacturing industry from 1992 to 2013. The patient, already under observation for past asbestos exposure, was diagnosed with asbestosis in 2012. By 2017, high-resolution computed tomography showed mild lung fibrosis progression. Inspiratory clicks were detected during auscultation, and pulmonary function tests revealed reduced CO diffusion capacity, leading to a diagnosis of OHP, believed to be triggered by exposure to terephthalic acid and dimethyl terephthalate (DMP), both precursors in PET production. The authors noted this as the first reported case of OHP linked to PET manufacturing (Sun et al., 2023). Another case of OHP, likely caused by exposure to DMP and styrene, involved a 46 year-old yacht factory worker who reported shortness of breath, chest tightness, and persistent coughing for two months (Kogevinas et al., 2018).

3.6 Blood Neoplasms

A cohort study examined cancer mortality among 5,201 workers from two boat-building plants. While no increase in mortality from hematological cancers was observed compared to the general population, there was a significant correlation between the duration and intensity of styrene exposure and the development of neoplasms (Banton et al., 2019). In 2020, a re-analysis was conducted on the relationship between styrene exposure and cancer mortality in a previously studied group of American boatbuilders who were occupationally exposed to styrene between 1959 and 1978. Styrene exposure estimates were based on workplace conditions and air monitoring data (Daniels & Bertke, 2020). The association was assessed using Cox proportional hazard models, which accounted for factors such as gender, race, age, socioeconomic status, and length of employment (exposure) (Christensen et al., 2017). A statistically significant link was found between styrene exposure and the development of leukemia and bladder cancer (Thakur & Pathania, 2019). Lifetime exposure to 0.05 ppm of styrene was associated with one additional death from blood cancer per 10,000 workers. The authors emphasized the need for further research to provide a more comprehensive understanding of these findings.

3.7 Biological Effects

The biological effects of microplastics and nanoplastics are complex and multifaceted, with potential implications for individual organisms, populations, and ecosystems. Microplastics and nanoplastics have a wide range of biological effects on organisms across the food chain, from marine invertebrates to mammals, including humans. Ingestion of these particles can lead to physical blockage of the digestive tract, internal injuries, reduced nutrient absorption, and starvation. They can also adsorb and concentrate toxic chemicals from the environment, such as persistent organic pollutants (POPs), which can be released into organisms' tissues, causing toxicity, cellular damage, and adverse health effects (Hu & Palić, 2020). Exposure to microplastics and nanoplastics particles can trigger inflammatory responses, leading to tissue damage, organ dysfunction, and increased susceptibility to diseases (González-Acedo et al., 2021; Subaramaniyam et al., 2023). Oxidative stress, caused by these particles, disrupts cellular redox balance, leading to damage to lipids, proteins, and DNA (Liu et al., 2024). This can lead to inflammation, aging, and an increased risk of diseases like cancer and neurodegenerative disorders (Huang et al., 2021; Yee et al., 2021; Yong et al., 2019). Microplastics and nanoplastics can interfere with physiological processes, disrupting feeding behavior, reproductive function, hormone regulation, and metabolism, impacting individual fitness, population dynamics, and ecosystem health. They can bioaccumulate in organisms at lower trophic levels of the food chain and biomagnify as consumed by predators, causing higher concentrations of microplastics and associated contaminants in top predators, including humans. Some studies suggest that microplastics and nanoplastics may exhibit genotoxic and mutagenic effects, causing DNA damage and increasing the risk of genetic mutations and cancer (Benson et al., 2022; Kaushik et al., 2018; Marchini & Gelain, 2022; Ribeiro et al., 2019; Rose et al., 2023).

3.8 Organoids Effect

Organoids, three-dimensional structures derived from stem cells, are widely used in biomedical research for studies on organ development, disease modeling, drug screening, and regenerative medicine (Bartfeld & Clevers, 2017; Das, 2023; Zhou et al., 2023). However, research on the effects of micro and nanoplastics on organoids is limited. Microplastics and nanoplastics can release toxic chemicals, causing cytotoxicity, apoptosis, or necrosis. They can also induce inflammatory responses in tissues and cells, leading to increased production of pro-inflammatory cytokines and disruption of tissue homeostasis. They may interfere with cellular signaling pathways, leading to aberrant cell behavior and impaired maturation. They can generate reactive oxygen species (ROS) and induce oxidative stress, damaging cellular components and impairing organoid viability and function (Yu et al., 2021). Exposure to these particles may induce epigenetic alterations in organoid cells, affecting gene expression patterns and cellular phenotype. Organoids derived from barrier tissues, such as the intestine or blood-brain barrier, may be sensitive to these effects. They can also interact with extracellular matrix components, altering the mechanical properties of the matrix, affecting cell-matrix interactions, and modulating organoid morphology and behavior (Sun et al., 2023).

3.9 Reproductive Toxicity

The effects of plastic as Microplastics and nanoplastics on human fertility and reproduction have not yet been studied. However, several animal experiments have been carried out. For instance, in the worm Caenorhabditis elegans, exposure to nanoplastics led to fewer progeny over successive generations (Qiang & Cheng, 2020). In another study using C. elegans, both unmodified nanoplastics and nanoplastics-NH_2 (amine group) were evaluated, and nanoplastics-NH_2 had a greater adverse impact on reproduction and gonadal development in later generations (Sun et al., 2021). The exposure of zebrafish to polystyrene (PS) has also been associated with gonadal injury (Fournier et al., 2020). As a result of exposure to nanoplastics, zebrafish embryos developed a significant amount of pericardial edema, over the no noticed unfavorable impact limit (NOAEL), nanoplastics emphatically brought down the cardiovascular result and blood stream speed (Ragusa et al., 2020). Mammalian progeny have been found to contain nanoplastics, suggesting that these materials may pass through the placenta and gather in fetal tissues. PS-nanoplastics (20 nm) were injected intratracheally into the lungs, heart, and spleen of pregnant Sprague Dawley rats, and the fetus's organs, such as the liver, kidneys, heart, and brain, were also shown to have these particles (Brachner et al., 2019). The placenta and fetus both saw a marked decrease in weight as a result of this exposure. It has been verified that plastic as nanoplastics and Microplastics exist in humans. Twelve microplastic fragments, measuring between five and ten mm, were discovered in four placentas out of six examined in a study. Three of the fragments were on the chorioamniotic membranes, four on the maternal side, and five on the foetal side (Park, 2020). A human placental perfusion ex vivo model assessed whether PS-nanoplastics with sizes of 50, 80, and 240 nm, Impact of plastics exposure on cells in vitro studies conducted on human cells have demonstrated that nano- and microplastics can cause a range of biological reactions, such as oxidative stress, apoptosis, inflammation, cytotoxicity, and genotoxicity. The most often utilized plastic polymer in research among those examined is PS. It is interesting to note that PS is not among the top five polymers manufactured worldwide. Data on other forms of plastic, specifically PVC and PET, are lacking. The sizes of the plastic particles range from nano to micro to even millimeters, and some of them exhibit unique surface changes that are now thought to be essential for the inflammatory and cytotoxic effects. It is difficult to determine how long an individual was exposed, and it is not apparent if the impacts are acute or long-term. Although plastic bioaccumulation in the human body lasts a lifetime, there is a dearth of research pertaining to long-term exposure. Extracting and separating plastic particles from biological and environmental samples is another challenge. This makes it difficult for studies to determine dose accurately (Ahsan et al., 2018). Moreover, no research has been done on the impact of extremely low quantities of plastic on cells. Once within the body, nanoplastics interact with a wide range of biomolecules, including carbohydrates, lipids, and proteins. It is proposed that these particles change from their initial biological qualities to new ones when they come into touch with biological fluids such as serum or plasma. The main cause of this is the development of a protein corona, or layer of proteins that have been adsorbed on the surface of the nanoplastic (Rai et al., 2022). Depending on the distinct plasma proteomes linked to various physiological conditions (stress, food, smoking, etc.) or illnesses (diabetes, cancer), the protein corona's composition may change (Kumar et al., 2022). This implies that under some circumstances, identical nanoplastics may display differing protein corona compositions. As of right now, no customized in vitro research has evaluated the exposure of cells to plastics. Therefore, it is still unknown how the development of the protein corona may affect possible harmful effects.

3.10 Carcinogenicity

The carcinogenic potential of micro and nanoplastics is still being explored, but evidence suggests they could contribute to cancer development and progression through multiple mechanisms. Micro and nanoplastics can adsorb and concentrate toxic chemicals from the environment, such as polycyclic aromatic hydrocarbons (PAHs), polychlorinated biphenyls (PCBs), and heavy metals (Real et al., 2021). When ingested or inhaled, these chemicals can be released into the body, damaging DNA, promoting mutations, and disrupting cellular processes involved in tumor suppression. They can induce inflammatory responses and oxidative stress in cells and tissues, which can promote DNA damage, cell proliferation, angiogenesis, and metastasis. Some studies suggest that micro and nanoplastics may exhibit genotoxic effects, leading to genetic alterations that contribute to carcinogenesis by disrupting key regulatory pathways involved in cell growth and differentiation (Suresh et al., 2023). Exposure to these particles can modulate immune responses, activating inflammatory pathways and suppressing immune surveillance mechanisms. They can interfere with cellular signaling pathways, disrupting normal tissue homeostasis and promoting uncontrolled cell growth. Exposure to these particles may induce epigenetic changes in cells, such as alterations in DNA methylation, histone modifications, and non-coding RNA expression, which can modulate gene expression patterns and cellular phenotypes, contributing to cancer development and progression (Zuo et al., 2023). Micro and nanoplastics can interact with cells in the tumor microenvironment, including cancer-associated fibroblasts, immune cells, and endothelial cells (Chen & Lin, 2024). These interactions can promote tumor growth, angiogenesis, and metastasis by enhancing tumor cell survival, migration, and invasion (Gopinath et al., 2022).

3.11 Translocation of Micro Plastics to the Circulatory System and Distant Tissue

Various studies affirm that microplastics can be transported from the gastrointestinal tract to the circulatory system and distant tissues in animals and humans. They can cross the gastrointestinal epithelial barrier, enter the bloodstream through passive diffusion, uptake by specialized cells, and transport via transcytosis or paracellular routes (Yin et al., 2021). They can also enter the lymphatic system, which drains into the bloodstream through lymphatic vessels in the intestinal mucosa. Once in the bloodstream, microplastics can be distributed throughout the body to organs and tissues via the circulatory system. They can travel to distant sites through arterial and venous circulation, potentially reaching organs like the liver, kidneys, lungs, brain, and spleen. Microplastics in the bloodstream can interact with endothelial cells lining blood vessels, facilitating their translocation across the endothelial barrier. Immune cells, such as macrophages and monocytes, can transport microplastics to secondary lymphoid organs or tissues, where they may accumulate and induce inflammatory responses. Some microplastics, particularly those with adsorbed lipophilic compounds, may be transported in association with lipoproteins or lipid droplets in the bloodstream (Mamun et al., 2023; Persiani et al., 2023). Lipid-mediated transport can facilitate the distribution of microplastics to distant tissues, including those rich in lipid content such as adipose tissue and the brain. The translocation of microplastics to the circulatory system and distant tissues may be influenced by their size, shape, surface properties, and chemical composition (Shetty et al., 2023). Smaller particles, including nanoplastics, may have a higher propensity for translocation due to their ability to penetrate biological barriers more easily.

3.12 Impact of Indian Population due to Microplastics

Microplastics pose a growing threat to the Indian population due to widespread plastic pollution in both terrestrial and aquatic environments. Contaminated water sources can lead to ingestion of microplastics through drinking water and contaminated seafood, exposing the population to potential health risks such as gastrointestinal irritation, inflammation, and harmful chemicals (Patil et al., 2021). Chronic exposure to microplastics, particularly in vulnerable groups like children and pregnant women, could also pose health risks. Microplastics can also cause environmental contamination, soil and water resource contamination, wildlife ingestion, and ecosystem disruption. In India, where agriculture and fisheries are major livelihoods, the environmental impact could have significant socio-economic implications. Microplastics can contribute to air pollution, exacerbate respiratory problems and other health issues in urban areas (Jeyasanta et al., 2023). India faces challenges in managing its plastic waste, with inadequate infrastructure for collection, recycling, and disposal. Improper management of plastic waste can lead to the accumulation of microplastics in the environment, increasing the risk of exposure to the Indian population (Henoy et al., 2022; Jeyasanta et al., 2023). In rural areas, the presence of microplastics could disrupt traditional livelihoods and cultural practices.

4 Toxicity Mechanism

Microplastics pose significant environmental and health risks, necessitating a multi-faceted approach involving governments, industry, academia, civil society, and international organizations (Ghosh et al., 2022). This includes investing in research to understand their sources, distribution, and effects, establishing comprehensive monitoring programs, developing regulations and policies to address microplastic pollution, investing in waste management infrastructure, and promoting the adoption of circular economy principles. Develop and implement regulations and policies to address microplastic pollution at the national and international levels (Usman et al., 2021). This could include measures such as bans or restrictions on certain types of plastic products, requirements for extended producer responsibility, and standards for wastewater treatment. Invest in infrastructure for effective waste management, including collection, recycling, and disposal facilities. Public awareness about the risks is also crucial, with education and outreach programs educating individuals and communities on reducing plastic consumption and waste. Research and innovation are encouraged to develop eco-friendly packaging, biodegradable polymers, and innovative waste management solutions. International cooperation is crucial to address the transboundary nature of microplastic pollution, with partnerships between governments, industry, academia, and non-governmental organizations sharing knowledge and resources. Risk assessments are conducted to evaluate potential hazards and develop risk management strategies to minimize exposure and mitigate adverse effects. Develop risk management strategies to minimize exposure and mitigate adverse effects, taking into account factors such as exposure pathways, toxicity, and vulnerability of ecosystems and populations. Stakeholders from diverse sectors, including government agencies, industry associations, environmental organizations, and community groups, are engaged in the development and implementation of microplastic risk mitigation strategies. This approach fosters dialogue, collaboration, and shared responsibility among stakeholders to achieve common goals.

5 Conclusions

In the modern era, there is a growing concern over the adverse health impacts of microplastics on the human wellbeing. However, more research is needed for better understand the mechanisms of toxicity, the routes of exposure, the long-term health consequences of microplastic exposure and to develop strategies for mitigating these risks. The translocation of microplastics to the circulatory system and distant tissues is well-documented in animal studies, while more research is needed to understand the kinetics, mechanisms, and potential health implications of this process in humans. In the mean time, efforts to reduce plastic pollution and mitigate human exposure to microplastics are crucial for protecting biodiversity and public health.

While a few drives address significant stages toward tending to microplastic contamination, there is a developing acknowledgment of the requirement for more thorough and composed activity at the worldwide level. That requires the implementation of effective waste management strategies, promoting public awareness and education about plastic pollution, and developing sustainable alternatives. Coordinated effort between government, industry, common society, and the scholarly world is fundamental to relieve the ecological and wellbeing gambles related to microplastics.

An effort, to develop a global framework specifically focused on microplastics is underway, with initiatives such as the Oslo Paris Convention for the Protection of the Marine Environment of the North-East Atlantic (OSPAR) and the United Nations Environment Assembly (UNEA) addressing the issue. Proceeded with coordinated effort and collaboration among nations and partners will be fundamental to successfully address the difficulties presented by microplastics on a worldwide scale. By implementing a comprehensive framework for risk mitigation, it is possible to address the challenges posed by microplastics and work towards a cleaner, healthier, and more sustainable future for all.

6 Recommendations

It is imperative to prioritize comprehensive research initiatives to effectively combat the multifaceted challenges posed by microplastics. Investigate the specific biological mechanisms through which microplastics exert toxic effects on human health and ecosystems. This can include elucidating the pathways of inflammation, oxidative stress, and potential genotoxicity associated with microplastic exposure. Conduct longitudinal studies to assess the long-term health consequences of microplastic exposure in human populations. Such studies should aim to establish causal relationships between microplastic exposure and various health outcomes, including respiratory disorders, reproductive toxicity, and cancer. Develop standardized methodologies for assessing the environmental impact of microplastics across different ecosystems. This includes evaluating the bioaccumulation and biomagnification of microplastics in food webs and their effects on biodiversity. National governments and international bodies must take decisive action to develop and enforce policies aimed at reducing plastic production and consumption. Implement strict regulations on the production and use of single-use plastics, including bans or restrictions on specific plastic products that contribute significantly to microplastic pollution, as similar actions have been taken by various countries. Mandate environmental protection policies that hold manufacturers accountable for the entire lifecycle of their products, including waste management and recycling. This encourages the development of sustainable packaging alternatives. Investment in robust waste management infrastructure is crucial to mitigate the release of microplastics into the environment. Develop and promote advanced recycling technologies that can effectively process a wider range of plastic materials, thereby reducing the volume of plastic waste that enters landfills and the environment

References

Agarwal, R., Robison, B., & Parks, R. (1977). [49] Nucleoside diphosphokinase from human erythrocytes. *Methods in Enzymology, 51*, 376–386. https://doi.org/10.1016/S0076-6879(78)51051-3

Ahsan, S. M., Rao, C. M., & Ahmad, M. F. (2018). Nanoparticle-protein interaction: The significance and role of protein corona. *Cellular and Molecular Toxicology of Nanoparticles*, 175–198.

Amato-Lourenço, L. F., Carvalho-Oliveira, R., Júnior, G. R., Dos Santos Galvão, L., Ando, R. A., & Mauad, T. (2021). Presence of airborne microplastics in human lung tissue. *Journal of Hazardous Materials, 416*, 126124. https://doi.org/10.1016/j.jhazmat.2021.126124

Bai, J., Chen, D., Zhang, Z., Huang, Y., Bai, Y., Hao, J., Song, J., Li, N., Ren, G., Wang, X., Yang, G., Yadav, S., Feng, Y., & Wang, X. (2024). Minimizing environment footprint through half-plastic film mulch and straw incorporation in maize-based system. *Agriculture, Ecosystems & Environment, 367*, 108957. https://doi.org/10.1016/j.agee.2024.108957

Banton, M. I., Bus, J. S., Collins, J. J., Delzell, E., Gelbke, H. P., Kester, J. E., Moore, M. M., Waites, R., & Sarang, S. S. (2019). Evaluation of potential health effects associated with occupational and environmental exposure to styrene—An update. *Journal of Toxicology and Environmental Health, Part B, 22*(1–4), 1–130. https://doi.org/10.1080/10937404.2019.1633718

Bartfeld, S., & Clevers, H. (2017). Stem cell-derived organoids and their application for medical research and patient treatment. *Journal of Molecular Medicine, 95*, 729–738. https://doi.org/10.1007/s00109-017-1531-7

Benson, N. U., Agboola, O. D., Fred-Ahmadu, O. H., De-la-Torre, G. E., Oluwalana, A., & Williams, A. B. (2022). Micro (nano) plastics prevalence, food web interactions, and toxicity assessment in aquatic organisms: A review. *Frontiers in Marine Science, 9*, 851281. https://doi.org/10.3389/fmars.2022.851281

Brachner, A., Fragouli, D., Duarte, I. F., Farias, P. M., Dembski, S., Ghosh, M., Barisic, I., Zdzieblo, D., Vanoirbeek, J., Schwabl, P., & Neuhaus, W. (2019). Assessment of human health risks posed by nano- and microplastics is currently not feasible. *International Journal of Environmental Research and Public Health, 17*(23), 8832. https://doi.org/10.3390/ijerph17238832

Broadwell, D. K., Darcey, D. J., Hudnell, H. K., Otto, D. A., & Boyes, W. K. (1995). Work-site clinical and neurobehavioral assessment of solvent-exposed microelectronics workers. *American Journal of Industrial Medicine, 27*(5), 677–698. https://doi.org/10.1002/ajim.4700270505

Brown, D., Wilson, M., MacNee, W., Stone, V., & Donaldson, K. (2001). Size-dependent proinflammatory effects of ultrafine polystyrene particles: A role for surface area and oxidative stress in the enhanced activity of ultrafines. *Toxicology and Applied Pharmacology, 175*(3), 191–199. https://doi.org/10.1006/taap.2001.9240

Chen, C., & Lin, Z. (2024). Exploring the potential and challenges of developing physiologically-based toxicokinetic models to support human health risk assessment of microplastic and nanoplastic particles. *Environment International, 186*, 108617. https://doi.org/10.1016/j.envint.2024.108617

Christensen, M. S., Hansen, J., Ramlau-Hansen, C. H., Toft, G., & Kolstad, H. (2017). Cancer incidence in workers exposed to styrene in the Danish-reinforced plastics industry, 1968–2012. *Epidemiology, 28*(2), 300–310. https://doi.org/10.1097/EDE.0000000000000608

Coggon, D., Ntani, G., Harris, E. C., & Palmer, K. T. (2015). Risk of cancer in workers exposed to styrene at eight British companies making glass-reinforced plastics. *Occupational and Environmental Medicine, 72*(3), 165–170. https://doi.org/10.1136/oemed-2014-102382

Cohen, W. M., Nelson, R. R., & Walsh, J. P. (2002). Links and impacts: The influence of public research on industrial R&D. *Management Science, 48*(1), 1–23. https://doi.org/10.1287/mnsc.48.1.1.14273

Collins, J. J., Bodner, K. M., & Bus, J. S. (2013) Cancer mortality of workers exposed to styrene in the U.S. reinforced plastics and composite industry. *Epidemiology, 24*(2), 195–203. https://doi.org/10.1097/EDE.0b013e318281a30f

Cox, K. D., Covernton, G. A., Davies, H. L., Dower, J. F., Juanes, F., & Dudas, S. E. (2019). Human consumption of microplastics. *Environmental science & technology, 53*(12), 7068–7074. https://doi.org/10.1021/acs.est.9b01517

Crawford, C. B., & Quinn, B. (2017). The interactions of microplastics and chemical pollutants. *Microplastic Pollutants*, 131–157.

Daniels, R. D., & Bertke, S. J. (2020). Exposure–response assessment of cancer mortality in styrene-exposed boatbuilders. *Occupational and Environmental Medicine, 77*(10), 706–712. https://doi.org/10.1136/oemed-2020-106445

Das, A. (2023). The emerging role of microplastics in systemic toxicity: Involvement of reactive oxygen species (ROS). *Science of the Total Environment, 895*, 165076. https://doi.org/10.1016/j.scitotenv.2023.165076

Dris, R., Gasperi, J., Mirande, C., Mandin, C., Guerrouache, M., Langlois, V., & Tassin, B. (2017). A first overview of textile fibers, including microplastics, in indoor and outdoor environments. *Environmental Pollution, 221*, 453–458. https://doi.org/10.1016/j.envpol.2016.12.013

Fournier, S. B., D'Errico, J. N., Adler, D. S., Kollontzi, S., Goedken, M. J., Fabris, L., Yurkow, E. J., & Stapleton, P. A. (2020). Nanopolystyrene translocation and fetal deposition after acute lung exposure during late-stage pregnancy. *Particle and Fibre Toxicology, 17*, 1–11. https://doi.org/10.1186/s12989-020-00385-9

Galloway, T. S. (2015). Micro- and nano-plastics and human health. *Marine Anthropogenic Litter*, 343–366. https://doi.org/10.1007/978-3-319-16510-3_13

Gewert, B., Plassmann, M. M., & MacLeod, M. (2015). Pathways for degradation of plastic polymers floating in the marine environment. *Environmental Science: Processes & Impacts, 17*(9), 1513–1521. https://doi.org/10.1039/C5EM00207A

Ghosh, S., Sinha, J. K., Ghosh, S., Vashisth, K., Han, S., & Bhaskar, R. (2022). Microplastics as an emerging threat to the global environment and human health. *Sustainability, 15*(14), 10821. https://doi.org/10.3390/su151410821

González-Acedo, A., García-Recio, E., Illescas-Montes, R., Ramos-Torrecillas, J., Melguizo-Rodríguez, L., & Costela-Ruiz, V. J. (2021). Evidence from in vitro and in vivo studies on the potential health repercussions of micro- and nanoplastics. *Chemosphere, 280*, 130826. https://doi.org/10.1016/j.chemosphere.2021.130826

Gopinath, P. M., Parvathi, V. D., Yoghalakshmi, N., Kumar, S. M., Athulya, P. A., Mukherjee, A., & Chandrasekaran, N. (2022). Plastic particles in medicine: A systematic review of exposure and effects to human health. *Chemosphere, 303*, 135227. https://doi.org/10.1016/j.chemosphere.2022.135227

Haldar, S., Muralidaran, Y., Míguez, D., Mulla, S. I., & Mishra, P. (2023). Eco-toxicity of nano-plastics and its implication on human metabolism: Current and future perspective. *Science of The Total Environment, 861*, 160571. https://doi.org/10.1016/j.scitotenv.2022.160571

Hartmann, N. B., Rist, S., Bodin, J., et al. (2017). Microplastics as vectors for environmental contaminants: Exploring sorption, desorption and transfer to biota. *Integrative Environmental Assessment and Management, 13*(3), 488–493. https://doi.org/10.1002/ieam.1904

Henoy, A., Haritash, A. K., Singh, S. K. (2022). Microplastics: Environmental issues and their management. In V. S. Kanwar, S. K. Sharma, & C. Prakasam (Eds.), *Proceedings of international conference on innovative technologies for clean and sustainable development (ICITCSD—2021)*. Springer. https://doi.org/10.1007/978-3-030-93936-6_9

Hu, M., & Palić, D. (2020). Micro- and nano-plastics activation of oxidative and inflammatory adverse outcome pathways. *Redox Biology, 37*, 101620. https://doi.org/10.1016/j.redox.2020.101620

Huang, W., Song, B., Liang, J., Niu, Q., Zeng, G., Shen, M., Deng, J., Luo, Y., Wen, X., & Zhang, Y. (2021). Microplastics and associated contaminants in the aquatic environment: A review on their ecotoxicological effects, trophic transfer, and potential impacts to human health. *Journal of Hazardous Materials, 405*, 124187. https://doi.org/10.1016/j.jhazmat.2020.124187

Iheanacho, S., Ogbu, M., Bhuyan, M. S., & Ogunji, J. (2023). Microplastic pollution: An emerging contaminant in aquaculture. *Aquaculture and Fisheries, 8*(6), 603–616. https://doi.org/10.1016/j.aaf.2023.01.007

Jeyasanta, K. I., Laju, R., Patterson, J., Jayanthi, M., Bilgi, D. S., Sathish, N., & Edward, J. P. (2023). Microplastic pollution and its implicated risks in the estuarine environment of Tamil Nadu, India. *Science of the Total Environment, 861*, 160572. https://doi.org/10.1016/j.scitotenv.2022.160572

Kalb, R. E., Fiorentino, D. F., Lebwohl, M. G., Toole, J., Poulin, Y., Cohen, A. D., Goyal, K., Fakharzadeh, S., Calabro, S., Chevrier, M., & Langholff, W. (2015). Risk of serious infection with biologic and systemic treatment of psoriasis: Results from the Psoriasis longitudinal assessment and registry (PSOLAR). *JAMA dermatology, 151*(9), 961–969. https://doi.org/10.1001/jamadermatol.2015.0718

Kaushik, G., Ponnusamy, M. P., & Batra, S. K. (2018). Concise review: Current status of three-dimensional organoids as preclinical models. *Stem Cells, 36*(9), 1329–1340. https://doi.org/10.1002/stem.2852

Kazi, A. H., Romanova, S. Y., Ilhami O., Dong, Z., Jesse, K., Xi, H., Bing, W., Lucia, F.-B., Yongfeng, L., Mathias, S., & Yusong, L. (2023). Assessing the release of microplastics and nanoplastics from plastic containers and reusable food pouches: Implications for human health. *Environmental Science & Technology*. https://doi.org/10.1021/acs.est.3c01942

Kelly, F. J., & Fussell, J. C. (2012). Size, source and chemical composition as determinants of toxicity attributable to ambient particulate matter. *Atmospheric Environment, 60*, 504–526. https://doi.org/10.1016/j.atmosenv.2012.06.039

Kirstein, I. V., Kirmizi, S., Wichels, A., Garin-Fernandez, A., Erler, R., Löder, M., & Gerdts, G. (2016). Dangerous hitchhikers? Evidence for potentially pathogenic Vibrio spp. on microplastic particles. *Marine Environmental Research, 120*, 1–8. https://doi.org/10.1016/j.marenvres.2016.07.004

Koelmans, A. A., Bakir, A., Burton, G. A., et al. (2016). Microplastics as a vector for chemicals in the aquatic environment: Critical review and model-supported reinterpretation of emphirical studies. *Environmental Science & Technology, 50*(7), 3315–3326. https://doi.org/10.1021/acs.est.5b06069

Kogevinas, M., Gwinn, W. M., Kriebel, D., Phillips, D. H., Sim, M., Bertke, S. J., Calaf, G. M., Colosio, C., Fritz, J. M., Fukushima, S., & Hemminki, K. (2018). Carcinogenicity of quinoline, styrene, and styrene-7, 8-oxide. *The Lancet Oncology, 19*(6), 728–729. https://doi.org/10.1016/S1470-2045(18)30316-4

Kolstad, H. A., Juel, K., Olsen, J., & Lynge, E. (1995). Exposure to styrene and chronic health effects: Mortality and incidence of solid cancers in the Danish reinforced plastics industry. *Occupational and Environmental Medicine, 52*(5), 320–327. https://doi.org/10.1136/oem.52.5.320

Kumar, R., Manna, C., Padha, S., Verma, A., Sharma, P., Dhar, A., Ghosh, A., & Bhattacharya, P. (2022). Micro(nano)plastics pollution and human health: How plastics can induce carcinogenesis to humans? *Chemosphere, 298*, 134267. https://doi.org/10.1016/j.chemosphere.2022.134267

Liu, S., He, Y., Yin, J., Zhu, Q., Liao, C., & Jiang, G. (2024). Neurotoxicities induced by micro/nanoplastics: A review focusing on the risks of neurological diseases. *Journal of Hazardous Materials, 469*, 134054. https://doi.org/10.1016/j.jhazmat.2024.134054

Mamun, A. A., Prasetya, T. A. E., Dewi, I. R., & Ahmad, M. (2023). Microplastics in human food chains: Food becoming a threat to health safety. *Science of the Total Environment, 858*, 159834. https://doi.org/10.1016/j.scitotenv.2022.159834

Marchini, A., & Gelain, F. (2022). Synthetic scaffolds for 3D cell cultures and organoids: Applications in regenerative medicine. *Critical Reviews in Biotechnology, 42*(3), 468–486. https://doi.org/10.1080/07388551.2021.1932716

Mazzei, M. A., Sartorelli, P., Bagnacci, G., Gentili, F., Sisinni, A. G., Fausto, A., Mazzei, F. G., & Volterrani, L. (2019). Occupational lung diseases: Underreported diagnosis in radiological practice. *Seminars in Ultrasound, CT and MRI, 40*(1), 36–50. https://doi.org/10.1053/j.sult.2018.10.019

Naomi, R., Yazid, M. D., Bahari, H., Keong, Y. Y., Rajandram, R., Embong, H., Teoh, S. H., Halim, S., & Othman, F. (2021). Bisphenol A (BPA) leading to obesity and cardiovascular complications: A compilation of current in vivo study. *International Journal of Molecular Sciences, 23*(6), 2969. https://doi.org/10.3390/ijms23062969

Nel, A., Xia, T., & Li, N. (2006). Toxic potential of materials at the nanolevel. *Science*, 0622.

Nurk, S., Koren, S., Rhie, A., Rautiainen, M., Bzikadze, A. V., Mikheenko, A., Vollger, M. R., Altemose, N., Uralsky, L., Gershman, A., Aganezov, S., Hoyt, S. J., Diekhans, M., Logsdon, G. A., Alonge, M., Antonarakis, S. E., Borchers, M., Bouffard, G. G., Brooks, S. Y., … Phillippy, A. M. (2022). The complete sequence of a human genome. *Science*, abj6987.

Oddone, E., Modonesi, C., & Gatta, G. (2014). Occupational exposures and colorectal cancers: A quantitative overview of epidemiological evidence. *World Journal of Gastroenterology: WJG, 20*(35), 12431–12444. https://doi.org/10.3748/wjg.v20.i35.12431

Park, S. J. (2020). Protein-nanoparticle interaction: Corona formation and conformational changes in proteins on nanoparticles. *International Journal of Nanomedicine, 15*, 5783–5802. https://doi.org/10.2147/IJN.S254808

Patil, S., Bafana, A., Naoghare, P. K., Krishnamurthi, K., & Sivanesan, S. (2021). Environmental prevalence, fate, impacts, and mitigation of microplastics—A critical review on present understanding and future research scope. *Environmental Science and Pollution Research, 28*, 4951–4974. https://doi.org/10.1007/s11356-020-11700-4

Patwa, R., Das, D., & Kaushik, G. (2023). Fate and occurrence of microplastic and nanoplastic pollution in industrial wastewater. In M. Shah (Ed.), *Microbial degradation and detoxification of pollutants* (pp. 251–276). De Gruyter. https://doi.org/10.1515/9783110743623-013

Pauly, D., Christensen, V., Dalsgaard, J., Froese, R., & Francisco Torres, J. (1998). Fishing down marine food webs. *Science*. https://doi.org/10.1126/science.279.5352.860

Persiani, E., Cecchettini, A., Ceccherini, E., Gisone, I., Morales, M. A., & Vozzi, F. (2023). Microplastics: A matter of the heart (and vascular system). *Biomedicines, 11*(2), 264. https://doi.org/10.3390/biomedicines11020264

Pletz, M. (2022). Ingested microplastics: Do humans eat one credit card per week? *Journal of Hazardous Materials Letters, 3*, 100071. https://doi.org/10.1016/j.hazl.2022.100071

Poma, A. M., Morciano, P., & Aloisi, M. (2023). Beyond genetics: Can micro and nanoplastics induce epigenetic and gene-expression modifications? *Frontiers in Epigenetics and Epigenomics, 1*, 1241583. https://doi.org/10.3389/freae.2023.1241583

Prata, J. C. (2018). Microplastics in wastewater: State of the knowledge on sources, fate and solutions. *Marine Pollution Bulletin, 129*(1), 262–265. https://doi.org/10.1016/j.marpolbul.2018.02.046

Qiang, L., & Cheng, J. (2020). Exposure to polystyrene microplastics impairs gonads of zebrafish (Danio rerio). *Chemosphere, 263*, 128161. https://doi.org/10.1016/j.chemosphere.2020.128161

Ragusa, A., Svelato, A., Santacroce, C., Catalano, P., Notarstefano, V., Carnevali, O., Papa, F., Rongioletti, M. C. A., Baiocco, F., Draghi, S., D'Amore, E., Rinaldo, D., Matta, M., &

Giorgini, E. (2020). Plasticenta: First evidence of microplastics in human placenta. *Environment International, 146*, 106274. https://doi.org/10.1016/j.envint.2020.106274
Rai, P. K., Sonne, C., Brown, R. J., Younis, S. A., & Kim, K. (2022). Adsorption of environmental contaminants on micro- and nano-scale plastic polymers and the influence of weathering processes on their adsorptive attributes. *Journal of Hazardous Materials, 427*, 127903. https://doi.org/10.1016/j.jhazmat.2021.127903
Real, Á. D., Santurtún, A., & Teresa Zarrabeitia, M. (2021). Epigenetic related changes on air quality. *Environmental Research, 197*, 111155. https://doi.org/10.1016/j.envres.2021.111155
Revel, M., Châtel, A., & Mouneyrac, C. (2018). Micro(nano)plastics: A threat to human health? *Current Opinion in Environmental Science & Health, 1*, 17–23. https://doi.org/10.1016/j.coesh.2017.10.003
Ribeiro, F., O'Brien, J. W., Galloway, T., & Thomas, K. V. (2019). Accumulation and fate of nano- and micro-plastics and associated contaminants in organisms. *TrAC Trends in Analytical Chemistry, 111*, 139–147. https://doi.org/10.1016/j.trac.2018.12.010
Rochman, C. M., Hoh, E., Kurobe, T., et al. (2013). Ingested plastic transfers hazardous chemicals to fish and induces hepatic stress. *Scientific Reports, 3*, 3263. https://doi.org/10.1038/srep03263
Rose, P. K., Yadav, S., Kataria, N., & Khoo, K. S. (2023). Microplastics and nanoplastics in the terrestrial food chain: Uptake, translocation, trophic transfer, ecotoxicology, and human health risk. *TrAC Trends in Analytical Chemistry, 167*, 117249. https://doi.org/10.1016/j.trac.2023.117249
Rubio, L., Marcos, R., & Hernández, A. (2020). Potential adverse health effects of ingested micro- and nanoplastics on humans. Lessons learned from in vivo and in vitro mammalian models. *Journal of Toxicology and Environmental Health, Part B, 23*(2), 51–68. https://doi.org/10.1080/10937404.2019.1700598
Sajadimajd, S., Bahramsoltani, R., Iranpanah, A., Kumar Patra, J., Das, G., Gouda, S., Rahimi, R., Rezaeiamiri, E., Cao, H., Giampieri, F., Battino, M., Tundis, R., Campos, M. G., Farzaei, M. H., & Xiao, J. (2019). Advances on natural polyphenols as anticancer agents for skin cancer. *Pharmacological Research, 151*, 104584. https://doi.org/10.1016/j.phrs.2019.104584
Salthouse, T. N., & Matlaga, B. F. (1975). Significance of cellular enzyme activity at nonabsorbable suture implant sites: Silk, polyester, and polypropylene. *Journal of Surgical Research, 19*(2), 127–132. https://doi.org/10.1016/0022-4804(75)90117-1
Schirinzi, G. F., Pérez-Pomeda, I., Sanchís, J., Rossini, C., Farré, M., & Barceló, D. (2017). Cytotoxic effects of commonly used nanomaterials and microplastics on cerebral and epithelial human cells. *Environmental Research, 159*, 579–587. https://doi.org/10.1016/j.envres.2017.08.043
Schmidt, C., Lautenschlaeger, C., Collnot, E., Schumann, M., Bojarski, C., Schulzke, J., Lehr, C., & Stallmach, A. (2013). Nano- and microscaled particles for drug targeting to inflamed intestinal mucosa—A first in vivo study in human patients. *Journal of Controlled Release, 165*(2), 139–145. https://doi.org/10.1016/j.jconrel.2012.10.019
Shetty, S. S., Sonkusare, S., Naik, P. B., & Madhyastha, H. (2023). Environmental pollutants and their effects on human health. *Heliyon*. https://doi.org/10.1016/j.heliyon.2023.e19496
Shi, Q., Tang, J., Liu, R., & Wang, L. (2021). Toxicity in vitro reveals potential impacts of microplastics and nanoplastics on human health: A review. *Critical Reviews in Environmental Science and Technology, 52*(21), 3863–3895. https://doi.org/10.1080/10643389.2021.1951528
Smith, L. K., Boukhaled, G. M., Condotta, S. A., Mazouz, S., Guthmiller, J. J., Vijay, R., Butler, N. S., Bruneau, J., Shoukry, N. H., Krawczyk, C. M., & Richer, M. J. (2018). Interleukin-10 directly inhibits CD8+ T cell function by enhancing N-glycan branching to decrease antigen sensitivity. *Immunity, 48*(2), 299–312. https://doi.org/10.1016/j.immuni.2018.01.006
Subaramaniyam, U., Allimuthu, R. S., Vappu, S., Ramalingam, D., Balan, R., Paital, B., Panda, N., Rath, P. K., Ramalingam, N., & Sahoo, D. K. (2023). Effects of microplastics, pesticides and nano-materials on fish health, oxidative stress and antioxidant defense mechanism. *Frontiers in Physiology, 14*, 1217666. https://doi.org/10.3389/fphys.2023.1217666

Sun, H., Jiao, R., Yu, J., & Wang, D. (2023). Combined effects of particle size and humic acid corona on the aggregation kinetics of nanoplastics in aquatic environments. *Science of the Total Environment, 901*, 165987. https://doi.org/10.1016/j.scitotenv.2023.165987

Sun, M., Ding, R., Ma, Y., Sun, Q., Ren, X., Sun, Z., & Duan, J. (2021). Cardiovascular toxicity assessment of polyethylene nanoplastics on developing zebrafish embryos. *Chemosphere, 282*, 131124. https://doi.org/10.1016/j.chemosphere.2021.131124

Suresh, P. K., Banerjee, A., & Singh, P. (2023). Promises and challenges for targeting the immunological players in the tumor micro-environment—Critical determinants for NP-based therapy. *OpenNano, 10*, 100134. https://doi.org/10.1016/j.onano.2023.100134

Thakur, M., & Pathania, D. (2019). Environmental fate of organic pollutants and effect on human health. *Abatement of Environmental Pollutants*, 245–262. https://doi.org/10.1016/B978-0-12-818095-2.00012-6

Van Tienhoven, E. A., Korbee, D., Schipper, L., Verharen, H. W., & De Jong, W. H. (2006). In vitro and in vivo (cyto)toxicity assays using PVC and LDPE as model materials. *Journal of Biomedical Materials Research Part A, 78A*(1), 175–182. https://doi.org/10.1002/jbm.a.30679

Tran, C. L., Buchanan, D., Cullen, R. T., Searl, A., Jones, A. D., & Donaldson, K. (2000). Inhalation of poorly soluble particles. II. Influence of particle surface area on inflammation and clearance. *Inhalation Toxicology, 12*(12), 1113–1126. https://doi.org/10.1080/08958370050166796

Usman, S., Abdull Razis, A. F., Shaari, K., Azmai, M. N., Saad, M. Z., Mat Isa, N., & Nazarudin, M. F. (2021). The burden of microplastics pollution and contending policies and regulations. *International Journal of Environmental Research and Public Health, 19*(11), 6773. https://doi.org/10.3390/ijerph19116773

Vianello, A., Jensen, R. L., Liu, L., & Vollertsen, J. (2019). Simulating human exposure to indoor airborne microplastics using a breathing thermal manikin. *Scientific Reports, 9*(1), 1–11. https://doi.org/10.1038/s41598-019-45054-w

Werder, E. J., Engel, L. S., Richardson, D. B., Emch, M. E., Gerr, F. E., Kwok, R. K., & Sandler, D. P. (2018). Environmental styrene exposure and neurologic symptoms in U.S. Gulf coast residents. *Environment International, 121*, 480–490. https://doi.org/10.1016/j.envint.2018.09.025

Winiarska, E., Jutel, M., & Zemelka-Wiacek, M. (2024). The potential impact of nano- and microplastics on human health: Understanding human health risks. *Environmental Research, 251*, 118535. https://doi.org/10.1016/j.envres.2024.118535

Wright, S. L., Thomicroplasticson, R. C., Galloway, T. S. (2013). The physical impacts of microplasticon marine organisms: A review. *Environmental Pollution, 178*, 483–492. https://doi.org/10.1016/j.envpol.2013.02.031

Wright, S. L., & Kelly, F. J. (2017). Plastic and human health: A micro issue? *Environmental Science & Technology, 51*(12), 6634–6647. https://doi.org/10.1021/acs.est.7b00423

Xu, H., Verbeken, E., Vanhooren, H. M., Nemery, B., & Hoet, P. H. (2004). Pulmonary toxicity of polyvinyl chloride particles after a single intratracheal instillation in rats. Time course and comparison with silica. *Toxicology and Applied Pharmacology, 194*(2), 111–121. https://doi.org/10.1016/j.taap.2003.09.018

Xu, H., Hoet, P. H. M., & Nemery, B. (2002). In vitro toxicity assessment of polyvinyl chloride particles and comparison of six cellular systems. *Journal of Toxicology and Environmental Health, Part A, 65*(16), 1141–1159. https://doi.org/10.1080/152873902760125372

Yang, Y., Yang, J., Wu, W.-M., et al. (2015). Biodegradation and mineralization of polystyrene by plastic-eating mealworms: Part 1. Chemical and physical characterization and isotopic tests. *Environmental Science & Technology, 49*(20), 12080–12086. https://doi.org/10.1021/acs.est.5b02661

Yasmin, B., Viviane, A. S. L., André H. B. d. O., Carla, B. V., & Rivelino, M. C. (2023). Microplastics and nanoplastics in the aquatic environment: Contamination, determination and interaction with other contaminants. *Periódico Tchê Química, 20*(43), 58–92. https://doi.org/10.52571/ptq.v20.n43.2023_05_cavalcante_pgs_58_92.pdf

Yee, M. S., Hii, L., Looi, C. K., Lim, W., Wong, S., Kok, Y., Tan, B., Wong, C., & Leong, C. (2021). Impact of microplastics and nanoplastics on human health. *Nanomaterials, 11*(2), 496. https://doi.org/10.3390/nano11020496

Yin, K., Wang, Y., Zhao, H., Wang, D., Guo, M., Mu, M., Liu, Y., Nie, X., Li, B., Li, J., & Xing, M. (2021). A comparative review of microplastics and nanoplastics: Toxicity hazards on digestive, reproductive and nervous system. *Science of the Total Environment, 774*, 145758. https://doi.org/10.1016/j.scitotenv.2021.145758

Yong, C. Q., Valiyaveettil, S., & Tang, B. L. (2019). Toxicity of microplastics and nanoplastics in mammalian systems. *International Journal of Environmental Research and Public Health, 17*(5), 1509. https://doi.org/10.3390/ijerph17051509

Yu, C., Luk, T. C., & Liao, V. H. C. (2021). Long-term nanoplastics exposure results in multi and trans-generational reproduction decline associated with germline toxicity and epigenetic regulation in Caenorhabditis elegans. *Journal of Hazardous Materials, 412*, 125173. https://doi.org/10.1016/j.jhazmat.2021.125173

Zhou, Y., Wu, Q., Li, Y., Feng, Y., Wang, Y., & Cheng, W. (2023). Low-dose of polystyrene microplastics induce cardiotoxicity in mice and human-originated cardiac organoids. *Environment International, 179*, 108171. https://doi.org/10.1016/j.envint.2023.108171

Zuo, Z., Liu, J., Sun, Z., Cheng, Y., Ewing, M., Bugge, T. H., Finkel, T., Leppla, S. H., & Liu, S. (2023). ERK and c-Myc signaling in host-derived tumor endothelial cells is essential for solid tumor growth. *Proceedings of the National Academy of Sciences, 120*(1), e2211927120. https://doi.org/10.1073/pnas.2211927120

Potential Bioindicators of Microplastics for Monitoring Microplastic Pollution

Vikram Mor, Wazir Singh, and Pradeep Khyalia

Abstract Microplastics (MPs) are ubiquitous contaminants that pose serious dangers to human health and the environment. Therefore, reliable bioindicator species that may collect and reflect the presence of MPs in diverse ecosystems are necessary for the efficient monitoring of MPs pollution. There are numerous studies on the spread and buildup of MPs in the food chain. Smaller MPs make it easier for aquatic and terrestrial biota to absorb and subsequently accumulate along the food chain. This study examines possible terrestrial and aquatic bioindicator species that can be used to track microplastic pollution. Additionally, the important selection factors for bioindicators, including sensitivity to environmental pollutants, ecological significance, broad distribution, and simplicity of collecting were also discussed.

Keywords Microplastics · Pollution · Bioindicators · Monitoring

1 Introduction

Plastic is a commonly used material in daily life, packaging, industries and agriculture due to its low cost and durability (Wright & Kelly, 2017; Zalasiewicz et al., 2016). But, excessive use of plastic has significantly altered the earth since the middle of the 20th century and is gradually leaving an enduring human imprint on air, land and water bodies (Porta, 2019). Thompson, (2009) refered to this current period of human

V. Mor (✉)
Department of Environmental Science, Faculty of Applied and Basic Sciences, SGT University, Gurugram, Haryana, India
e-mail: vikram_fps@sgtuniversity.org

W. Singh
Department of Environmental Science, GL Bajaj Group of Institution Akbarpur, Uttar Pradesh, Mathura, India

P. Khyalia
Department of Environmental Science, Maharshi Maharshi Dayanand University, Rohtak, Haryana, India

P. Khyalia and S. Yadav (eds.), *Microplastic Pollution in India and Its Environmental Impacts*, Earth and Environmental Sciences Library,
https://doi.org/10.1007/978-3-031-92530-6_10

history as the "Plastic Age." In recent years, both the production and consumption of plastics has grown exponentially going from 365.5 million tons in 2018 to 390.7 million tons in 2021 (Plastics Europe, 2022).

Globally, a total of 10 billion metric tons of plastics is present, which is twenty-five times beyond the total population (Bostan et al., 2023; Shekhar et al., 2022). It has consequently turned into a significant urban garbage due to its slow decomposition and inadequate management (Geyer et al., 2017; Wang et al., 2018). However, the mismanagement of plastic waste led to the indiscriminate release of uncontrolled disposal of countless small plastic debris into the environment (Eriksen et al., 2014; Jambeck, 2015). When this plastic waste is subjected to UV radiation and mechanical wear, it undergo degradation and fragmentation, changing from macroplastics (which measure 5–50 cm) to microplastic (<5 mm) and nanoplastics (<0.1 μm) (Cheng et al., 2021) .

Microplastics (MPs) can be released from various sources such as the degradation of big plastic items, microbeads from personal care products and the release of microfibers from clothes and fabrics (Kumari et al., 2025). Eventually, these tiny particles are transported to the soil and water by the wind and water currents. Various scientific studies have revealed that MPs are present in all the ecosystems, air (Wright et al., 2020; Amato-Lourenco, 2020), water (Abelouah et al., 2022), and soil (Chang et al., 2022). MPs can enter the soil directly from the plastic mulch and indirectly due to the use of untreated sewage and wastewater for irrigation (Fuller & Gautam, 2016; Horton et al., 2017).

Over the last decade, MPs pollution has become an environmental concerns because once released, they can live for hundreds to thousands of years, accumulating in all types of environments, from high alpine regions to the deep ocean (Esposito et al., 2022; Hu et al., 2019; Pastorino et al., 2024; Wang et al., 2018). MPs can cause a wide range of effects on the environment because they can be consumed by different forms of life planktons to top predators, which can result in physical injury, digestive obstructions, negative effects on reproductively and energy metabolism of (Wang et al., 2020a; Ali et al., 2021; Benson et al., 2022). In addition, MPs can disrupt the soil microbial community (Liu et al., 2019b) and alter the physical and chemical characteristics of soils, including pH, porosity, structure, and nutrient availability (Wang et al., 2022a). Furthermore, as pollutants pass through the food chain, these particles have the ability to absorb and transfer contaminants including heavy metals and persistent organic pollutants, perhaps amplifying their harmful effects (Chen et al., 2023). Contaminated fish or other aquatic species can introduce MPs into the human food chain which can cause a direct risk to human health. According to a recent study, humans ingest between 39,000 and 52,000 MPs particles annually on average. If the estimates for MPs particles inhaled are taken into account, the total could potentially exceed 74,000 particles. The total might reach up to 74,000 particles if the inhaled MPs are also included (Cox et al., 2019). Schirinzi et al. (2017) reported the cytotoxic effects of MPs on human brain. In addition, because MPs can be released as plastic waste material by humans and animals, the cycle of MPs in the environment never stops.

Given the pervasiveness of MPs pollution and its possible consequences, there is an urgent need of effective monitoring techniques because monitoring MPs is significant in figuring out the long-term effects and mitigating techniques (Multisanti et al., 2022). Biomonitoring can be a crucil approach to monitor to monitor MPs due to its economic, environmental, and user-friendly advantages (L´opez-Pedrouso et al., 2020).

2 Sources and Types of MPs

MPs exist in various forms in different ecosystems such as oceans, rivers, and other aquatic environments. Moreover, MPs are also present in the soil and air. Generally, sources of microplastic are classified into two major categories: Primary MPs and Secondary MPs.

Primary MPs are purposely manufactured for use in household items like exfoliating scrubs, toothpaste, and cosmetics (Barnes et al., 2009) as well as industrial pellets called nurdles that may be accidentally released during production or shipping (Rochman et al., 2013). Furthermore, synthetic fibers released from textiles and carpets when washed also play a role in primary microplastic pollution (Napper & Thompson, 2016). Primary MPs is most common form found in marine environments, freshwater lakes, and rivers (McCormick et al., 2014; Su et al., 2016; Zhang et al., 2017). These tiny pieces of plastic are frequently released into the environment where they may be consumed by sea creatures, become part of the food web or may be inhaled by the terrestrial organisms.

On the other hand, secondary MPs are produced when larger plastic objects such as plastic bottles, bags, and packaging break down from exposure to UV rays mechanical forces, and microorganisms, leading to the emission of smaller plastic particles into the surroundings (Andrady, 2011). This group also covers the breakdown of bigger items such as tires and plastic mulch utilized in farming activities (Kumar et al., 2020).

2.1 Microplastic: shape and structure

MPs exist in different forms such as fibers, fragments, spheres, particles etc. (Hartmann et al., 2019) and each arises from different sources and processes. Shape and size of the MPs could affect their fate in the environment and interactions with living beings (Frank et al., 2020; Schwarzer et al., 2022). Therefore, accurately assessing the ecological and health impacts of MPs requires a thorough understanding of their shape, size, and structure.

2.2 Classification of MPs on shape and structure

1. Sphericals: These are frequently discovered in toothpaste and exfoliating scrubs, among other personal care items. They are generally created during the production process. Although their uniform shape makes them easier to employ in industrial settings, it also increases the risk of their dispersing into the surrounding area.
2. Fibers: During washing, synthetic textiles like nylon and polyester lose fibers that subsequently turn into MPs contaminants. These long, thin strands, which frequently resemble threads, can build up in aquatic systems and endanger marine life by being ingested.
3. Fragments: The physical deterioration of bigger plastic objects produces these asymmetrically formed bits. Angular and uneven MPs result from the fragmentation process, which is fueled by mechanical weathering, UV radiation, and chemical deterioration. The fact that various creatures may consume these particles makes them very dangerous.
4. Films: Microplastic films are created when thin, flexible plastic fragments break apart. These fragments are frequently made of shopping bags or packaging materials. They are extremely vulnerable to environmental weathering processes because to their huge surface area relative to their volume.
5. Pellets: Also referred to as nurdles, they are tiny, spherical or rounded plastic beads that are utilized as raw materials while making plastic. They are easy to handle during manufacture because of their uniform size, but they are also easily spilled.

3 Bio-monitoring

Monitoring of MPs is crucial for understanding their long-term effects and determining mitigation strategies. The economic, environmental, and user-friendly benefits of biomonitoring make it one of the most popular monitoring techniques.

Since MPs are ubiquitous in ecosystems and food chains, monitoring their presence in living things aids in evaluating possible hazards such toxicity, inflammation, human endocrine disruption. In addition to its detrimental impacts on biodiversity. It also helps in determining the sources of contamination and comprehending the environmental movement of MPs.

3.1 Bio-indicators selection criteria

Choosing the right bioindicators is a crucial step for ensuring that the species selected offer significant information on the sources, impacts, and spread of microplastic contamination. Selecting the right sentinel species for each ecosystem is crucial for monitoring MP pollution in both terrestrial and aquatic settings. Because of ecological features, species habits, and exposure pathways, the selection criteria for sentinel species in these habitats may overlap but also differ. Key factors in terrestrial ecosystems include a species' susceptibility to MPs exposure depending on how it uses its habitat and how it feeds. For instance, herbivorous species, burrowing mammals, and urban-dwelling birds may be more prone to come into contact with MPs in soil or plants. Sampling viability is another essential factor, considering elements such as species accessibility, population density, and ease of monitoring in different habitats. According to Hilty and Merenlender (2000) and Goodsell et al. (2009), a comprehensive bioindicator for environmental monitoring must meet the following requirements: a broad range of distribution, a well-established biology, immobility, the capacity to provide an early warning, a crucial role in the ecosystem, a uniform reaction to contaminants, and the presence of observable toxic effects correlated with the level of pollution. The major factors for selection are as follows:

1. Trophic Position:

   The selection of species for biomonitoring MPs pollution heavily depends on the trophic position of species, which affects the way MPs are consumed, collected, and moved along the food chain. Lower trophic level organisms are effective for identifying the amount of MPs in the environment since they are likely to consume them straight from the water or sediment. These species can offer important information on the level of microplastic contamination in their environments. Top predators can help in understanding the broader ecological effects of plastic pollution, underscoring the possible threats that microplastic buildup poses to ecosystem health and food security.

2. Sampling ease and accessibility:

   It is a critical factor for the selection of bioindicator species, because it affects the viability, affordability, and accuracy of long-term environmental assessments. Since, assessing environmental changes over time requires regular and repeated monitoring, which is made possible by species that are abundant, widely distributed, and easy to sample.

3. Sensitivity to Pollutants:

   The best species for identifying the effects of microplastic contamination are those that show distinct biological reactions to environmental contaminants. For example, animals that have sensitive immune or physiological reactions to stresses, such as toxins or foreign particles, can offer early indicators of environmental damage brought on by MPs.

4. Life Cycle and Longevity:

   Species' life cycle and longevity are other critical factors for biomonitoring of MPs. Species with shorter life cycles can be helpful for tracking acute contamination events, while long-lived species can offer insights into the long-term impacts of microplastic exposure. Species with longer life cycles and multiple stages can also be useful for investigating the persistence and buildup of MPs over time.

5. Ecological significance:

Ecologically important species (like top predators and keystone species) and economically valuable species (like commercially fished species like tuna or shellfish) are given priority for biomonitoring because of the direct impact that their health and contamination levels have on ecosystem function and human livelihoods. Furthermore, since they are often linked to important ecological processes like the nutrient cycle, bioindicator species with ecological relevance can reveal long-term ecosystem trends.

6. Distribution of species:

An important consideration for choosing bioindicator species for microplastic monitoring is their wide distribution, which ensures that the species may be utilized for measuring microplastic pollution in various ecosystems and geographical areas. Widely dispersed species, such as filter-feeding invertebrates, tiny fish, or specific seabirds, are exposed to MPs from various environmental sources, providing important information about the degree of contamination in freshwater, marine, and terrestrial habitats.

4 Bio-indicators for microplastics

4.1 Aquatic organisms

Several aquatic organisms have been found to absorb MPs through various processes, such as feeding, drinking, respiration, and absorption in both inland and marine environments. These organisms include algae, macrophytes, zooplankton, insects, crustaceans, mollusks, fish, amphibians, birds, and mammals (Multisanti et al., 2022).

Vertebrates

Fishes are the common vertebrates used for biomonitoring of MPs. Most likely due to their increased exposure to MPs. MPs can indeed enter fish bodies through a variety of routes, including the gills and food and water consumption. The initial route by which MPs enter fish bodies is by feeding, and the kind of feeding has an impact on the capture of MPs. According to the findings of a study by Li et al. (2021), fish inadvertently consume MPs regardless of how they feed. Bivalves and crustaceans, which are members of the Mollusca and Arthropoda groups, are the primary organisms utilized for MPs biomonitoring (Proki´c et al., 2021). Because of their biological traits, including their broad distribution, high tolerance to various environmental factors and feeding type (filtration), these two classes of organisms are frequently employed as bioindicators for tracking pollution trends in waters (Beyer et al., 2017).

In this regard, Mytilus species are thought to be the best sentinel for marine biomonitoring research.. Their presence in temperate coastal waters worldwide and their sessile nature, which allows them to provide location-specific information, are the main factors contributing to their appropriateness. Additionally, they are hardy organisms that are simple to maintain in culture, forming (often large) mussel beds in shallow waters from which they can be readily collected, and they are medium-sized (one individual provides an adequate amount of tissue for analysis). These characteristics make them appropriate for both translocation and caging exposure in situ and laboratory exposure studies (Li et al., 2019). They primarily use their huge ciliated gills to pump and filter large amounts of water to filter-feed on phytoplankton. They effectively absorb pollutant compounds from the seawater due to their seawater filtration characteristic, which also gives them an integrative assessment of the concentration and bioavailability of seawater pollutants in situ. Moreover, the blue mussels serve as primary consumers and act as conduits for the movement of MPs from the abiotic phase to the higher trophic levels in the marine food chain, which further contributes to their ecological significance.

Macroinvertebrates

Macroinvertebrates are highly significant as bioindicators for evaluating MP pollution in the water column and freshwater sediment due to their prevalence in many ecological niches. Apart from the fact that aquatic macroinvertebrates consume MPs, a number of epibenthic insects, including caddis fly (*Trichoptera*) species in freshwater settings, may incorporate MPs into their larval cases or biological structures (Ehlers et al., 2019). Thus, larval cases produced by aquatic insects could be used as bioindicators for MPs and in their evaluation in freshwater systems.

Benthic macroinvertebrates may be able to consume MPs from sludge, reflecting the MP loads in their bodies. Chironomus sp. larvae are a perfect organism for evaluating MP contamination in various aquatic environments, (Bere et al., 2016; Nhiwatiwa et al., 2017). Khdre et al. (2023) studied MPs in the upper Egypt wastewater stream and found that *Chironomus* sp. has an abundance of blue-coloured MPs. This could indicate that larvae like this color, which could be similar to what they eat. This indicates that this species can be used as bioindicator for MPs in polluted freshwater ecosystems. In another study, Álvarez Troncoso et al. (20244) found microplastics of different sizes, polymer types and shapes in freshwater aquatic invertebrates such as Odonata bodies (*Gomphidae*) and Trichoptera cases (*Leptoceridae* and *Limnephilidae*), therefore they might be used as MP bioindicators in freshwater ecosystem.

Invertebrates

The crustaceans *Ocypode quadrata*, *Atlantorchestoidea brasiliensis*, *Excirolana braziliensis*, and *Emerita brasiliensis* can be utilized as bioindicators for monitoring the pollutants such as heavy metals, MPs because they are susceptible to anthropogenic effect (Cardoso et al., 2016; Schlacher et al., 2016). According to Cau et al. (2020), *Norwegian langoustine* is a reliable bioindicator of deep-sea MP pollution. MPs of relatively larger size are found in the stomach as compared to the intestine. Additionally, it was noted that the stomach functions as a size bottleneck for MPs that are consumed, improving the accumulation of bigger particles there and encouraging their disintegration into tiny plastic debris that is subsequently expelled in the intestine.

The red swamp crayfish Procambarus clarkii is another organism which has the potential to act as the bioindicators of MPs. It may colonize a variety of settings, including brackish and lagoon ecosystems, due to its biological and behavioral adaptability and dispersal tendencies (Dörr et al., 2020). Due to its ability to withstand extremely contaminated habitats, this invasive species has been used as a bioindicator of environmental contamination from organic compounds and heavy metals (Goretti et al., 2016; Mistri et al., 2020). Pastorino et al. (2023) studied the sediment, water and crayfish samples and detected the same microplastic objects across all three evaluations, suggesting that red swamp crayfish could be useful markers of microplastic contamination. Because of their feeding patterns, the crayfish with the highest weight had the fewest MPs, indicating that smaller members of the species make better indicators than larger ones. Therefore, Crayfish may be a viable option to study MPs incidence and temporal variations in a variety of aquatic habitats.

The crustaceans, known as barnacles, are a dominant group in some rocky coastal areas. They are regarded as excellent biomonitoring organisms because of their easy accessibility, sessile adult life, and broad distribution (Vaezzadeh et al., 2021). Additionally, they can accumulate MPs; therefore, these may be used as bioindicators for MPs, which could help identify the sources of MPs (Xu et al., 2020).

4.2 Terrestrial organisms

MPs are becoming a global emerging pollutant and a threat to biodiversity in both terrestrial and aquatic ecosystems (Zang et al., 2020). They can also cause negative ecological surprises (Baho et al., 2021) by altering basic physiological and ecosystem processes in terrestrial vertebrates exposed to plastic pollution levels (de Souza Machado et al., 2018). A growing number of terrestrial species are being employed as bioindicators of MPs pollution, revealing important information about the prevalence and impacts of MPs in terrestrial ecosystems. Terrstrial speces such as small mammals, birds and plants can be excellent choices for MPs monitoring because these species are exposed to MPs through contaminated air, soil and food. Therefore, these can serve as indicators of the wider environmental effects of plastic pollution on terrestrial habitats and aid in monitoring the presence and buildup of MPs in soil, air, and plants.

Since earthworms are frequently found in soil and are sensitive to some external pollutants, they are considered to be bioindicators to assess the level of soil pollution (Fourie et al., 2007). Recent research has demonstrated that MPs cause ecotoxicity effects on earthworms, such as immunological stress response (Zhao et al., 2023), spermatogenesis inhibition (Kwak & An, 2021). Trakic et al. (2024) have demonstrated that MPs were able to enter earthworm bodies and, consequently, food webs through non-selective feeding. These studies clearly indicate that earthworm has the potential to act as bioindicators for MPs pollution in the soil.

Birds have been employed as bioindicators of several important pollutants, including flame retardants (Tongue et al., 2019), and heavy metals (Rashid et al., 2021). These are useful bioindicators due to their wide range, participation in a variety of trophic levels, and intricate food webs that frequently contain bioaccumulated contaminants (Egwumah et al., 2017).

Another significant bioindicator species of the effects of MPs on terrestrial flora is wheat (*Triticum aestivum* L.). Qi et al. (2018), have reported that wheat exposed to sandy soil containing low-density polyethylene and a kind of biodegradable plastic film based on starch showed decreased root growth, plant height, and seed germination. Furthermore, MPs were found in the plant's aerial portions, suggesting that this species has absorbed them.

Similar to this, a variety of terrestrial biota were employed as bioindicators for microplastic contamination, including *Lumbricus terrestris, Porcellio, Armadillo, Eobania vermiculata, Rumina decollata,* and *Scolopendra.*

5 Conclusion

Microplastic (MPs) pollution is a serious and expanding environmental hazard that has broad effects on ecosystem health and human welfare. Considering the pervasiveness of MNP contamination and its possible effects, efficient monitoring techniques are required. Conventional approaches of MPs contamination monitoring, are useful in some situations, but fall short in of offering a comprehensive, economical, and environmentally sound way to keep tabs on this ubiquitous contaminant. Therefore, investigating biological systems—bioindicators—as instruments for tracking microplastic pollution may be a promising way to close this gap. This paper highlights the potential of a variety of species, from terrestrial species like earthworms and birds to marine invertebrates, to function as bioindicators of MPs contamination. These creatures offer a vital way to measure pollution levels in both aquatic and terrestrial ecosystems because they interact with MPs through ingestion, filtration, or absorption. Nevertheless, there are certain difficulties with using bioindicators. Therefore, finding appropriate bioindicators for monitoring MPs contamination in various ecosystems will require more extensive research.

6 Recommendations

Future research should focus on identifying and standardizing bioindicators across diverse ecosystems to ensure reliable microplastic monitoring. Studies should assess the bioaccumulation and toxicological effects of microplastics in different organisms, considering factors like size, polymer type, and trophic level. Advanced analytical techniques, such as spectroscopy and molecular biomarkers, should be explored for accurate detection and quantification. Policy planners and decision-makers should establish guidelines for selecting bioindicator species based on ecological relevance and sensitivity to microplastic exposure. Stakeholders should promote interdisciplinary collaborations between ecologists, toxicologists, and policymakers to integrate bioindicator-based monitoring into national pollution control programs. Additionally, public awareness campaigns should emphasize the role of bioindicators in assessing microplastic pollution, fostering community participation in data collection and mitigation efforts.

Acknowledgement VM and WS thank Vice-Chancellor SGT University for providing financial support and facilities for the work.

References

Abelouah, M. R., Ben-Haddad, M., Hajji, S., De-la-Torre, G. E., Aziz, T., Abou Oualid, J., & Alla, A. A. (2022a). Floating microplastics pollution in the Central Atlantic Ocean of Morocco: Insights into the occurrence, characterization, and fate. *Marine Pollution Bulletin, 182*, 113969.

Ali, I., Cheng, Q., Ding, T., Yiguang, Q., Yuechao, Z., Sun, H., & Liu, J. (2021). Micro-and nanoplastics in the environment: Occurrence, detection, characterization and toxicity–a critical review. *Journal Cleaner Production, 313*, Article 127863.

Álvarez, Troncoso., M. R., Gutierrez Rial, D., Villar Comesaña, I., Ehlers, S. M., Soto González, B., Mato de la Iglesia, S., Garrido González, J. (2024). Microplastics in water, sediments and macroinvertebrates in a small river of NW Spain. *Limnetica.*

Amato-Lourenço, L. F., dos Santos Galvão, L., de Weger, L. A., Hiemstra, P. S., Vijver, M. G., & Mauad, T. (2020). An emerging class of air pollutants: potential effects of microplastics to respiratory human health? *Science of the Total Environment, 749*, Article 141676.

Andrady, A. L. (2011). Microplastics in the marine environment. *Marine Pollution Bulletin, 62*(8), 1596–1605.

Baho, D. L., Bundschuh, M., & Futter, M. N. (2021). Microplastics in terrestrial ecosystems: Moving beyond the state of the art to minimize the risk of ecological surprise. *Global Change Biology, 27*(17), 3969–3986.

Barnes, D. K., Galgani, F., Thompson, R. C., & Barlaz, M. (2009). Accumulation and fragmentation of plastic debris in global environments. *Philosophical Transactions of the Royal Society b: Biological Sciences, 364*(1526), 1985–1998.

Benson, N. U., Agboola, O. D., Fred-Ahmadu, O. H., De-la-Torre, G. E., Oluwalana, A., & Williams, A. B. (2022). Micro (nano) plastics prevalence, food web interactions, and toxicity assessment in aquatic organisms: a review. *Frontiers in Marine Science, 9*, Article 851281.

Bere, T., Dalu, T., & Mwedzi, T. (2016). Detecting the impact of heavy metal contaminated sediment on benthic macroinvertebrate communities in tropical streams. *Science of the Total Environment, 572*, 147–156.

Bermúdez, J. R., & Swarzenski, P. W. (2021). A microplastic size classification scheme aligned with universal plankton survey methods. *MethodsX, 8*, Article 101516.

Bermúdez, J. R., Metian, M., Oberhänsli, F., Taylor, A., & Swarzenski, P. W. (2021). Preferential grazing and repackaging of small polyethylene microplastic particles ($\leq$ 5 μm) by the ciliate Sterkiella sp. *Marine Environmental Research, 166*, Article 105260.

Besseling, E., Quik, J. T., Sun, M., & Koelmans, A. A. (2017). Fate of nano-and microplastic in freshwater systems: A modeling study. *Environmental Pollution, 220*, 540–548.

Beyer, J., Green, N. W., Brooks, S., Allan, I. J., Ruus, A., Gomes, T., & Schøyen, M. (2017). Blue mussels (Mytilus edulis spp.) as sentinel organisms in coastal pollution monitoring: a review. *Marine Environmental Research, 130*, 338–365.

Bostan, N., Ilyas, N., Akhtar, N., Mehmood, S., Saman, R. U., Sayyed, R. Z., & Pandiaraj, S. (2023). Toxicity assessment of microplastic (MPs); a threat to the ecosystem. *Environmental Research*, 116523.

Cardoso, R. S., Barboza, C. A., Skinner, V. B., & Cabrini, T. M. (2016). Crustaceans as ecological indicators of metropolitan sandy beaches health. *Ecological Indicators, 62*, 154–162.

Cau, A., Avio, C. G., Dessì, C., Moccia, D., Pusceddu, A., Regoli, F., & Follesa, M. C. (2020). Benthic crustacean digestion can modulate the environmental fate of microplastics in the deep sea. *Environmental Science & Technology, 54*(8), 4886–4892.

Chang, X., Fang, Y., Wang, Y., Wang, F., Shang, L., & Zhong, R. (2022). Microplastic pollution in soils, plants, and animals: a review of distributions, effects and potential mechanisms. *Science of the Total Environment, 850*, Article 157857.

Chen, X., Yu, X., Zhang, L., Zhao, W., & Sui, Q. (2023). Organic pollutants adsorbed on microplastics: Potential indicators for source appointment of microplastics. *Journal of hazardous materials*, 133225.

Cheng, Y., Song, W., Tian, H., Zhang, K., Li, B., Du, Z., & Zhu, L. (2021). The effects of high-density polyethylene and polypropylene microplastics on the soil and earthworm Metaphire guillelmi gut microbiota. *Chemosphere, 267*, Article 129219.

Cox, K. D., Covernton, G. A., Davies, H. L., Dower, J. F., Juanes, F., & Dudas, S. E. (2019). Human consumption of microplastics. *Environmental Science & Technology, 53*(12), 7068–7074.

de Souza Machado, A. A., Kloas, W., Zarfl, C., Hempel, S., & Rillig, M. C. (2018). Microplastics as an emerging threat to terrestrial ecosystems. *Global Change Biology, 24*(4), 1405–1416.

Dörr, A. J. M., Scalici, M., Caldaroni, B., Magara, G., Scoparo, M., Goretti, E., & Elia, A. C. (2020). Salinity tolerance of the invasive red swamp crayfish Procambarus clarkii (Girard, 1852). *Hydrobiologia, 847*, 2065–2081.

Egwumah, F. A., Egwumah, P. O., & Edet, D. I. (2017). Paramount roles of wild birds as bioindicators of contamination. *Int J Avian & Wildlife Biol, 2*(6), 00041.

Ehlers, S. M., Manz, W., & Koop, J. H. (2019). Microplastics of different characteristics are incorporated into the larval cases of the freshwater caddisfly Lepidostoma basale. *Aquatic Biology, 28*, 67–77.

Eriksen, M., Lebreton, L. C., Carson, H. S., Thiel, M., Moore, C. J., Borerro, J. C., & Reisser, J. (2014). Plastic pollution in the world's oceans: more than 5 trillion plastic pieces weighing over 250,000 tons afloat at sea. *PLoS ONE, 9*(12), Article e111913.

Esposito, G., Prearo, M., Renzi, M., Anselmi, S., Cesarani, A., Barcelò, D., & Pastorino, P. (2022). Occurrence of microplastics in the gastrointestinal tract of benthic by–catches from an eastern Mediterranean deep–sea environment. *Marine Pollution Bulletin, 174*, Article 113231.

Fourie, F., Reinecke, S. A., & Reinecke, A. J. (2007). The determination of earthworm species sensitivity differences to cadmium genotoxicity using the comet assay. *Ecotoxicology and Environmental Safety, 67*(3), 361–368.

Frank, Y., Vorobiev, E., Babkina, I., Antsiferov, D., & Vorobiev, D. (2020). Microplastics in fish gut, first records from the tom river in west siberia, russia. *Vestnik Tomskogo Gosudarstvennogo Universiteta Biologiya*, (52), 130–139. https://doi.org/10.17223/19988591/52/7

Fuller, S., & Gautam, A. (2016). A procedure for measuring microplastics using pressurized fluid extraction. *Environmental Science & Technology, 50*(11), 5774–5780.

Geyer, R., Jambeck, J. R., & Law, K. L. (2017). Production, use, and fate of all plastics ever made. *Science Advances, 3*(7), Article e1700782.

Goodsell, P. J., Underwood, A. J., & Chapman, M. G. (2009). Evidence necessary for taxa to be reliable indicators of environmental conditions or impacts. *Marine Pollution Bulletin, 58*(3), 323–331.

Goretti, E., Pallottini, M., Ricciarini, M. I., Selvaggi, R., & Cappelletti, D. (2016). Heavy metals bioaccumulation in selected tissues of red swamp crayfish: An easy tool for monitoring environmental contamination levels. *Science of the Total Environment, 559*, 339–346.

Hartmann, N. B., Huffer, T., Thompson, R. C., Hassellov, M., Verschoor, A., Daugaard, A. E., & Wagner, M. (2019). Are we speaking the same language? Recommendations for a definition and categorization framework for plastic debris.

Hilty, J., & Merenlender, A. (2000). Faunal indicator taxa selection for monitoring ecosystem health. *Biological Conservation, 92*(2), 185–197.

Horton, A. A., Walton, A., Spurgeon, D. J., Lahive, E., & Svendsen, C. (2017). Microplastics in freshwater and terrestrial environments: Evaluating the current understanding to identify the knowledge gaps and future research priorities. *Science of the Total Environment, 586*, 127–141.

Hu, Y., Gong, M., Wang, J., & Bassi, A. (2019). Current research trends on microplastic pollution from wastewater systems: a critical review. *Reviews in Environmental Science and Bio/technology, 18*, 207–230.

Jambeck, J. R., Geyer, R., Wilcox, C., Siegler, T. R., Perryman, M., Andrady, A., & Law, K. L. (2015). Plastic waste inputs from land into the ocean. *science, 347*(6223), 768–771.

Khdre, A. M., Ramadan, S. A., Ashry, A., & Alaraby, M. (2023). Chironomus sp. As a bioindicator for assessing microplastic contamination and the heavy metals associated with it in the sediment of wastewater in Sohag Governorate, Egypt. *Water, Air, & Soil Pollution, 234*(3), 161.

Kumar, M., Xiong, X., He, M., Tsang, D. C., Gupta, J., Khan, E., & Bolan, N. S. (2020). Microplastics as pollutants in agricultural soils. *Environmental Pollution, 265*, Article 114980.

Kumari, S., Yadav, D., Yadav, S., Selvaraj, M., Sharma, G., Karnwal, A., & Yadav, S. (2025). From macro to micro: The key parameters influencing the degradation mechanism and the toxicity of microplastics in the environment. *Polymer Degradation and Stability*, 111174.

Kwak, J. I., & An, Y. J. (2021). Microplastic digestion generates fragmented nanoplastics in soils and damages earthworm spermatogenesis and coelomocyte viability. *Journal of Hazardous Materials, 402*, Article 124034.

Li, J., Lusher, A. L., Rotchell, J. M., Deudero, S., Turra, A., Bråte, I. L. N., & Shi, H. (2019). Using mussel as a global bioindicator of coastal microplastic pollution. *Environmental Pollution, 244*, 522–533.

Li, B., Liang, W., Liu, Q. X., Fu, S., Ma, C., Chen, Q., & Shi, H. (2021). Fish ingest microplastics unintentionally. *Environmental Science & Technology, 55*(15), 10471–10479.

Liu, S. L., Jian, M. F., Zhou, L. Y., Li, W. H., Wu, X. E., & Rao, D. (2019). Pollution characteristics of microplastics in migratory bird habitats located within Poyang Lake wetlands. *Huan jing ke xue= Huanjing kexue, 40*(6), 2639–2646.

López-Pedrouso, M., Varela, Z., Franco, D., Fernández, J. A., & Aboal, J. R. (2020). Can proteomics contribute to biomonitoring of aquatic pollution? *A Critical Review. Environmental Pollution, 267*, Article 115473.

McCormick, A., Hoellein, T. J., Mason, S. A., Schluep, J., & Kelly, J. J. (2014). Microplastic is an abundant and distinct microbial habitat in an urban river. *Environmental Science & Technology, 48*(20), 11863–11871.

Mistri, M., Munari, C., Pagnoni, A., Chenet, T., Pasti, L., & Cavazzini, A. (2020). Accumulation of trace metals in crayfish tissues: is Procambarus clarkii a vector of pollutants in Po Delta inland waters? *The European Zoological Journal, 87*(1), 46–57.

Multisanti, C. R., Merola, C., Perugini, M., Aliko, V., & Faggio, C. (2022a). Sentinel species selection for monitoring microplastic pollution: A review on one health approach. *Ecological Indicators, 145*, Article 109587.

Napper, I. E., & Thompson, R. C. (2016). Release of synthetic microplastic fibres from domestic washing machines: A case study of domestic laundry. *Environmental Science and Technology, 50*(21), 11532–11538.

Nhiwatiwa, T., Brendonck, L., & Dalu, T. (2017). Understanding factors structuring zooplankton and macroinvertebrate assemblages in ephemeral pans. *Limnologica, 64*, 11–19.

Pastorino, P., Anselmi, S., Zanoli, A., Esposito, G., Bondavalli, F., Dondo, A., & Prearo, M. (2023). The invasive red swamp crayfish (Procambarus clarkii) as a bioindicator of microplastic pollution: Insights from Lake Candia (northwestern Italy). *Ecological Indicators, 150*, Article 110200.

Pastorino, P., Elia, A. C., Pizzul, E., Bertoli, M., Renzi, M., & Prearo, M. (2024). The old and the new on threats to high-mountain lakes in the Alps: A comprehensive examination with future research directions. *Ecological Indicators, 160*, Article 111812.

Porta, R. (2019). The plastics sunset and the bio-plastics sunrise. *Coatings, 9*(8), 526.

Prokić, M. D., Gavrilović, B. R., Radovanović, T. B., Gavrić, J. P., Petrović, T. G., Despotović, S. G., & Faggio, C. (2021). Studying microplastics: Lessons from evaluated literature on animal model organisms and experimental approaches. *Journal of Hazardous Materials, 414*, Article 125476.

Qi, Y., Yang, X., Pelaez, A. M., Lwanga, E. H., Beriot, N., Gertsen, H., & Geissen, V. (2018). Macro-and micro-plastics in soil-plant system: effects of plastic mulch film residues on wheat (Triticum aestivum) growth. *Science of the Total Environment, 645*, 1048–1056.

Rashid, B., Zubair, M., Irshad, N., Ahmad, K. S., Mahmood, M., Tahir, M. M., & Shaheen, A. (2021). Correction to: biomonitoring of toxic metals in feathers of birds from North-Eastern Pakistan. *Bulletin of Environmental Contamination and Toxicology, 106*(6), 909.

Rochman, C. M., Hoh, E., Kurobe, T., & Teh, S. J. (2013). Ingested plastic transfers hazardous chemicals to fish and induces hepatic stress. *Scientific Reports, 3*(1), 1–7.

Schirinzi, G. F., Pérez-Pomeda, I., Sanchís, J., Rossini, C., Farré, M., & Barceló, D. (2017). Cytotoxic effects of commonly used nanomaterials and microplastics on cerebral and epithelial human cells. *Environmental Research, 159*, 579–587.

Schlacher, T. A., Lucrezi, S., Connolly, R. M., Peterson, C. H., Gilby, B. L., Maslo, B., & Schoeman, D. S. (2016). Human threats to sandy beaches: A meta-analysis of ghost crabs illustrates global anthropogenic impacts. *Estuarine, Coastal and Shelf Science, 169*, 56–73.

Schwarzer, M., Brehm, J., Vollmer, M., Jasinski, J., Xu, C., Zainuddin, S., & Laforsch, C. (2022). Shape, size, and polymer dependent effects of microplastics on Daphnia magna. *Journal of Hazardous Materials, 426*, Article 128136.

Shekhar, A. R., Kumar, A., Syamsai, R., Cai, X., & Pol, V. G. (2022). Is the plastic pandemic a greater threat to humankind than COVID-19? *ACS Sustainable Chemistry & Engineering, 10*(10), 3150–3154.

Su, L., Xue, Y., Li, L., Yang, D., Kolandhasamy, P., Li, D., & Shi, H. (2016). Microplastics in taihu lake, China. *Environmental Pollution, 216*, 711–719.

Thompson, R. C. (2009). Structuring policy problems for plastics, the environment and human health. *Our Plastic Age, 364*(1526), 1973–1976.

Tongue, A. D., Reynolds, S. J., Fernie, K. J., & Harrad, S. (2019). Flame retardant concentrations and profiles in wild birds associated with landfill: A critical review. *Environmental Pollution, 248*, 646–658.

Trakić, T., Popović, F., Sekulić, J., & Hackenberger, D. K. (2024). Ecotoxicological Effects of Commercial Microplastics on Earthworm Eisenia fetida (Savigny, 1826)(Clitellata; Lumbricidae). *Agriculture, 14*(2), 267.

Vaezzadeh, V., Thomes, M. W., Kunisue, T., Tue, N. M., Zhang, G., Zakaria, M. P., & Bong, C. W. (2021). Examination of barnacles' potential to be used as bioindicators of persistent organic pollutants in coastal ecosystem: A Malaysia case study. *Chemosphere, 263*, Article 128272.

Wang, W., Yuan, W., Chen, Y., & Wang, J. (2018). Microplastics in surface waters of Dongting lake and Hong lake, China. *Science of the Total Environment, 633*, 539–545.

Wang, S., Zhang, C., Pan, Z., Sun, D., Zhou, A., Xie, S., & Zou, J. (2020). Microplastics in wild freshwater fish of different feeding habits from Beijiang and Pearl River Delta regions, south China. *Chemosphere, 258*, Article 127345.

Wang, C., Wang, L., Ok, Y. S., Tsang, D. C., & Hou, D. (2022). Soil plastisphere: exploration methods, influencing factors, and ecological insights. *Journal of Hazardous Materials, 430*, Article 128503.

Wright, S. L., & Kelly, F. J. (2017). Plastic and human health: a micro issue? *Environmental Science & Technology, 51*(12), 6634–6647.
Wright, S. L., Ulke, J., Font, A., Chan, K. L. A., & Kelly, F. J. (2020). Atmospheric microplastic deposition in an urban environment and an evaluation of transport. *Environment International, 136*, Article 105411.
Xu, X. Y., Wong, C. Y., Tam, N. F. Y., Liu, H. M., & Cheung, S. G. (2020). Barnacles as potential bioindicator of microplastic pollution in Hong Kong. *Marine Pollution Bulletin, 154*, Article 111081.
Zalasiewicz, J., Waters, C. N., Do Sul, J. A. I., Corcoran, P. L., Barnosky, A. D., Cearreta, A., & Yonan, Y. (2016). The geological cycle of plastics and their use as a stratigraphic indicator of the Anthropocene. *Anthropocene, 13*, 4–17.
Zang, H., Zhou, J., Marshall, M. R., Chadwick, D. R., Wen, Y., & Jones, D. L. (2020). Microplastics in the agroecosystem: are they an emerging threat to the plant-soil system? *Soil Biology and Biochemistry, 148*, Article 107926.
Zhang, W., Zhang, S., Wang, J., Wang, Y., Mu, J., Wang, P., & Ma, D. (2017). Microplastic pollution in the surface waters of the Bohai Sea, China. *Environmental Pollution, 231*, 541–548.
Zhao, Y., Jia, H., Deng, H., Xing, W., Feng, D., Li, J., & Chen, H. (2023). Response of earthworms to microplastics in soil under biogas slurry irrigation: Toxicity comparison of conventional and biodegradable microplastics. *Science of the Total Environment, 858*, Article 160092.

Assessment of Potential Ecological Risk of Microplastics in India

Manish Pant, Naveen Singhal, and Jabrinder Singh

Abstract The prevalence of microplastics (MPs), identified as plastic particles of smaller- size (<5 mm), poses a significant global environmental concern, particularly for aquatic ecosystems. Amidst India's ongoing battle with plastic pollution, addressing the ecological impact of MPs becomes imperative. This chapter provides a comprehensive exploration of MPs pollution in India, encompassing both terrestrial and marine environments. Although India is widely recognized as a major source of marine plastic pollution, research on MPs contamination in its freshwater habitats remains sparse. Beginning with a detailed overview of MPs, including their origins, diversity, and persistence, the chapter scrutinizes the current state of MPs contamination across India's diverse environmental landscapes. With a focus on understanding MPs pollution in freshwater systems, the chapter offers cutting-edge insights into MPs abundance, composition, and associated ecological risks. Furthermore, it presents a detailed review of MPs studies conducted in India over the past two decades, highlighting the potential genetic effects of indirect MPs intake on human health, like sterility, obesity, and chronic cancer. To mitigate environmental pollution risks, the chapter emphasizes the necessity of controlling the overuse of plastics and the implementation of laws and strategies aimed at curtailing the sources of plastic waste. In conclusion, the chapter advocates for the adoption of sustainable management practices to effectively mitigate the ecological risks posed by MPs in India.

Keywords Micro-plastics · India · Freshwater · Pollution · FTIR · Ecological risk

M. Pant (✉) · N. Singhal · J. Singh
Department of Chemistry- SoPS, DIT University Dehradun, Dehradun Uttarakhand, India
e-mail: manishpant66@gmail.com

N. Singhal
e-mail: drnaveen.singhal@dituniversity.edu.in

J. Singh
e-mail: drjabrinder.singh@dituniversity.edu.in

P. Khyalia and S. Yadav (eds.), *Microplastic Pollution in India and Its Environmental Impacts*, Earth and Environmental Sciences Library,
https://doi.org/10.1007/978-3-031-92530-6_11

1 Introduction

The surge in plastic production has surged significantly due to various factors such as affordability, resistance to corrosion, robustness, and insulation properties (Gao et al., 2021). In 2021, Europe reported that the yearly output of plastic worldwide has reached approximately 367 million tonnes. Terrestrial sources are the primary contributors responsible for up to 80% of plastic litter. Additionally, plastic waste in marine environments is exacerbated by the aquaculture and fishing sectors (Radhakrishnan et al., 2021). Poor management of waste, in conjunction with human activities in marine and coastal regions, results in the infiltration of plastic particles into marine ecosystems. This poses a substantial risk to the survival of marine creatures (Veerasingam et al., 2016).

Plastic waste is divided into four primary size groups on a global scale: macroplastics (>25 mm), microplastics (<5 mm), mesoplastics (5–25 mm), and nanoplastics (<100 nm) (Robin et al., 2020; Yadav et al., 2023). Microplastics (MPs) have been detected across various ecosystems, from deep ocean environments to sediment in coastal regions like beaches and shorelines. In close proximity to densely populated zones, coastal areas commonly display notably increased concentrations of microplastics compared to rural counterparts. These microplastics exist in various polymer forms, like polypropylene (PP), polystyrene (PS), polyethylene (PE), polyvinyl chloride (PVC), polyethylene terephthalate (PET) and similar substances, originating from diverse sources and present in a variety of products (Koutnik et al., 2021).

MPs known for their diverse physical and chemical compositions, have been documented in various consumer goods beyond just soil and water sources. Studies have identified their presence in everyday items such as table salts (Syamsu et al., 2024), cold tea, and beer, soft drinks, energy drinks as well as in commonly consumed seafood like commercial fishes (Harikrishnan et al., 2023) and mussels, and packaged meat (Kedzierski et al., 2020). Furthermore, the presence of MPs particles has been detected in drinking water as well as beverages, alongwith several food sources, like commercial seaweed-nori and food containers (Guo et al., 2024). Additionally, personal care and cosmetic items have been identified as potential contributors to MPs pollution (Sun et al., 2020).

In India, research has primarily focused on microplastics on coastal regions, with less attention given to terrestrial water bodies (Amrutha et al., 2021). Nonetheless, multiple studies have revealed microplastic pollution in the sediment of rivers across the country (Ayyamperumal et al., 2022; Chauhan et al., 2021; Tsering et al., 2021). Furthermore, our comprehension of how microplastics interact with sedimentary and organic characteristics within river systems remains constrained. It is vital to bridge this knowledge gap, especially considering the susceptibility of Indian river systems to south-west monsoon heavy rains, which lead to flooding in catchment areas. The occurrence of floods aids in the transportation of a broad spectrum of materials, plastics of differing sizes, among other materials, into the Arabian Sea.

The widespread consequences of MPs contamination also endanger heritage sites (Khaleel et al., 2022).

Hence, it is crucial to prioritize efforts towards conducting an eco-toxicological menace assessment of microplastics (MPs) to fully grasp the potential threat they pose upon ingestion by marine life. In recent decades, the swift pace of industrialization, economic expansion, and climate shifts in India has resulted in the pollution of estuarine sediments by an array of contaminants, both organic and inorganic, resulting in significant environmental health challenges in coastline and oceanic ecosystems. In the past three years, several studies have delved into the abundance and polymer composition of MPs in diverse environmental settings across India (Veerasingam et al., 2020). However, there has been a notable oversight in conducting ecological risk assessments of MPs in sediments despite the wealth of available data. This research initiative embarks on a meta-analysis of existing studies to evaluate the potential ecological threats posed by MPs in both terrestrial and marine sediments throughout India.

2 Sources and Pathways of Microplastics

Microplastics (MPs) exhibit a diverse array of characteristics including different shapes, sizes, morphologies, polymer compositions, and densities (Wang et al., 2019). Based on their source, classified into two main groups: primary and secondary.

2.1 Primary Microplastics

Primary MPs originate primarily from two sources: industrial manufacturing processes and household activities, both of which directly introduce micro-sized plastic into the environment. Industrial operations such as pharmaceutical production, plastic blasting (such as thermal cutting), tire wear (such as from vehicles and aircraft), plastic pellet manufacturing, cosmetics production. Across terrestrial soils and freshwater ecosystems, the escalation of primary MPs levels is notably influenced by insect repellents (Guo & Wang, 2019). Likewise, household items which includes hand cleaners, shower gels, scrubs, nail polish, facial cleansers, shaving lotion, hair dyes, baby products, insect repellent, sunscreens, and toothpaste, among others, release microplastics which enter into the surroundings via wastewater/sewage treatment (Auta et al., 2017; Pant et al., 2025a). Another route for primary MPs to infiltrate the environment is through leaks from small manufacturing facilities. Plastic manufacturers utilize artificial resin pellets of uniform shape and 3–5 mm in diameter in plastic production, significantly contributing to microplastic pollution. However, MPs in freshwater sediment rarely contain microbeads or resin pellets, potentially due to the inherent preservation of microbeads in biosolids (Covernton et al., 2019).

2.2 Secondary Microplastics

Secondary microplastics consist of fibers produced from the breakdown of larger plastics on both land and in the ocean. These substances degrade over time, continuously releasing MPs into the environment through mechanisms like photodegradation induced by solar radiation, mechanical action from wind and water waves, chemical degradation under acidic or alkaline conditions, and biodegradation by microorganisms (Malankowska et al., 2021). Secondary MPs arise from numerous sources, each exhibiting diversity and abundance. A variety of factors, including climatic conditions (such as temperature, sunlight, and precipitation), as well as plastic characteristics (density & size), and chemical composition, influence the degradation of debris of microplastic. Exposure to UV light expedites the natural degradation of plastics by initiating polymer structural oxidation, ultimately breaking down plastic bonds (Wagner et al., 2014a, b). Alongside UV exposure, physical forces like turbulence and wind-induced abrasion is another mechanism for microplastic production (Barnes et al., 2009). Moreover, factors like the chemical composition of precipitation, soil as well as water pH levels, and substances released by micro-organisms all contribute to the oxidative breakdown of plastic polymers (Shim & Thompson, 2015). In the decomposition of large plastics into MPs, microorganisms play a vital role through various biochemical processes. Within terrestrial ecosystems, these processes entail surface abrasion, fragmentation, and oxidative deterioration of MPs, complicating efforts to trace their origins in different environmental settings (Zhou et al., 2020a). Additionally, the use of low-density plastics in agricultural practices such as mulching and greenhouse operations, aimed at enhancing soil quality and crop yields, contributes to the buildup of MPs in terrestrial eosystem (Huang et al., 2020). MPs originating from municipal solid waste and landfill sites function as secondary outlets, leading to soil and groundwater system contamination (Zhou et al., 2020b). The existence of MPs in aquatic ecosystems presents a significant threat to aquatic organisms due to their enduring properties, facilitating leaching, accumulation, and adsorption in the environmental framework, leading to adverse outcomes (Shim & Thompson, 2015). Upon entering the environment, MPs have the potential to degrade into nanoparticles, potentially heightening risks to ecosystems (Huang et al., 2020). Figure 1 shows sources and pathways of microplastics enters into environment.

3 Destiny and Pathway of Microplastics in Indian Environments

Microplastics (MPs) journey across vast expanses propelled by their distinctive characteristics, encompassing their shape, diminutive size, lightweight properties, buoyancy, and enduring nature, all influenced by the interplay of wind and water dynamics. Numerous scientific investigations have highlighted the pivotal role played by wastewater treatment plant (WWTP) effluent as a major conduit for the direct

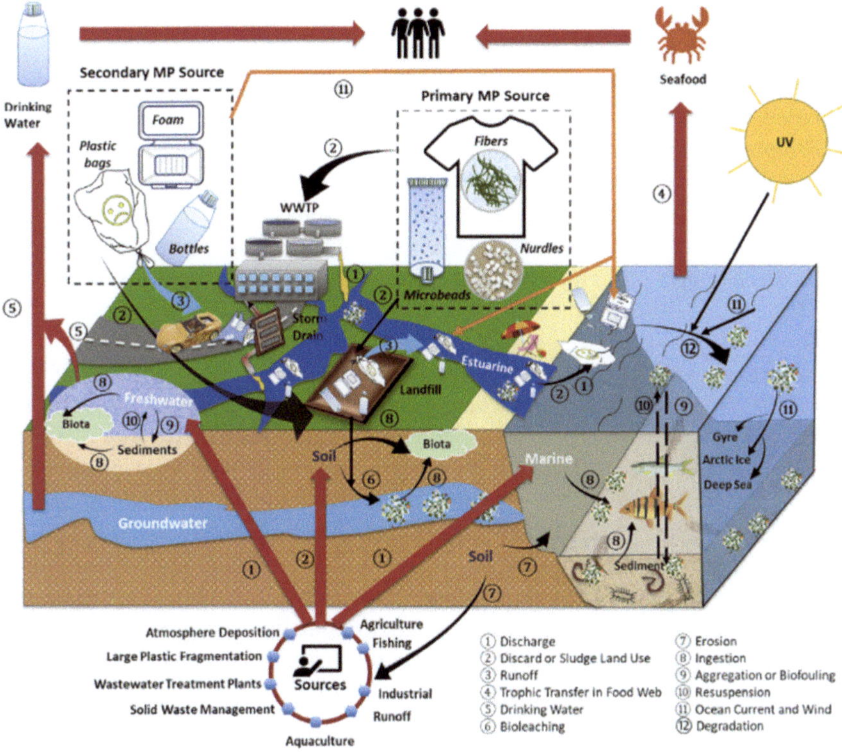

Fig. 1 Destiny and pathways of microplastics in various environmental systems (Reprinted from original article by Wu et al. (2019) with permission from Elsevier)

dissemination of MPs into both aquatic and terrestrial ecosystems (Karbalaei et al., 2018). Initially, within the WWTP process, larger MPs as well as other particles may become ensnared in mechanical screens and grit chambers. Subsequently, during the initial treatment phase, microplastic fibers settle due to the gravitational force, aided by processes such as coagulation and flocculation (Zhou et al., 2021). During the transition to the secondary phase of treatment, MPs interact with microbial communities, potentially impacting microbial activity and becoming ensnared in activated sludge. Consequently, the sludge produced by WWTPs emerges as yet another conceivable route for the transportation of MPs from WWTPs to soil ecosystems (Auta et al., 2017).

Plastic debris plays a crucial role in generating secondary MPs and facilitating their movement within soil as well as aquatic environment (Obbard et al., 2014). Over time, plastic waste decomposes into MPs, building up on the soil's surface. MPs penetrate deeper into the soil through agricultural practices, rainfall, soil fissures, and the activities of soil organisms (He et al., 2018). These routes facilitate the downward movement of MPs with water, eventually reaching groundwater. Soil organisms like earthworms, insects, nematodes, bacteria, fungi, and algae play crucial roles in the movement and fate of through processes such as ingestion, chemical reactions, and excretion. According to specific research findings, MPs might move from the digestive systems of soil organisms to their waste, potentially serving as an indirect origin of marine microplastics. Via processes like surface runoff and wind dispersal, traces of microplastics originating from soil infiltrate freshwater streams and marine ecosystems. MPs, in a range of forms, continually settle in soil or sediment after entering the atmosphere, potentially endangering human respiratory health. These MPs are widespread in marine environments, including beaches, surface waters, seabed sediments, and among various marine organisms like seabirds, fish, bivalves, mammals, and crustaceans (Rose et al., 2023). Distribution of MPs in various ecosystems, see Fig. 2.

Numerous research efforts have pointed out the widespread distribution of MPs worldwide, even in remote locales where human influence is minimal (Guo & Wang, 2019). MPs, owing to their small size and lightweight, can traverse vast distances via marine currents and atmospheric deposition, resulting in their widespread distribution across ecosystems. This occurrence has resulted in the amassing of MPs in substantial repositories such as polar ice and marine sediments (Obbard et al., 2014). Approximations indicate that roughly 92% of the 269 million tonnes of particles worldwide are composed of MPs. This highlights a notable buildup of microplastics in marine sediments, serving as a reservoir, while their surface presence remains relatively scarce (Auta et al., 2017). Nonetheless, the specific extent of microplastic penetration and endurance in the marine ecosystem since the advent of human activity is still unclear.

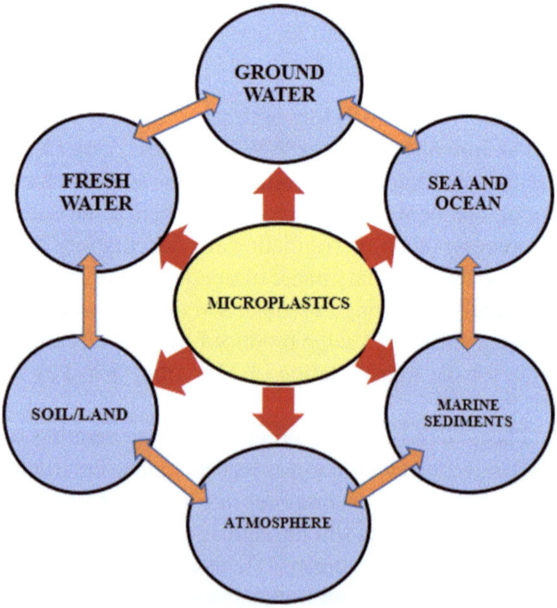

Fig. 2 transportation and presence of microplastics within the environment

4 Ecological Impacts of Microplastics in India

4.1 Impact of Microplastics on Terrestrial Species

Microplastic particles (MPs) can accumulate over time, persisting in ecosystems and exerting significant impacts on biodiversity. MPs have been shown to impact soil characteristics such as water retention capacity, bulk density, microbial activity, as well as soil aggregation (De Souza Machado et al., 2018). Plastic mulching practices also play a role in shaping the biogeochemistry of soil. The increasing prevalence of plastic consumption in everyday life leads to the deterioration and fragmentation of these materials, leading to the release of microplastics into terrestrial environments. Improperly managed plastic waste, tire abrasion, deterioration of construction materials, and the airborne dispersal of microplastics from diverse origins constitute sources of these particles. Terrestrial habitats bear a substantial burden of microplastic pollution, with contamination levels estimated to range from 4 to 23 times higher than those observed in oceans. Agricultural areas emerge as prominent repositories of microplastics, surpassing even oceanic basins in their accumulation (Horton et al., 2017; Bisen et al., 2025).

MPs make their way into the food chain and can eventually reach top-level consumers, potentially breaching the blood–brain barrier and causing behavioral changes or disorders depending on their size. MPs accumulation and distribution in tissues are shaped by their particle size. Animal studies have demonstrated that microplastics can trigger oxidative stress, disrupt energy and lipid metabolism, and induce neurotoxicity. The buildup of MPs carries various health risks, including cytotoxicity, hypersensitivity, adverse immune reactions, and acute responses such as hemolysis (Hwang et al., 2019).

4.2 Impact of Microplastics on Aquatic Environments

4.2.1 Freshwater Environment

Microplastic pollution is a pressing environmental issue recognized primarily in the world's oceans. However, research into freshwater microplastics (MPs) has been limited, despite freshwater being a vital drinking water source. While global efforts in this area are slowly advancing, there remains a notable lack of attention in India, as indicated by recent literature (Gopinath et al., 2020). This underscores the urgent need to evaluate microplastic contamination in India's freshwater systems. Through in-depth examinations of freshwater ecosystems, researchers can acquire essential knowledge about the origins, behaviors, and potential toxicities of MPs in these settings.

Freshwater lakes are fundamental as significant stores of freshwater, representing approximately 90% of the Earth's surface freshwater. However, escalating human activities have subjected these invaluable resources to direct threats such as heavy metal contamination, eutrophication, and several other forms of pollution. MPs, as emerging contaminants, represent a relatively recent addition to the array of pollutants affecting these water bodies. Recent studies of Indian lakes have reported that a significant contributor to these synthetic polymers resulted from the breakdown of larger plastic debris (Ganesan et al., 2019; Pant et al., 2024a).

Around 80% of plastic waste found in marine ecosystems is traced back to terrestrial origins, frequently carried by rivers (Wagner et al., 2014a, b). According to a recent investigation by Napper et al. (2021), it was approximated that the combined discharge of the Ganges, Meghna, and Brahmaputra rivers into the Bay of Bengal ranges from 1 to 3 billion microplastics per day. However, not all microplastics transported by rivers make their way to the oceans; a portion accumulates in river sediments, acting as significant reservoirs for these particles.

River water pollution presents a significant global issue, particularly pronounced in India, housing approximately 20 river basins (Central Water Commission, 2019). The river basins face numerous pollution challenges, encompassing heavy metals, pesticides as well as persistent organic pollutants (POPs), and an array of harmful biological and chemical substances. Ineffective waste management practices exacerbate the release of plastic waste into these rivers, with microplastics being of particular concern. The primary origins of microplastic contamination in Indian rivers arise from the fragmentation of larger plastic items, including packaging materials, ropes, fishing nets, wrappers, pipes, and synthetic textiles. Moreover, shifts in hydrodynamic and climatic factors, such as variations in wave height, flow velocity, and wind speed, play a role in dispersing these synthetic polymers across various sections of the water column (Amrutha & Warrier, 2020; Pant et al., 2024b; Singh et al., 2025).

In the current review of literature, only two studies have delved into the occurrence of MPs in groundwater within India (Ganesan et al., 2019; Selvam et al., 2020). These investigations employed distinct approaches for both sample collection and processing. Ganesan et al. (2019) opted for the direct collection of groundwater samples in glass bottles, followed by the isolation of MPs through laboratory-based filtration. Following this step, microplastics captured on the filter were analyzed utlilizing an optical microscope. In contrast, Selvam et al. (2020) opted for sample retrieval using a Teflon pump, conducting on-site filtration employing stainless steel sieves with a mesh size of 50 μm. Subsequently, microplastics were subjected to processing and extraction via the wet peroxide oxidation method, followed by the filtration of the oxidized solution. Afterwards, the filters were examined using a stereomicroscope to validate the presence of microplastics.

Additionally, within the same investigation conducted by Selvam et al. (2020), the study evaluated the adsorption capacities of various polymers, including polypropylene (PP), polyvinyl chloride (PVC), polyamide (PA), cellulose ester (CE), and polyethylene (PE), for heavy metals like arsenic (As), manganese (Mn), zinc (Zn), chromium (Cr), cadmium (Cd), copper (Cu), and lead (Pb). The results indicated that MPs composed of PP and PE polymers exhibited notable abilities to adsorb substantial amounts of metals, even at trace levels, posing significant risks if consumed by aquatic or terrestrial organisms.

4.2.2 Marine Environment

Microplastics (MPs), traditionally considered a concern mainly in ocean environments, are now increasingly recognized as prevalent in inland systems as well. They have been identified not solely along coastlines but also within surface waters, throughout the water column, and among various organisms inhabiting these ecosystems (Welden & Cowie, 2016).

At Caranzalem beach in Goa, Nigam (1982) documented an average of 50 to 300 plastic particles per square meter along the high-water mark. These particles are typically measured between 3–5 mm wide and 1 to 4 mm long. In the Alang-Sosiya shipbreaking yard located in Mumbai Metropolitan, India, Reddy et al. (2006) utilized FTIR and SEM techniques for analysis, discovering the presence of 81 plastic items per kilogram of sediment. Meanwhile, Jayasiri et al. (2013) detected 68.83 particles per square meter, ranging between 1–20 mm in size (with 40% of particles falling between 1 and 5 mm), on recreational beaches in Mumbai. Plastic pellets numbering from 1200 to 3000, sized between 2–5 mm, were found scattered along the coastlines of Goa and Chennai (Veerasingham et al., 2016). FTIR spectra examination has revealed that the primary plastic particles are mainly composed of polyethylene and polypropylene polymers. Currently, there are no restrictions on the detection of microplastics in organisms under natural conditions. Various methods have been utilized for the collection of microplastic samples from marine waters in India, including manual retrieval using glass bottles (Sathish et al., 2020), deploying boats equipped with different nets like manta trawl net, neuston net, and plankton net with varying mesh sizes (Goswami et al., 2023), or utilizing a Teflon pump followed by filtration through stainless steel sieves (Selvam et al., 2020).

Once samples are collected, they are preserved either by refrigeration or using a 4–5% formalin solution (Selvam et al., 2020) or a 5% formaldehyde solution (Sathish et al., 2020). After preservation, samples are subjected to microplastic separation techniques, including wet peroxide oxidation combined by density separation or density separation exclusively. Nonetheless, investigators' current choices of research sites and the limited exploration of Indian marine environments do not provide a comprehensive insight into the nation's impact on global marine microplastic pollution. Researchers have mainly concentrated on beaches due to their accessibility, while the water column has received less attention. Although extensive studies have been conducted along a portion of the southern coastline, other regions remain largely unexplored. Hence, the current research scenario in India appears fragmented, indicating the necessity for further studies to fill gaps and enhance comprehension of microplastic pollution in marine ecosystems.

4.3 Impacts on Human

Despite widespread recognition of microplastic (MP) contamination in the hydrosphere, atmosphere, and lithosphere, research efforts in these realms remain relatively scant in India. Organisms residing in these environments may ingest MPs, leading to a notable risk of human exposure through consumption or inhalation. Recent findings of MPs in diverse human consumables in India, like seafood, salt, drinking water, and tap water, have drawn attention to the issue of microplastic pollution. However, methods for collecting samples for MP analysis varied among studies; for example, in one study, salt samples were directly obtained from salt pans (Selvam et al., 2020). The increasing utilization of plastics in the food packaging sector is progressively contributing to the prevalence of MPs in human consumables. Furthermore, apprehensions are emerging regarding the contamination of sources from which these consumables are derived, potentially influencing the abundance of MPs in these products. A recent study led by (Nithin et al., 2021) proposed that changing the manufacturing process of consumer goods, such as table salts, to utilize groundwater rather than seawater, may offer a promising strategy to reduce microplastic pollution levels.

5 Assessment Methods of Microplastics

Various monitoring techniques are utilized Tt improve comprehension of the environmental consequences of microplastics (MPs). A thorough examination of MPs involves three main stages: collecting microplastic samples, preprocessing, and identifying and quantifying microplastics.

5.1 Sampling

Microplastic sampling techniques are broadly classified into three categories: Selective, Bulk and Volume-reduced sampling. Each method serves distinct purposes and employs specific tools (Hidalgo-Ruz et al., 2012). The direct selection method involves visually identifying and manually extracting MPs from environmental samples, primarily effective for surface sediment samples containing plastic particles sized between 1 to 6 mm (Karlsson et al., 2017). Conversely, the concentrated sample approach entails on-site filtering and sieving of large samples to isolate particular components for later analysis (Redondo-Hasselerharm et al., 2018). Meanwhile, the large sample technique preserves all samples, which is beneficial when detecting MPs is challenging without magnification. In aquatic environments, plankton nets of various mesh sizes are commonly utilized for microplastic collection. Large volumes of water are often to ensure sample accuracygathered. Surface water sampling typically employs trawl sampling devices like manta nets and neuston nets (Faure et al., 2015). Manta nets are particularly suitable for calm waters to prevent equipment damage from waves and to ensure precise measurement of sampled water volume.

Different types of nets are deployed depending on the environmental conditions and depth levels during microplastic sampling. The Neuston net is suitable for use even in heavy storms, while the Bongo net is typically chosen for sampling in middle water layers. For deep-sea sampling at the bottom, benthic trawling is preferred (Cózar et al., 2014). When deploying trawls, they are positioned in the windward direction to prevent interference from the vessel's hull and towed at speeds of approximately 2 knots for 15 to 30 min (Auta et al., 2017). The size of the trawl's mesh determines the range of microplastic sizes captured and the amount of filtered water volume. Currently, the most common mesh size is around 300 μm, balancing between capturing MPs effectively and minimizing clogging risks. For sediment sampling, tools encompassing box samplers and stainless steel shovels are used commonly (Hidalgo-Ruz et al., 2012). When examining MPs on beaches, researchers commonly use square quadrat sampling or transects parallel or perpendicular to the coastline. Multiple samples are often merged to provide a comprehensive evaluation of microplastic pollution levels. Initial screening might take place on-site, with collected samples stored for later laboratory analysis.

5.2 *Preprocessing*

The presence of various impurities in environmental samples makes direct microplastic (MPs) analysis challenging. Hence, it is imperative to separate and recover microplastic samples. Common pretreatment techniques include screening, filtration, density separation, and biochemical separation via digestion (Hidalgo-Ruz et al., 2012).

Filtration and screening techniques employ fine pores to capture MPs, while the sieving method eliminates larger particles and impurities by passing samples through sieves with various pore sizes (Roch & Brinker, 2017). Screens are typically made of stainless steel or copper, while filtration employs filter membranes with much smaller pore diameters (Hidalgo-Ruz et al., 2012). The principle behind density separation lies in the contrast in density between MPs and impurities found in environmental samples. According to the "Guidance on monitoring of marine litter in European seas," this process entails saturating the sample with brine, thoroughly stirring it, allowing heavier components to settle, while MPs stay suspended or float on the surface, and subsequently collecting the MPs from the upper layer of the solution (Schulz et al., 2013).

The technique of digestion is primarily employed to prepare biological samples and minimize substrate interference in environmental samples. This process typically involves alkaline digestion, acid digestion and enzyme digestion (Enders et al., 2017). In certain instances, researchers have used a 35% H_2O_2 solution for continuous sample digestion, effectively breaking down most bio-organic components and enabling further investigation (Nuelle et al., 2014). Alkaline and acid digestion entail treating samples with acid or alkaline solutions, respectively. However, the chemical resistance of different MPs varies, potentially limiting the effectiveness of strong acids or bases. Take, for instance, polyformaldehyde and polycarbonate, which can undergo reactions with potent acids or alkalis. Investigations utilizing acid digestion have exhibited substantial recovery rates for substances like styrene, yet recovery rates for nylon fibers usually remain low (Desforges et al., 2015). Enzyme digestion holds promise, with particular research indicating its capacity to digest more than 97% of plankton samples while safeguarding microplastic remnants. Nevertheless, both alkaline and acid digestion methods may influence the digestion outcomes to certain degree (Cole et al., 2014).

5.3 Identification and Quantification of Microplastics

Once microplastics (MPs) are separated, they need to be identified, typically through morphological and chemical analyses. Common methods include visual observation, staining, spectroscopy, and pyrolysis techniques.

5.3.1 Visual

Visual identification is a commonly employed method where particles ranging from 1 to 5 mm, as well as suspected particles, are directly chosen from mixed samples. This method allows for the rapid processing of numerous samples and enables the easy identification of physical properties such as color, shape, and size of MPs. However, varying standards among different observers and insufficient pretreatment can introduce bias in estimating microplastic abundance, a notable drawback. To mitigate this, strict adherence to recommended criteria is advised, aiming for a maximum difference of 10% in multiple counting results (Irfan et al., 2020).

The smaller MPs become, the higher the chances of misidentification. Optical microscopy, stereomicroscopy, and, scanning electron microscopy (SEM) are commonly employed techniques for detailed identification. Combining optical and stereomicroscopes provides intricate imaging characteristics of MPs (Rani et al., 2015). SEM offers superior resolution and magnification, even at the nanoscale, enabling observation of microplastic surface textures like grooves, pits, fragmentation, and fractures, aiding in the analysis of weathering progression (Cai et al., 2017). For example, SEM has effectively scrutinized polypropylene degradation in ocean environments (Khoironi et al., 2020). Additionally, the combination of SEM and energy-dispersive X-ray (EDX) allows for prompt evaluation of morphology and supports comprehensive qualitative and quantitative examination of microplastic composition components. This method is notably effective in pinpointing polymer additives (Zhang et al., 2020). Nevertheless, preparing samples for SEM entails significant time and effort.

5.3.2 Staining

In order to enhance identification precision, researchers have introduced Nile red, a lipophilic dye, and a 1% Rose Bengal stain for microplastic analysis. Nile red is applied to samples and examined using fluorescence microscopy or UV light photobox techniques (Maes et al., 2017). This approach effectively distinguishes various microplastic types, including polyethylene (PE), polypropylene (PP), polystyrene (PS), polycarbonate (PC), and polyurethane (PU), with PE identification comparable to results obtained through Fourier-transform infrared spectroscopy (FTIR) or infrared (IR) microscopy (Hermabessiere et al., 2018). Nevertheless, the existence of natural organic matter presents a limitation, potentially leading to an overestimation of microplastic quantities in environmental samples. Therefore, digestion becomes an essential procedure (Shim et al., 2016).

5.3.3 Spectrometry Methods

The chemical composition of microplastics (MPs) is frequently analyzed using Raman spectroscopy and Fourier transform infrared spectroscopy (FTIR) (Cabernard et al., 2018). FTIR, often implemented through attenuated total reflection (ATR), transmission, and reflection, is known for its effectiveness in microplastic identification (Song et al., 2015). ATR, providing stable surface spectral information, is suitable for detecting MPs with particle sizes equal to or larger than 300 μm, with analysis completed rapidly and precisely within 1 min. The transmission mode provides high-resolution spectra but necessitates infrared light penetration into the sample, while the reflection mode is suitable for opaque and thicker samples. FTIR accurately identifies plastic particle types without being affected by fluorescence interference, also revealing the weathering degree of MPs, aiding in source and input path determination (Käppler et al., 2016). However, FTIR has limitations: it's optimal for particles $\geq$ 20 μm, struggles with dark or opaque plastics, and is influenced by sample moisture, requiring individual particle analysis, a time-consuming process (Gong & Xie, 2020).

Raman spectroscopy is pivotal for gaining intricate understandings of high molecular weight polymers, thereby advancing the identification of MPs. Unlike Infrared Spectrophotometry, Raman spectroscopy can differentiate MPs with particle sizes smaller than 20 μm (Fries et al., 2013). However, this approach presents certain challenges: it requires extended imaging durations, limiting its practical utility, and is incompatible with samples showing fluorescence. Furthermore, accurate polymer identification of MPs may be impeded by the presence of additives and pigments within them (Loder et al., 2015a). Fluorescence generated by photosensitive materials in the sample also poses challenges to polymer identification (Song et al., 2015). Despite its broader wavelength range and greater sensitivity to nonpolar functional groups, Raman spectroscopy faces challenges due to interference from fluorescence emitted by organic matter or microplastic additives. Weak laser signals extend acquisition time, resulting in sample degradation (Loder & Gerdts, 2015). However, when

Raman spectroscopy is paired with automatic particle tracking technology for image analysis, it facilitates automated recognition, encompassing detailed point mapping. This procedure entails precision in directing the laser to pinpointed areas of particles for region-specific data gathering, thus painting a thorough picture of microplastic distribution (Chen et al., 2021).

5.3.4 Pyrolysis

Pyrolysis serves as a valuable adjunct to enhance the precision of various identification techniques, particularly in polymer analysis. One notably effective method is Pyrolysis gas chromatography-mass spectrometry (Py-GC-MS), which has gained prominence for its direct analysis of samples without prior treatment (Fischer et al., 2016). This approach provides comprehensive insights into polymer types and additives concurrently, regardless of microplastic characteristics such as shape, size, or color. However, there are limitations to consider. Pyrolysis can produce higher molecular weight products, leading to equipment blockages and damage, necessitating frequent maintenance and increased costs. Additionally, while adept at identifying polymer compositions, thermal decomposition methods lack the capacity to furnish detailed morphological information due to their sample-destructive nature.

This method operates with a minimal sample size (100–500 μg) per run and typically requires 30 to 100 min for a single measurement, which can limit its suitability for analyzing large sample numbers. To overcome these limitations, ongoing efforts have focused on refining and enhancing pyrolysis techniques. A notable advancement is the integration of Thermo-extraction and desorption with gas chromatography-mass spectroscopy (TED-GS-MS). This method integrates thermal extraction with thermogravimetric analysis (TGA) and thermal desorption-gas chromatography-mass spectrometry (TDS-GC-MS), enabling the effective identification of polymers like PP, PS, PE, PET, and PA (Dumichen et al., 2017). However, challenges persist, particularly in obtaining representative samples with good homogeneity, which often necessitates additional pretreatments encompassing screening and digestion. Efforts to enhance and optimize Py-GC-MS have been undertaken. For example, increasing the pyrolysis and injector temperatures to 700 °C and 300 °C, respectively, and adjusting the cracking ratio to 5 have resulted in a limit of detection (LOD) of less than 1 μg for common polymers. Comparative studies with Raman spectroscopy have demonstrated a 100% coincidence rate, suggesting the superiority of Py-GC-MS (Chen et al., 2021).

6 Ecological Risk Assessment of MPs in India

The ecological risk assessment of MP was performed using a combination of pollutant load index (PLI) and potential ecological risk index (PERI), following the methods proposed by Xu et al. (2018) and Peng et al. (2018). Several studies have been conducted on a wide range of habitats, encompassing rivers, lakes, groundwater, and the marine environment. Several researchers identified significant contaminants in Indian water bodies, impacting the environment and endangering human and aquatic health. Table 1 summarizes various key studies and environmental dangers.

7 Mitigation and Management Strategies

To halt the escalating influx of microplastic (MP) waste, it is essential to adopt preemptive measures targeting the underlying causes, rather than solely managing the resultant fallout (Löhr et al., 2017). Given that a substantial portion of MP originates from the deterioration of MP products, strategies geared towards curtailing plastic utilization and its ecological ramifications assume paramount significance in the battle against microplastic pollution. Such preventative measures may encompass the implementation of regulatory frameworks, stringent enforcement mechanisms, levies on plastic commodities, augmentation of public awareness campaigns, advancement of recycling initiatives, and embracement of circular economy paradigms. Effectively addressing this challenge necessitates synchronized endeavors from diverse sectors, encompassing both public as well as private domains, spanning global to local echelons (GESAMP, 2015). Educational campaigns, awareness-raising endeavors, alongwith community outreach programs emerge as efficacious instruments in reshaping individual behaviors and attitudes vis-à-vis littering (Pettipas et al., 2016). These endeavors should accord precedence to enduring strategies for fostering sustainable behavior modifications.

Source reduction and management have proven effective for restoration efforts. Emphasizing the 4R's—reduce, reuse, recycle, and recover—is crucial for enhancing waste disposal practices (Schneider & Ragossnig, 2015). Recycling stands out as the preferred waste management method due to its energy and resource efficiency, reduced pollutant emissions, diminished need for landfills, and positive impacts on both the economy and the environment (Prata et al, 2019). conversely, landfilling is considered the least desirable option due to its elevated risk of contamination, environmental consequences, and resource depletion. Moving from a linear economy to a circular one presents a sustainable, enduring solution by reducing waste generation, enhancing material value, and promoting an efficient post-use plastic economy. The concept of achieving zero waste production can be realized through circular economy strategies, which aim to eliminate the necessity for waste disposal or landfilling (Wagner & Lambert, 2018). Additionally, landfill mining emerges as a viable approach within circular economy frameworks to recover plastic waste that has been previously landfilled (Burlakovs et al., 2019).

The adoption of biobased plastics offers an appealing alternative to traditional plastics, primarily due to their renewable source, ability to decompose, and reduced environmental footprint (Rujnić-Sokele & Pilipović, 2017). Utilizing such bioplastics as substitutes can contribute to reducing plastic usage. Additionally, strategies like implementing extended producer responsibility (EPR) systems, enhancing

Table 1 Studies of microplastics in various ecosystem of India and their identified ecological risk

Study area	Microplastics detected	Detection method and size range	Sources	Ecological risk identified	References
Red Hills Lake, Tamil Nadu	High-density polyethylene (HDPE), low-density polyethylene (LDPE), polypropylene, and polystyrene	FTIR and EDX, <5 mm	Plastic items can break down owing to ageing, dry deposition, or fishing nets	Serious threat to plankton, fishes, and other organism	Gopinath et al. (2020)
River Sharavathi, Karnataka	Polyethylene, polyethylene terephthalate, and polypropylene	ATR-FTIR, <5 mm	Factors to consider include population distribution, land use and cover, geomorphology, and hydrodynamics	Harm to the biota and the ecosystem	Amrutha et al. (2023)
Ganga river, Bihar	Polyethylene terepthalate	Stereomicroscope (Nikon) and ATR F, 1mm to >5 mm	Direct entry of plastic waste into the river via sewage	Highly impact to aquatic biodiversity	Sarkar et al. (2019)
Indian Coast	Polyethylene, Polypropylene, Polyamide (PA) and polystyrene (PS)	–	Rain, flood and storm events, urban runoff	Contaminated seafood and the transfer of hazardous substances to humans	Ranjani et al. (2021)
Kodaikanal Lake, Tamil Nadu	Polyethylene, polypropylene, polyethylene terephthalate, polystyrene , cellophane (CP), poly vinyl acetate (PVA), and polyether urethane (PEU)	ATR-FTIR, SEM/ EDAX <5 mm	Tourism is the main source of MPs	The study identified a problem with freshwater, especially in tourist locations, that requires further attention	Laju et al. (2022)
Kerela coastal area	Polyethylene, polyamide, and polypropylene	FTIR	No systematic reports available on the distribution & source of microplastics	Extreme ecological risk due to a clear link among the toxicity score for individual polymers & the abundance of MPs	Radhakrishnan et al. (2023)

(continued)

Table 1 (continued)

Study area	Microplastics detected	Detection method and size range	Sources	Ecological risk identified	References
Maharastra coastal area	Polyethylene (PE), polypropylene (PP), polystyrene (PS)	FTIR, SEM/EDAX	A large amount of discarded microplastics, fishing ropes, and nets	Bioaccumulation, particularly in marine species, poses a significant risk to human health	Kumkar et al. (2023)
Kerela coastal area	Polyethylene, polypropylene	ATR-FTIR, Raman Spectrum analysis, <1 mm	Following, floods, the availability of MPs enhanced sevenfold	Potentially harmful to the food chain, plankton, fish, and other organism	Nikhil et al. (2023)

single-use plastic recycling rates, enacting plastic regulations, and exploring innovative technologies offer avenues for plastic waste reduction (Liu et al., 2018). The United Nations Environment Assembly, involving approximately 170 countries, has committed to a significant reduction in plastic consumption by 2030 (BBC News, 2019). Despite ongoing efforts, they are often considered inadequate in light of the extensive plastic pollution, its non-biodegradable characteristics, and its enduring presence in the environment. Major challenges include limited information regarding pollution sources, ineffective implementation, insufficient data on microplastic pollution, and a lack of prioritization in mitigation efforts. Overcoming these obstacles requires a more cohesive and coordinated approach to combat both plastic and microplastic pollution (Patil et al., 2021; Pant et al., 2025b).

8 Bottle Necks and Challenges

Understanding the diversity in plastic products and their varying lifespans is essential for grasping the plastic product lifecycle. As plastic exists till the end of its utility, it transforms into waste. Primary microplastics are deliberately diminished in size for incorporation into personal care products, medications, and various industries. Conversely, secondary microplastics arise as larger plastic items slowly break down under the effects of chemical, physical, mechanical and biological influences.

The linear economy model, particularly reliant on single-use plastics, significantly adds to the buildup of plastic in natural settings such as oceans and landfills. Cordier and Uehara's (2019) study estimates that the production, distribution, and consumption of commercial plastics and products lead to about $13 billion in annual damage to marine ecosystems. Nevertheless, despite this fact, an astonishing 95% of all packaging plastic material, totaling $80–120 billion, is disposed of, worsening environmental challenges. Research indicates that removing approximately 135 million metric tons of plastic waste from oceans would require around 0.7–1.0% of the total global GDP in 2017, which is equivalent to about 492–708 billion euros. However, this endeavor only addresses 15% of the anticipated total ocean plastics between 2020 and 2030, as projected by Watt et al. (2021). Consequently, the significant cost associated constitutes a notable barrier to the successful elimination of plastics as well as microplastics (MPs) from the natural environment.

Every aspect of MPs analysis, from sampling to identification, demands significant time, presenting a significant challenge to extensive monitoring endeavors. Moreover, variation in methodologies used in these processes, resulting in non-standardized results. Collecting a sufficient quantity of MPs particles, especially for chemical analysis, becomes increasingly challenging, as highlighted by Yu et al. (2018). This challenge leads to infrequent detections and elevated detection thresholds. Moreover, accurately measuring MPs in terrestrial environments is daunting due to the necessity of separating them from extensive amounts of biomass. Despite the adaptation of sediment and water sample characterization techniques for soil samples, a standardized approach remains elusive, as highlighted by Miri et al. (2022). While various extraction methods offer speed, cost-effectiveness, and efficiency, a universally accepted standard method is lacking.

Despite their capacity to identify particles as minute as 20 μm, alternative techniques like infrared spectroscopy, differential scanning calorimetry, thermogravimetric analysis, and Raman spectroscopy encounter hurdles in their extensive usage due to cost factors and restricted applicability. Recently, micro-FTIR (Fourier Transform Infrared Spectroscopy linked to a microscope) has attracted attention for its capability to map samples, characterize multiple polymers, and identify irregularly shaped MPs. However, it continues to be a costly and process that requires a considerable amount of time to complete, as highlighted by Miri et al. (2022).

In wastewater treatment plants (WWTPs), MPs are primarily removed through various processes, encompassing screening, grease removal, grit removal, skimming, and sedimentation (both preliminary and primary treatment). Secondary treatment, utilizing activated sludge, is followed by tertiary treatment methods such as membrane bioreactors (MBR), rapid sand filters (RSF), disc filters (DF), ultrafiltration (UF), reverse osmosis (RO), dissolved air flotation (DAF) and gravity filters. However, it's essential to note that these technologies are not specifically designed to effectively eliminate MPs, resulting in a significant portion being discharged into water bodies. Since 2015, microplastic pollution has risen as a critical environmental issue, paralleling global challenges encompassing climate change, ozone depletion, and ocean acidification (Yu et al., 2018). Despite attempts to tackle plastic pollution through legislation and policies, these measures often prove insufficient in managing plastic production, consumption, and disposal effectively. As a result, there has been a discernible rise in MPs in natural settings. Moreover, the dynamics of human exposure to MPs have been altered by the COVID-19 pandemic, introducing a notable new source of MPs into the environment within a brief period. This raises concerns regarding heightened interaction, ingestion, and potential adverse effects within food chains. Therefore, gaining a thorough understanding of the considerable implications of reusable face masks on both the human health and the environment presents an unprecedented challenge (Rose et al., 2023; Pant et al., 2025c).

9 Conclusion and Future Perspectives

India, a significant contributor to plastic waste, is increasingly focusing on research into microplastics (MPs). However, the extent of India's role in global MPs pollution remains unclear. This study identified knowledge gaps concerning microplastic pollution across different compartments of the Indian environment. Although considerable research has focused on marine environments, particularly along the southeast coast where Tamil Nadu is the most studied state, freshwater systems, notably rivers, have garnered more attention compared to lakes and groundwater. However, there's still a necessity for a thorough comprehension of the origins, routes, and destiny of MPs in aquatic environments.

The degradation of plastics presents a significant concern, leading to the generation of MPs and posing risks to ecosystems, humans, animals, and plants. In the future, research efforts may aim to assess how advanced treatment approaches could alter the levels of MPs observed in treated water. The MPs detection occurrences in wastewater treatment plants (WWTPs) hold crucial importance for effective control, given that these facilities release MPs into the ecosystem. It is imperative to target smaller microplastics (<20 μm) prevalent in water, as they can infiltrate the circulatory systems of aquatic organisms. Channeling efforts towards microplastic removal from wastewater is crucial, given the role of WWTPs in facilitating microplastic entry into natural aquatic systems. Furthermore, it is imperative that future research endeavors prioritize understanding how plastics behave across different environments and developing technologies to mitigate plastic pollution. It is urgent to investigate the creation and decomposition of nano-plastics to assess their environmental destiny and related risks.

Effective detection and mitigation of MPs pollution heavily rely on employing proper sampling, identification, and removal techniques. Greater research efforts are warranted to fully understand the potential consequences of MPs on human health. It is crucial to adopt strategies such as refuse, reduce, reuse, recycle, recover and rethink to effectively address MPs pollution. Implementing bans or taxes on environmentally harmful plastic products can offer comprehensive protection against microbead pollution. Efforts aimed at reducing MPs production and consumption should focus on improving product design and embracing alternative materials that offer durability while replacing plastics. It is imperative for scientific endeavors to prioritize the identification and mitigation of plastic pollution hotspots in water bodies to reduce their discharge into rivers. Developing standardized scientific protocols for sampling and extracting MPs can greatly streamline risk assessment procedures.

Investigating the potential toxicity of pollutants attached to MPs in the ecosystem is critical. It is essential to comprehend both the direct and indirect sources of pollution, as well as particle interactions and transport mechanisms. This comprehension is critical for accurately modeling contaminant sources and their accumulation in river ecosystems, enabling the assessment of their impact on river health and biodiversity. Strategies such as reducing consumption, plastics production and implementing effective waste management practices are imperative for addressing MPs pollution. Delving into the dynamics of MPs transport within sediment layers can significantly bolster risk assessment methodologies in aquatic ecosystems. Vigilant monitoring of MPs presence, especially in regions dependent on seafood, remains imperative for formulating robust management protocols. Efforts in industries like cosmetics and beauty to phase out MPs in favor of eco-friendly alternatives indicate progress. Some nations have even implemented bans on cosmetics containing MPs. Encouraging positive shifts in human behavior is strongly advocated, offering a sustainable solution to the microplastic challenge.

10 Recommendation

Sensitizing local residents (rural & tribal communities) and tourists about eco-friendly behaviour, averting contagion of the ecosystems, watershed & land management alongwith proper waste water treatment must be at core planning. Continous monitoring and assessment must be conducted by pollution control boards and people participation must be ensured to achieve SDG's and smart city projects. Municipal Authorities and government policy-makers must take cognizance to formulate effective plans (treatment of sewage and watershed development) to control plastic pollution and ecosystem management. Research pertaining to adverse health and ecological impacts of microplastics and other pollutants needs to organised in multidisciplinary mode by nodal agencies of sustainability.

References

Amrutha, K., & Warrier, A. K. (2020). The first report on the source-to-sink characterization of microplastic pollution from a riverine environment in tropical India. *Science of the Total Environment, 739*, Article 140377. https://doi.org/10.1016/j.scitotenv.2020.140377.

Amrutha, K., Unnikrishnan, V., Shajikumar, S., & Warrier, A. K. (2021). Current state of microplastics research in SAARC countries—a review. *Microplastic Pollution*, 27–63. https://doi.org/10.1007/978-981-16-0297-9_2.

Amrutha, K., Shajikumar, S., Warrier, A. K., Sebastian, J. G., Sali, Y. A., Chandran, T., & Unnikrishnan, V. (2023). Assessment of pollution and risks associated with microplastics in the riverine sediments of the Western Ghats: A heritage site in southern India. *Environmental Science and Pollution Research, 30*(12), 32301–32319. https://doi.org/10.1007/s11356-022-24437-z.

Auta, H. S., Emenike, C. U., & Fauziah, S. H. (2017). Distribution and importance of microplastics in the marine environment: A review of the sources, fate, effects, and potential solutions. *Environment International, 102*, 165–176. https://doi.org/10.1016/j.envint.2017.02.013.

Ayyamperumal, R., Huang, X., Li, F., Chengjun, Z., Chellaiah, G., Gopalakrishnan, G., & Antony, J. K. (2022). Investigation of microplastic contamination in the sediments of Noyyal River-Southern India. *Journal of Hazardous Materials Advances, 8*, Article 100198. https://doi.org/10.1016/j.hazadv.2022.100198.

Barnes, D. K., Galgani, F., Thompson, R. C., & Barlaz, M. (2009). Accumulation and fragmentation of plastic debris in global environments. *Philosophical Transactions of the Royal Society b: Biological Sciences, 364*(1526), 1985–1998. https://doi.org/10.1098/rstb.2008.0205.

BBC News, 2019 UN resolution pledges to plastic reduction by 2030. https://www.bbc.co.uk/news/science-environment-47592111. Accessed 22 May 2020.

Bisen, D., Chouhan, A. P. S., Pant, M., & Chakma, S. (2025). Advancement of thermochemical conversion and the potential of biomasses for production of clean energy: A review. *Renewable and Sustainable Energy Reviews, 208*, 115016.

Burlakovs, J., Kriipsalu, M., Porshnov, D., Jani, Y., Ozols, V., Pehme, K. M., & Klavins, M. (2019). Gateway of landfilled plastic waste towards circular economy in Europe. *Separations, 6*(2), 25. https://doi.org/10.3390/separations6020025.

Cabernard, L., Roscher, L., Lorenz, C., Gerdts, G., & Primpke, S. (2018). Comparison of Raman and Fourier transform infrared spectroscopy for the quantification of microplastics in the aquatic environment. *Environmental Science & Technology, 52*(22), 13279–13288. https://doi.org/10.1021/acs.est.8b03438.

Cai, L., Wang, J., Peng, J., Tan, Z., Zhan, Z., Tan, X., & Chen, Q. (2017). Characteristic of microplastics in the atmospheric fallout from Dongguan city, China: Preliminary research and first evidence. *Environmental Science and Pollution Research, 24*, 24928–24935. https://doi.org/10.1007/s11356-017-0116-x.

Central Water Commission. (2019). Reassessment of water availability in India using space inputs. http://old.cwc.gov.in/main/downloads/ReassessmentMainReport.pdf. Accessed 15 Oct 2020.

Chauhan, J. S., Semwal, D., Nainwal, M., Badola, N., & Thapliyal, P. (2021). Investigation of microplastic pollution in river Alaknanda stretch of Uttarakhand. *Environment, Development and Sustainability*, 1–15. https://doi.org/10.1007/s10668-021-01388-y.

Chen, J., Wang, W., Liu, H., Xu, X., & Xia, J. (2021). A review on the occurrence, distribution, characteristics, and analysis methods of microplastic pollution in ecosystems. *Environmental Pollutants and Bioavailability, 33*(1), 227–246. https://doi.org/10.1080/26395940.2021.1960198.

Cole, M., Webb, H., Lindeque, P. K., Fileman, E. S., Halsband, C., & Galloway, T. S. (2014). Isolation of microplastics in biota-rich seawater samples and marine organisms. *Scientific Reports, 4*(1), 4528. https://doi.org/10.1038/srep04528.

Cordier, M., & Uehara, T. (2019). How much innovation is needed to protect the ocean from plastic contamination? *Science of the Total Environment, 670*, 789–799. https://doi.org/10.1016/j.scitotenv.2019.03.258.

Covernton, G. A., Pearce, C. M., Gurney-Smith, H. J., Chastain, S. G., Ross, P. S., Dower, J. F., & Dudas, S. E. (2019). Size and shape matter: A preliminary analysis of microplastic sampling technique in seawater studies with implications for ecological risk assessment. *Science of the Total Environment, 667*, 124–132. https://doi.org/10.1016/j.scitotenv.2019.02.346.

Cózar, A., Echevarría, F., González-Gordillo, J. I., Irigoien, X., Úbeda, B., Hernández-León, S., & Duarte, C. M. (2014). Plastic debris in the open ocean. *Proceedings of the National Academy of Sciences, 111*(28), 10239–10244. https://doi.org/10.1073/pnas.1314705111.

de Souza Machado, A. A., Kloas, W., Zarfl, C., Hempel, S., & Rillig, M. C. (2018). Microplastics as an emerging threat to terrestrial ecosystems. *Global Change Biology, 24*(4), 1405–1416. https://doi.org/10.1111/gcb.14020.

Desforges, J. P. W., Galbraith, M., & Ross, P. S. (2015). Ingestion of microplastics by zooplankton in the Northeast Pacific Ocean. *Archives of Environmental Contamination and Toxicology, 69*, 320–330. https://doi.org/10.1007/s00244-015-0172-5.

Dümichen, E., Eisentraut, P., Bannick, C. G., Barthel, A. K., Senz, R., & Braun, U. (2017). Fast identification of microplastics in complex environmental samples by a thermal degradation method. *Chemosphere, 174*, 572–584. https://doi.org/10.1016/j.chemosphere.2017.02.010.

Enders, K., Lenz, R., Beer, S., & Stedmon, C. A. (2017). Extraction of microplastic from biota: Recommended acidic digestion destroys common plastic polymers. *ICES Journal of Marine Science, 74*(1), 326–331. https://doi.org/10.1093/icesjms/fsw173.

Europe, P. (2021). Plastics—the facts 2021. An analysis of European latest plastics production, demand and waste data.

Faure, F., Saini, C., Potter, G., Galgani, F., De Alencastro, L. F., & Hagmann, P. (2015). An evaluation of surface micro-and mesoplastic pollution in pelagic ecosystems of the Western Mediterranean Sea. *Environmental Science and Pollution Research, 22*, 12190–12197. https://doi.org/10.1007/s11356-015-4453-3.

Fischer, E. K., Paglialonga, L., Czech, E., & Tamminga, M. (2016). Microplastic pollution in lakes and lake shoreline sediments–a case study on Lake Bolsena and Lake Chiusi (central Italy). *Environmental Pollution, 213*, 648–657. https://doi.org/10.1016/j.envpol.2016.03.012.

Fries, E., Dekiff, J. H., Willmeyer, J., Nuelle, M. T., Ebert, M., & Remy, D. (2013). Identification of polymer types and additives in marine microplastic particles using pyrolysis-GC/MS and scanning electron microscopy. *Environmental Science: Processes & Impacts, 15*(10), 1949–1956. https://doi.org/10.1039/C3EM00214D.

Ganesan, M., Nallathambi, G., & Srinivasalu, S. (2019). Fate and transport of microplastics from water sources. *Current Science, 117*, 1874–1879. https://wwwops.currentscience.ac.in/Volumes/117/11/1879.pdf. Accessed 17 May 2020.

Gao, F., Li, J., Hu, J., Sui, B., Wang, C., Sun, C., & Ju, P. (2021). The seasonal distribution characteristics of microplastics on bathing beaches along the coast of Qingdao, China. *Science of the Total Environment, 783*, Article 146969. https://doi.org/10.1016/j.scitotenv.2021.146969.

GESAMP. (2015). Sources, fate and effects of microplastics in the marine environment: a global assessment. In: Kershaw, P. J. (ed.). (IMO/ FAO/UNESCO-IOC/UNIDO/WMO/IAEA/UN/UNEP/UNDP Joint Group of Experts on the Scientific Aspects of Marine Environmental Protection). Rep. Stud. GESAMP No. 90, p. 96.

Gong, J., & Xie, P. (2020). Research progress in sources, analytical methods, eco-environmental effects, and control measures of microplastics. *Chemosphere, 254*, Article 126790. https://doi.org/10.1016/j.chemosphere.2020.126790.

Gopinath, K., Seshachalam, S., Neelavannan, K., Anburaj, V., Rachel, M., Ravi, S., & Achyuthan, H. (2020). Quantification of microplastic in red hills lake of Chennai city, Tamil Nadu, India. *Environmental Science and Pollution Research, 27*, 33297–33306. https://doi.org/10.1007/s11356-020-09622-2.

Goswami, P., Selvakumar, N., Verma, P., Saha, M., Suneel, V., Vinithkumar, N. V., & Nayak, J. (2023). Microplastic intrusion into the zooplankton, the base of the marine food chain: Evidence from the Arabian Sea, Indian Ocean. *Science of the Total Environment, 864*, Article 160876. https://doi.org/10.1016/j.scitotenv.2022.160876.

Guo, X., & Wang, J. (2019). The chemical behaviors of microplastics in marine environment: A review. *Marine Pollution Bulletin, 142*, 1–14. https://doi.org/10.1016/j.marpolbul.2019.03.019.

Guo, X., Dai, H., & He, L. (2024). Migration testing of microplastics from selected water and food containers by Raman microscopy. *Journal of Hazardous Materials, 462*, Article 132798. https://doi.org/10.1016/j.jhazmat.2023.132798.

Harikrishnan, T., Janardhanam, M., Sivakumar, P., Sivakumar, R., Rajamanickam, K., Raman, T., & Singaram, G. (2023). Microplastic contamination in commercial fish species in southern coastal region of India. *Chemosphere, 313*, Article 137486. https://doi.org/10.1016/j.chemosphere.2022.137486.

He, D., Luo, Y., Lu, S., Liu, M., Song, Y., & Lei, L. (2018). Microplastics in soils: Analytical methods, pollution characteristics and ecological risks. *TrAC Trends in Analytical Chemistry, 109*, 163–172. https://doi.org/10.1016/j.trac.2018.10.006.

Hermabessiere, L., Himber, C., Boricaud, B., Kazour, M., Amara, R., Cassone, A. L., & Duflos, G. (2018). Optimization, performance, and application of a pyrolysis-GC/MS method for the identification of microplastics. *Analytical and Bioanalytical Chemistry, 410*, 6663–6676. https://doi.org/10.1007/s00216-018-1279-0.

Hidalgo-Ruz, V., Gutow, L., Thompson, R. C., & Thiel, M. (2012). Microplastics in the marine environment: A review of the methods used for identification and quantification. *Environmental Science & Technology, 46*(6), 3060–3075. https://doi.org/10.1021/es2031505.

Horton, A. A., Walton, A., Spurgeon, D. J., Lahive, E., & Svendsen, C. (2017). Microplastics in freshwater and terrestrial environments: Evaluating the current understanding to identify the knowledge gaps and future research priorities. *Science of the Total Environment, 586*, 127–141. https://doi.org/10.1016/j.scitotenv.2017.01.190.

Huang, Y., Liu, Q., Jia, W., Yan, C., & Wang, J. (2020). Agricultural plastic mulching as a source of microplastics in the terrestrial environment. *Environmental Pollution, 260*, Article 114096. https://doi.org/10.1016/j.envpol.2020.114096.

Hwang, J., Choi, D., Han, S., Choi, J., & Hong, J. (2019). An assessment of the toxicity of polypropylene microplastics in human derived cells. *Science of the Total Environment, 684*, 657–669. https://doi.org/10.1016/j.scitotenv.2019.05.071.

Irfan, T., Khalid, S., Taneez, M., & Hashmi, M. Z. (2020). Plastic driven pollution in Pakistan: The first evidence of environmental exposure to microplastic in sediments and water of Rawal Lake.

Environmental Science and Pollution Research, 27, 15083–15092. https://doi.org/10.1007/s11356-020-07833-1.
Jayasiri, H. B., Purushothaman, C. S., & Vennila, A. (2013). Quantitative analysis of plastic debris on recreational beaches in Mumbai, India. *Marine Pollution Bulletin, 77*(1–2), 107–112. https://doi.org/10.1016/j.marpolbul.2013.10.024.
Käppler, A., Fischer, D., Oberbeckmann, S., Schernewski, G., Labrenz, M., Eichhorn, K. J., & Voit, B. (2016). Analysis of environmental microplastics by vibrational microspectroscopy: FTIR, Raman or both? *Analytical and Bioanalytical Chemistry, 408*, 8377–8391. https://doi.org/10.1007/s00216-016-9956-3.
Karbalaei, S., Hanachi, P., Walker, T. R., & Cole, M. (2018). Occurrence, sources, human health impacts and mitigation of microplastic pollution. *Environmental Science and Pollution Research, 25*, 36046–36063. https://doi.org/10.1007/s11356-018-3508-7.
Karlsson, T. M., Vethaak, A. D., Almroth, B. C., Ariese, F., van Velzen, M., Hassellöv, M., & Leslie, H. A. (2017). Screening for microplastics in sediment, water, marine invertebrates and fish: Method development and microplastic accumulation. *Marine Pollution Bulletin, 122*(1–2), 403–408. https://doi.org/10.1016/j.marpolbul.2017.06.081.
Kedzierski, M., Lechat, B., Sire, O., Le Maguer, G., Le Tilly, V., & Bruzaud, S. (2020). Microplastic contamination of packaged meat: Occurrence and associated risks. *Food Packaging and Shelf Life, 24*, Article 100489. https://doi.org/10.1016/j.fpsl.2020.100489.
Khaleel, R., Valsan, G., Rangel-Buitrago, N., & Warrier, A. K. (2022). Hidden problems in geological heritage sites: The microplastic issue on Saint Mary's Island, India, Southeast Arabian Sea. *Marine Pollution Bulletin, 182*, Article 114043. https://doi.org/10.1016/j.marpolbul.2022.114043.
Khoironi, A., Hadiyanto, H., Anggoro, S., & Sudarno, S. (2020). Evaluation of polypropylene plastic degradation and microplastic identification in sediments at Tambak Lorok coastal area, Semarang, Indonesia. *Marine Pollution Bulletin, 151*, Article 110868. https://doi.org/10.1016/j.marpolbul.2019.110868.
Koutnik, V. S., Leonard, J., Alkidim, S., DePrima, F. J., Ravi, S., Hoek, E. M., & Mohanty, S. K. (2021). Distribution of microplastics in soil and freshwater environments: Global analysis and framework for transport modeling. *Environmental Pollution, 274*, Article 116552. https://doi.org/10.1016/j.envpol.2021.116552.
Kumkar, P., Verma, C. R., Hýsek, Š, Pise, M., Źółtowska, S., Gosavi, S. M., & Kalous, L. (2023). Contaminants and their ecological risk assessment in beach sediments and water along the Maharashtra coast of India: A comprehensive approach using microplastics, heavy metal (loid) s, pharmaceuticals, personal care products and plasticisers. *Science of the Total Environment, 892*, Article 164712. https://doi.org/10.1016/j.scitotenv.2023.164712.
Laju, R. L., Jayanthi, M., Jeyasanta, K. I., Patterson, J., Asir, N. G. G., Sathish, M. N., & Edward, J. P. (2022). Spatial and vertical distribution of microplastics and their ecological risk in an Indian freshwater lake ecosystem. *Science of the Total Environment, 820*, Article 153337. https://doi.org/10.1016/j.scitotenv.2022.153337.
Liu, Z., Adams, M., & Walker, T. R. (2018). Are exports of recyclables from developed to developing countries waste pollution transfer or part of the global circular economy? *Resources, Conservation and Recycling, 136*, 22–23. https://doi.org/10.1016/j.resconrec.2018.04.005.
Löder, M. G., & Gerdts, G. (2015). Methodology used for the detection and identification of microplastics—a critical appraisal. *Marine anthropogenic litter*, 201–227. https://doi.org/10.1007/978-3-319-16510-3_8.
Löder, M. G. J., Kuczera, M., Mintenig, S., Lorenz, C., & Gerdts, G. (2015). Focal plane array detector-based micro-fourier-transform infrared imaging for the analysis of microplastics in environmental samples. *Environmental Chemistry, 12*(5), 563–581. https://doi.org/10.1071/EN14205.
Löhr, A., Savelli, H., Beunen, R., Kalz, M., Ragas, A., & Van Belleghem, F. (2017). Solutions for global marine litter pollution. *Current Opinion in Environmental Sustainability, 28*, 90–99. https://doi.org/10.1016/j.cosust.2017.08.009.

Maes, T., Jessop, R., Wellner, N., Haupt, K., & Mayes, A. G. (2017). A rapid-screening approach to detect and quantify microplastics based on fluorescent tagging with Nile Red. *Scientific Reports, 7*(1), 44501. https://doi.org/10.1038/srep44501.

Malankowska, M., Echaide-Gorriz, C., & Coronas, J. (2021). Microplastics in marine environment: A review on sources, classification, and potential remediation by membrane technology. *Environmental Science: Water Research & Technology, 7*(2), 243–258. https://doi.org/10.1039/D0EW00802H.

Miri, S., Saini, R., Davoodi, S. M., Pulicharla, R., Brar, S. K., & Magdouli, S. (2022). Biodegradation of microplastics: Better late than never. *Chemosphere, 286*, Article 131670. https://doi.org/10.1016/j.chemosphere.2021.131670.

Napper, I. E., Baroth, A., Barrett, A. C., Bhola, S., Chowdhury, G. W., Davies, B. F., & Koldewey, H. (2021). The abundance and characteristics of microplastics in surface water in the transboundary Ganges River. *Environmental Pollution, 274*, Article 116348. https://doi.org/10.1016/j.envpol.2020.116348.

Nigam, R. A. I. I. V. (1982). Plastic pellets on the Caranzalem beach sands, Goa, India. Mahasagar. http://drs.nio.org/drs/handle/2264/6650.

Nikhil, V. G., Ranjeet, K., & Varghese, G. K. (2023). Spatio-temporal evaluation and risk assessment of microplastics in nearshore surface waters post-2018 Kerala deluge along the southwest coast of India. *Marine Pollution Bulletin, 192*, Article 115058. https://doi.org/10.1016/j.marpolbul.2023.115058.

Nithin, A., Sundaramanickam, A., Surya, P., Sathish, M., Soundharapandiyan, B., & Balachandar, K. (2021). Microplastic contamination in salt pans and commercial salts–a baseline study on the salt pans of Marakkanam and Parangipettai, Tamil Nadu, India. *Marine Pollution Bulletin, 165*, Article 112101. https://doi.org/10.1016/j.marpolbul.2021.112101.

Nuelle, M. T., Dekiff, J. H., Remy, D., & Fries, E. (2014). A new analytical approach for monitoring microplastics in marine sediments. *Environmental Pollution, 184*, 161–169. https://doi.org/10.1016/j.envpol.2013.07.027.

Obbard, R. W., Sadri, S., Wong, Y. Q., Khitun, A. A., Baker, I., & Thompson, R. C. (2014). Global warming releases microplastic legacy frozen in Arctic Sea ice. *Earth's Future, 2*(6), 315–320. https://doi.org/10.1002/2014EF000240.

Pant, M., & Singh, J. (2024a). Distribution of heavy metals contamination in the environment and their toxicological impacts. In *Heavy metal contamination in the environment* (pp. 95–113). CRC Press.

Pant, M., Singh, S., & Singh, J. (2024b). Seasonal behavior and spatial variations of water quality index and micro-biological changes in the springs of Indian Himalayan Region. *Environment, Development and Sustainability*, 1–22.

Pant, M., Singh, J., Bisen, D., Manori, S., & Shukla, R. K. (2025a). Nanocomposite-based photocatalysis: Tackling pharmaceutical and personal care products (PPCPs) pollutants in environmental remediation. In *Sustainable development goals towards environmental toxicity and green chemistry* (pp. 83–105). Springer, Cham.

Pant, M., Singh, J., Singhal, N., & Bisen, D. (2025b). Bioremediation strategies for pharmaceutical and personal care products contaminants: Microbial and plant-based approaches. In *Emerging contaminants in water and wastewater: sources and substances* (pp. 331–354). Cham, Springer Nature Switzerland.

Pant, M., Naik, S. C. S., Singhal, N., & Singh, J. (2025c). Bioremediation approaches for mitigation of emerging water contaminants in wastewater. In *Biotechnologies for wastewater treatment and resource recovery* (pp. 93–108). Elsevier.

Patil, S., Bafana, A., Naoghare, P. K., Krishnamurthi, K., & Sivanesan, S. (2021). Environmental prevalence, fate, impacts, and mitigation of microplastics—a critical review on present understanding and future research scope. *Environmental Science and Pollution Research, 28*(5), 4951–4974. https://doi.org/10.1007/s11356-020-11700-4.

Peng, G., Xu, P., Zhu, B., Bai, M., & Li, D. (2018). Microplastics in freshwater river sediments in Shanghai, China: A case study of risk assessment in mega-cities. *Environmental Pollution, 234*, 448–456. https://doi.org/10.1016/j.envpol.2017.11.034.

Pettipas, S., Bernier, M., & Walker, T. R. (2016). A Canadian policy framework to mitigate plastic marine pollution. *Marine Policy, 68*, 117–122. https://doi.org/10.1016/j.marpol.2016.02.025.

Prata, J. C., Silva, A. L. P., Da Costa, J. P., Mouneyrac, C., Walker, T. R., Duarte, A. C., & Rocha-Santos, T. (2019). Solutions and integrated strategies for the control and mitigation of plastic and microplastic pollution. *International Journal of Environmental Research and Public Health, 16*(13), 2411. https://doi.org/10.3390/ijerph16132411.

Radhakrishnan, K., Sivapriya, V., Rajkumar, A., Akramkhan, N., Prakasheswar, P., Krishnakumar, S., & Hussain, S. M. (2021). Characterization and distribution of microplastics in estuarine surface sediments, Kayamkulam estuary, southwest coast of India. *Marine Pollution Bulletin, 168*, Article 112389. https://doi.org/10.1016/j.marpolbul.2021.112389.

Radhakrishnan, K., Krishnakumar, S., Prakasheswar, P., Pradhap, D., Akramkhan, N., Gomathi, S., & Hussain, S. M. (2023). Potential ecological risk assessment studies based on source and distribution of microplastics from the surface sediments of tropical backwaters, kerala, India. *Total Environment Research Themes, 7*, Article 100063. https://doi.org/10.1016/j.totert.2023.100063.

Rani, M., Shim, W. J., Han, G. M., Jang, M., Al-Odaini, N. A., Song, Y. K., & Hong, S. H. (2015). Qualitative analysis of additives in plastic marine debris and its new products. *Archives of Environmental Contamination and Toxicology, 69*, 352–366. https://doi.org/10.1007/s00244-015-0224-x.

Ranjani, M., Veerasingam, S., Venkatachalapathy, R., Mugilarasan, M., Bagaev, A., Mukhanov, V., & Vethamony, P. J. M. P. B. (2021). Assessment of potential ecological risk of microplastics in the coastal sediments of India: A meta-analysis. *Marine Pollution Bulletin, 163*, Article 111969. https://doi.org/10.1016/j.marpolbul.2021.111969.

Reddy, M. S., Basha, S., Adimurthy, S., & Ramachandraiah, G. (2006). Description of the small plastics fragments in marine sediments along the Alang-Sosiya ship-breaking yard, India. *Estuarine, Coastal and Shelf Science, 68*(3–4), 656–660. https://doi.org/10.1016/j.ecss.2006.03.018.

Redondo-Hasselerharm, P. E., Falahudin, D., Peeters, E. T., & Koelmans, A. A. (2018). Microplastic effect thresholds for freshwater benthic macroinvertebrates. *Environmental Science & Technology, 52*(4), 2278–2286.

Robin, R. S., Karthik, R., Purvaja, R., Ganguly, D., Anandavelu, I., Mugilarasan, M., & Ramesh, R. (2020). Holistic assessment of microplastics in various coastal environmental matrices, southwest coast of India. *Science of the Total Environment, 703*, Article 134947. https://doi.org/10.1016/j.scitotenv.2019.134947.

Roch, S., & Brinker, A. (2017). Rapid and efficient method for the detection of microplastic in the gastrointestinal tract of fishes. *Environmental Science & Technology, 51*(8), 4522–4530. https://doi.org/10.1021/acs.est.7b00364.

Rose, P. K., Jain, M., Kataria, N., Sahoo, P. K., Garg, V. K., & Yadav, A. (2023). Microplastics in multimedia environment: A systematic review on its fate, transport, quantification, health risk, and remedial measures. *Groundwater for Sustainable Development, 20*, Article 100889. https://doi.org/10.1016/j.gsd.2022.100889.

Rujnić-Sokele, M., & Pilipović, A. (2017). Challenges and opportunities of biodegradable plastics: A mini review. *Waste Management & Research, 35*(2), 132–140. https://doi.org/10.1177/0734242X16683272.

Sarkar, D. J., Sarkar, S. D., Das, B. K., Manna, R. K., Behera, B. K., & Samanta, S. (2019). Spatial distribution of meso and microplastics in the sediments of river Ganga at eastern India. *Science of the Total Environment, 694*, Article 133712. https://doi.org/10.1016/j.scitotenv.2019.133712.

Sathish, M. N., Jeyasanta, K. I., & Patterson, J. (2020). Monitoring of microplastics in the clam Donax cuneatus and its habitat in Tuticorin coast of Gulf of Mannar (GoM), India. *Environmental Pollution, 266*, 115219. https://doi.org/10.1016/j.envpol.2020.115219.

Schneider, D. R., & Ragossnig, A. (2015). Recycling and incineration, contradiction or coexistence? *Waste Management & Research, 33*(8), 693–695. https://doi.org/10.1177/0734242X15593421.
Schulz, M., Neumann, D., Fleet, D. M., & Matthies, M. (2013). A multi-criteria evaluation system for marine litter pollution based on statistical analyses of OSPAR beach litter monitoring time series. *Marine Environmental Research, 92*, 61–70. https://doi.org/10.1016/j.marenvres.2013.08.013.
Selvam, S., Jesuraja, K., Venkatramamam, S., Roy, P. D., & Kumari, V. J. (2020). Hazardous microplastic characteristics and its role as heavy metal in groundwater and surface water of coastal south India. *Journal of Hazardous Materials, 402*, Article 123786. https://doi.org/10.1016/j.jhazmat.2020.123786.
Shim, W. J., & Thomposon, R. C. (2015). Microplastics in the ocean. *Archives of Environmental Contamination and Toxicology, 69*, 265–268. https://doi.org/10.1007/s00244-015-0216-x.
Shim, W. J., Song, Y. K., Hong, S. H., & Jang, M. (2016). Identification and quantification of microplastics using Nile Red staining. *Marine Pollution Bulletin, 113*(1–2), 469–476. https://doi.org/10.1016/j.marpolbul.2016.10.049.
Singh, J., Pant, M., Ghoshal, T., Srivastava, A.L., & Singh, S. (2025). Spatio-seasonal observations of water quality index, heavy-metal pollution index and microbial contagion in springs of lower Indian Himalayan zone. *Environmental Claims Journal*, 1–36.
Song, Y. K., Hong, S. H., Jang, M., Han, G. M., Rani, M., Lee, J., & Shim, W. J. (2015). A comparison of microscopic and spectroscopic identification methods for analysis of microplastics in environmental samples. *Marine Pollution Bulletin, 93*(1–2), 202–209. https://doi.org/10.1016/j.marpolbul.2015.01.015.
Sun, Q., Ren, S. Y., & Ni, H. G. (2020). Incidence of microplastics in personal care products: An appreciable part of plastic pollution. *Science of the Total Environment, 742*, Article 140218. https://doi.org/10.1016/j.scitotenv.2020.140218.
Syamsu, D. A., Deswati, D., Syafrizayanti, S., Putra, A., & Suteja, Y. (2024). Presence of microplastics contamination in table salt and estimated exposure in humans. *Global Journal of Environmental Science and Management, 10*(1), 205–224. https://doi.org/10.22034/gjesm.2024.01.14.
Tsering, T., Sillanpää, M., Sillanpää, M., Viitala, M., & Reinikainen, S. P. (2021). Microplastics pollution in the Brahmaputra River and the Indus River of the Indian Himalaya. *Science of the Total Environment, 789*, Article 147968. https://doi.org/10.1016/j.scitotenv.2021.147968.
Veerasingam, S., Saha, M., Suneel, V., Vethamony, P., Rodrigues, A. C., Bhattacharyya, S., & Naik, B. G. (2016). Characteristics, seasonal distribution and surface degradation features of microplastic pellets along the Goa coast, India. *Chemosphere, 159*, 496–505. https://doi.org/10.1016/j.chemosphere.2016.06.056.
Veerasingam, S., Ranjani, M., Venkatachalapathy, R., Bagaev, A., Mukhanov, V., Litvinyuk, D., & Vethamony, P. (2020). Microplastics in different environmental compartments in India: Analytical methods, distribution, associated contaminants and research needs. *TrAC Trends in Analytical Chemistry, 133*, Article 116071. https://doi.org/10.1016/j.trac.2020.116071.
Wagner, M., Scherer, C., Alvarez-Muñoz, D., Brennholt, N., Bourrain, X., Buchinger, S., & Reifferscheid, G. (2014a). Microplastics in freshwater ecosystems: What we know and what we need to know. *Environmental Sciences Europe, 26*(1), 1–9. https://doi.org/10.1186/s12302-014-0012-7.
Wagner, M., Scherer, C., Alvarez-Muñoz, D., Brennholt, N., Bourrain, X., Buchinger, S., Fries, E., Grosbois, C., Klasmeier, J., Marti, T., Rodriguez-Mozaz, S., Urbatzka, R., Vethaak, A. D., Winther-Nielsen, M., & Reiferscheid, G. (2014b). Microplastics in freshwater ecosystems: What we know and what we need to know. *Environmental Sciences Europe, 26*, 12. https://doi.org/10.1186/s12302-014-0012-7.
Wagner, M., & Lambert, S. (2018). *Freshwater microplastics: Emerging environmental contaminants?* (p. 303). Springer Nature.
Wang, H., Liu, X., Wang, H. (2019). The threats faced by the Yangtze River and the flood plain ecosystem and the overall protection countermeasures. Acta Hydrobiol. Sin. 43. Suppl.

Watt, E., Picard, M., Maldonado, B., Abdelwahab, M. A., Mielewski, D. F., Drzal, L. T., & Mohanty, A. K. (2021). Ocean plastics: Environmental implications and potential routes for mitigation–a perspective. *RSC Advances, 11*(35), 21447–21462. https://doi.org/10.1039/D1RA00353D.

Welden, N. A., & Cowie, P. R. (2016). Long-term microplastic retention causes reduced body condition in the langoustine, Nephrops norvegicus. *Environmental Pollution, 218*, 895–900. https://doi.org/10.1016/j.envpol.2016.08.020.

Wu, P., Huang, J., Zheng, Y., Yang, Y., Zhang, Y., He, F., Chen, H., Quan, G., Yan, J., Li, T., & Gao, B. (2019). Environmental occurrences, fate, and impacts of microplastics. *Ecotoxicology and Environmental Safety, 184*, 109612.

Xu, P., Peng, G., Su, L., Gao, Y., Gao, L., & Li, D. (2018). Microplastic risk assessment in surface waters: A case study in the Changjiang Estuary, China. *Marine Pollution Bulletin, 133*, 647–654. https://doi.org/10.1016/j.marpolbul.2018.06.020.

Yadav, S., Kataria, N., Khyalia, P., Rose, P. K., Mukherjee, S., Sabherwal, H., & Khoo, K. S. (2023). Recent analytical techniques, and potential eco-toxicological impacts of textile fibrous microplastics (FMPs) and associated contaminates: A review. *Chemosphere, 326*, Article 138495. https://doi.org/10.1016/j.chemosphere.2023.138495.

Yu, Y., Zhou, D., Li, Z., & Zhu, C. (2018). Advancement and challenges of microplastic pollution in the aquatic environment: A review. *Water, Air, & Soil Pollution, 229*, 1–18. https://doi.org/10.1007/s11270-018-3788-z.

Zhang, Y., Kang, S., Allen, S., Allen, D., Gao, T., & Sillanpää, M. (2020). Atmospheric microplastics: A review on current status and perspectives. *Earth-Science Reviews, 203*, Article 103118. https://doi.org/10.1016/j.earscirev.2020.103118.

Zhou, R., Lu, G., Yan, Z., Jiang, R., Bao, X., & Lu, P. (2020a). A review of the influences of microplastics on toxicity and transgenerational effects of pharmaceutical and personal care products in aquatic environment. *Science of the Total Environment, 732*, Article 139222. https://doi.org/10.1016/j.scitotenv.2020.139222.

Zhou, Y., Wang, J., Zou, M., Jia, Z., Zhou, S., & Li, Y. (2020b). Microplastics in soils: A review of methods, occurrence, fate, transport, ecological and environmental risks. *Science of the Total Environment, 748*, Article 141368. https://doi.org/10.1016/j.scitotenv.2020.141368.

Zhou, G., Wang, Q., Li, J., Li, Q., Xu, H., Ye, Q., & Zhang, J. (2021). Removal of polystyrene and polyethylene microplastics using PAC and FeCl3 coagulation: Performance and mechanism. *Science of the Total Environment, 752*, Article 141837. https://doi.org/10.1016/j.scitotenv.2020.141837.

MIX
Papier aus verantwortungsvollen Quellen
Paper from responsible sources
FSC® C105338

If you have any concerns about our products,
you can contact us on
ProductSafety@springernature.com

In case Publisher is established outside the EU,
the EU authorized representative is:
Springer Nature Customer Service Center GmbH
Europaplatz 3, 69115 Heidelberg, Germany

Printed by Libri Plureos GmbH
in Hamburg, Germany